T0271137

Logic Colloquium '96

Since their inception, the Perspectives in Logic and Lecture Notes in Logic series have published seminal works by leading logicians. Many of the original books in the series have been unavailable for years, but they are now in print once again.

This volume, the 12th publication in the Lecture Notes in Logic series, collects the proceedings of the European Summer Meeting of the Association of Symbolic Logic, held at the University of the Basque Country, at Donostia (San Sebastian) Spain, in July 1996. The main topics were model theory, proof theory, recursion and complexity theory, models of arithmetic, logic for artificial intelligence, formal semantics of natural language, and philosophy of contemporary logic. It includes 11 papers from preeminent researchers in mathematical logic.

J. M. LARRAZABAL works in the Department of Logic and Philosophy of Science at the University of the Basque Country, San Sebastian.

D. LASCAR works in the Faculty of Mathematics at Université de Paris VII.

G. MINTS works in the Department of Philosophy at Stanford University, California.

LECTURE NOTES IN LOGIC 12

Logic Colloquium '96

Proceedings of the Colloquium held in San Sebastián,
Spain, July 9–15, 1996

Edited by

J. M. LARRAZABAL
University of the Basque Country, San Sebastián

D. LASCAR
Université de Paris VII

G. MINTS
Stanford University, California

ASSOCIATION FOR SYMBOLIC LOGIC

CAMBRIDGE
UNIVERSITY PRESS

CAMBRIDGE
UNIVERSITY PRESS

University Printing House, Cambridge CB2 8BS, United Kingdom

One Liberty Plaza, 20th Floor, New York, NY 10006, USA

477 Williamstown Road, Port Melbourne, VIC 3207, Australia

4843/24, 2nd Floor, Ansari Road, Daryaganj, Delhi – 110002, India

79 Anson Road, #06–04/06, Singapore 079906

Cambridge University Press is part of the University of Cambridge.

It furthers the University's mission by disseminating knowledge in the pursuit of education, learning, and research at the highest international levels of excellence.

www.cambridge.org
Information on this title: www.cambridge.org/9781107166080
10.1017/9781316716816

First edition © 1998 Springer-Verlag Berlin Heidelberg
This edition © 2016 Association for Symbolic Logic under license to Cambridge University Press.

Association for Symbolic Logic
Richard A. Shore, Publisher
Department of Mathematics, Cornell University, Ithaca, NY 14853
http://www.aslonline.org

A catalogue record for this publication is available from the British Library.

ISBN 978-1-107-16608-0 Hardback

Preface

The 1996 European Summer Meeting of the Association of Symbolic Logic was held held the University of the Basque Country, at Donostia (San Sebastián) Spain, on July 9-15, 1996. It was organised by the Institute for Logic, Cognition, Language and Information (ILCLI) and the Department of Logic and Philosophy of Sciences of the University of the Basque Country. It was supported by: the University of Pais Vasco/Euskal Herriko Unibertsitatea, the Ministerio de Educatión y Ciencia (DGCYT), Hezkuntza Saila (Eusko Jaurlaritza), Gipuzkoako Foru Aldundia, and Kuxta Fundazioa.

The main topics of the meeting were Model Theory, Proof Theory, Recursion and Complexity Theory, Models of Arithmetic, Logic for Artificial Intelligence, Formal Semantics of Natural Language and Philosophy of Contemporary Logic.

The Program Committee consisted of K. Ambos Spies (Heidelberg), J.L. Balcázar (Barcelona), J.E. Fenstad (Oslo), D. Israel (Stanford), H. Kamp (Stuttgart), R. Kaye (Birmingham), J.M. Larrazabal (San Sebastián), D. Lascar (Paris, chairman), A. Marcja (Firenze), G. Mints (Stanford), M. Otero (Madrid), S. Ronchi della Rocca (Torino), K. Segerberg (Uppsala) and L. Vega (Madrid).

The organizing Committee consisted of X. Arrazola (San Sebastián), A. Arrieta (San Sebastián), R. Beneyeto (Valencia), B. Carrascal (San Sebastián), K. Korta (San Sebastián), J.M. Larrazabal (San Sebastián, chairman), J.C. Martínez (Barcelona), J.M. Méndez (Salamanca), F. Migura (Victoria) and J. Pérez (Victoria).

Twenty-three one hour lectures were given at the meeting by: N. Asher (University of Texas), M.L. Bonet (Universidad Politécnica de Cataluña), J. Etchmendy (Stanford University), D. Evans (University of East Anglia), L. Gordeev (Tuebingen), M. Hyland (University of Cambridge), V. Kanovei (Moscow TEI), U. Kohlenbach (Frankfurt University), R. Kossak (City University of New York), M. Kummer (Universität Karlsruhe), S. Lempp (University of Wisconsin), B. Nebel (University of Ulm), A. Nesin (Tehran), H.J. Ohlbach (Imperial College, London), M. Pentus (Moscow State University), M. Pinkal (Universität des Saarlandes), G. Priest (University of Queensland), M. Reynolds (King's College, London), E. Sandewall (Linkoping University), J. Schmerl (University of Connecticut), S. Starchenko (Vanderbilt University), G. Takeuti (University of Illinois) and L. Torentliev (University of Amsterdam).

Tutorials were given by: A. Berarducci (University of Pisa), M. Dezani-

Ciancaglini (University of Torino), M. Hyland (University of Cambridge), J. Lutz (Iowa State University), J.J.C. Meyer (Utrecht University), C. Steinhorn (Vassar College).

Abstracts of invited lectures, tutorials and contributed talks given by members of the Association for Symbolic Logic can be found in the *Bulletin of Symbolic Logic*, vol.3, number 2, June 1997, pages 243-277.

The present volumes contains the papers of the invited talks as they were made available by the authors. All papers have been refereed.

We wish to thank the above mentioned institutions for their generous support and the referees of the papers contributed in this volume.

Table of contents

The Logical Foundations of Discourse Interpretation

Nicholas Asher

A central but still unsettled question in formal theories about discourse interpretation is: What are the key theoretical structures on which discourse interpretation should depend? If we take our cue from theories that analyze the meanings of individual sentences, the meaning of the discourse's parts should determine the meaning of the whole; some sort of principle of compositionality of meaning must hold at the level of discourse interpretation. So a theory of discourse interpretation must develop from an account of discourse structure.

Unlike the syntactic structure of a sentence, the discourse structure of a text is not a structure studied by syntacticians or governed principally by syntactic concerns. It has to be inferred from a variety of knowledge sources. Recent work on discourse structure in AI, philosophy and linguistics has shown that discourse structure depends on numerous information sources—compositional semantic principles, lexical semantics, pragmatic principles, and information about the speaker's and interpreter's mental states. So a theory of discourse interpretation must in fact also be a theory of semantics and pragmatics and their interaction—a theory of the pragmatics-semantics interface.

Such a theory linking together pragmatics and semantics brings up a foundational question about frameworks. Pragmatics, though not often formalized, has often made appeal to different types of logical principles than semantics. While semantic theories have typically used a classical, monotonic, logical framework, pragmatic theories appear to best couched within nonmonotonic logic. How should we model the interaction of these multiple knowledge sources needed to construct discourse structure, or the interaction between defeasible pragmatic principles and nondefeasible semantic principles?

---

[1]Received June 20, 1996; revised version Marsh 19, 1997.

Another fundamental difference between pragmatics and semantics also requires resolution. The model-theoretic approach to semantics is thoroughly entrenched, whereas in pragmatics appeals are often made to representations of information, beliefs and other mental states of the participants. How should a discourse context be thought of—as a structured representation, or model-theoretically? How should we model in a logically precise fashion the updating of discourse contexts with new information?

One of the good things about recent developments in theories of discourse interpretation is that we know at least some of the answers to these questions. I will try to provide a guide to some strategies for building such theories. I will begin with a simple and now standard semantic approach to discourse interpretation, Discourse Representation Theory (DRT) without eventualities. I will then add in steps components needed for pragmatic interpretation. The approaches become more complex, but each bit of additional complexity lets us analyze some phonenomenon or represent some bit of pragmatic reasoning that we were not able to before. The end result will be a theory of the semantic-pragmatic interface, Segmented Discourse Representation Theory or SDRT. Like DRT, SDRT exploits representations to model discourse contexts and to determine the semantic effects of discourse structure.

Finally, a question that should be asked in this context, is why should any of this be of interest to logicians? Like formal semantics, discourse semantics is an area of application for logical techniques. The few theorems mentioned in this text are not terribly surprising nor do they introduce any new formal techniques unfamiliar to logicians. But discourse semantics has uncovered some new logical problems that were not part of standard formal semantics and could not easily be formulated within the framework of higher order intensional logic, the logical framework of Montague Grammar. It is my hope that some of these problems and the attempts that semanticists have made to formalize them will be of interest to logicians and may prompt them to investigations in these areas.

1   Motivations for Dynamic Semantics

To understand the development of dynamic semantics, one must understand the problems that motivated this development and to do that, one must say something about the interpretation of the constituents of discourse —its constituent sentences and their constituent phrases— i.e., standard formal semantics. In the 1960s Montague developed a very influential theory for analyzing the truth conditions of sentences, Montague Grammar. The question then arose: How do we now analyze the truth conditions of a discourse? For just as one can only understand the meanings of expresions within the context of a sentence, the meaning of a sentence is not something that occurs in isolation—there is always a discourse context.

One might think that there is nothing more to the interpretation of discourse than simply building up the meanings of its constituent sentences and then combining them together. If meanings are taken to be sets of possible worlds or other indices, the operation of combination is particularly simple; it's intersection. But once a clear and precise proposal for the semantics of sentences was given by Montague Grammar, several problems emerged that showed that the interpretation of discourse would not be simple. Below I discuss two: pronominal anaphora and the interpretation of tense in discourse.

1.1 Pronominal Anaphora

Pronominal anaphora occurs when an anaphoric pronoun refers back to some word or phrase in the preceding discourse. In the second sentence of the discourse (1) below, the anaphoric pronouns *he* and *it* refer back respectively to the farmer and the donkey.

(1)   A farmer owns a donkey. He beats it.

In translating each sentence of (1) into logical formulae using the tools of Montague Grammar, we encounter a problem in the translation of anaphoric pronouns. The pronouns in the second sentence are anaphorically bound to the noun phrases in the first sentence, and perhaps the most natural way to try to represent this linkage in the logical form of (1) is to translate the pronouns as variables that are to be bound by the quantifiers introduced by the noun phrases in the first sentence. But the problem is that by the time we attempt this "bound variable" translation of the pronouns, we have finished the translation of the first sentence and so closed off the rightward scope of the quantifiers. Conjoining the translation of the second sentence does not produce a bound variable reading of the variables introduced by the pronouns. In fact it produces the open sentence:

(1') $\exists x(farmer(x) \land \exists y(donkey(y) \land owns(x,y))) \land beats(x,y)$

The Montague Grammarian's approach to quantifying-in offers a partial solution to this problem (Gamut 1991). However, if we wish to use the procedure of quantifying in to deal with discourses in which anaphoric linkage to an antecedently occurring noun phrase exists over multiple sentences as in (2) below, then we must suppose a complete syntactic analysis of the discourse prior to the interpretation of any of its constituent sentences.

(2)   A farmer owned a donkey. He beat it. It ran away.

This conclusion is not cognitively plausible and prompts us to look for a different solution to the problem of intersentential anaphora.

1.2   Temporal Anaphora

Observations of Barbara Partee (1973) and of Kamp and Rohrer working on the analysis of French tenses in discourse during the seventies brought to attention a facet of the meaning of tenses that was missing from the best analyses of tense of the day. Those analyses took tenses to be tense operators of the sort found in tense logic (cf. Montague 1974, and some of the references in Dowty 1979). In a French discourse like

(3)   Pierre entra dans le salon. Il s'assit sur la banquette. Il s'endormit.
         Pierre entered the room. He sat down on the sofa. He fell asleep.

the three events introduced, Pierre's entering, his sitting down and his falling asleep, all occur in the past if (3) is true, but they also occur in a definite sequence—the sequence in which they are introduced in the discourse. That is, we naturally understand the story as telling us that his entering the room occurred prior to his sitting down which in turn occurred prior to his falling asleep. The operator view of tenses, on which a past-tensed sentence the logical form of which is $P\phi$ is true iff ϕ holds of some time prior to the moment of speech, is incapable of capturing this contextual sensitivity. Even views on which verbs are treated as predicates of events and the past tense introduces a relation of earlier than between the event and the moment of speech are not by themselves equipped to capture this context sensitive interpretation.

2   Dynamic Semantics and Basic DRT

The solution that Kamp and Heim independently proposed to the problem of anaphoric pronoun interpretation was to redefine the semantic contribution of a sentence and its constituents to a discourse. This was the first attempt at discourse semantics, using rigorous, formal methods similar to those found in Montague Grammar. In Montague Grammar and on other accounts of discourse interpretation (Stalnaker 1978), the contribution of a sentence is a proposition, or, formally, a set of possible worlds in which the sentence was true. Such a proposition contributes to the content of a discourse in a simple way: the meaning of a discourse is just those possible worlds that are in the intersection of all the propositions that are the meanings of the discourse's constituent sentences. For Kamp and Heim, a sentence S, when interpreted in a discourse D no longer simply yields a set of worlds; the meaning of S is rather a relation, a relation between contexts. This new relational conception of the meaning of S was dubbed its context change potential.[2]

To define the context change potential of a sentence, Kamp uses a representational theory of discourse semantics, Discourse Representation Theory (DRT). DRT assigns a truth conditional meaning to a natural language discourse in two steps: the DRS construction procedure and the correctness definition. In the first step, we construct a representation of the content of the discourse known as a discourse representation structure or DRS. DRT uses DRSs to define context change potential. I won't detail here how DRSs for clauses and their interpretation can be built up compositionally (for details see Asher 1993 or Muskens 1996); I want to focus on discourse aspects of dynamic semantics.

The basic fragment of DRT is defined by the following definition of DRSs and DRS conditions.

Discourse Referents is a set of objects denoted by x, y, z, with or without subscripts. Predicates is a set of predicate constants associated with various natural language nouns, verbs and adjectives. Supose $U \subseteq$ Discourse Referents; we then define DRSs K and conditions γ recursively:

$$K \quad := \quad \langle U, 0 \rangle \mid K^{\cap} \gamma$$

Let $R \in$ Predicates be an n-ary predicate and $x_1, \cdots, x_n$ be discourse referents.

$$\gamma \quad := \quad R(x_1, \cdots, x_n) \mid \neg K \mid K_1 \Rightarrow K_2 \mid K_1 \vee K_2.$$

The truth definition for discourses that give rise to DRSs described by the definition above is given by embedding the DRSs they generate into a standard Tarskian model. Given the semantics for DRSs, any DRS in the fragment above has a first order translation. E.g.

---

[2]Many other formalisms have adopted the Kamp-Heim approach in a different guise. See Barwise (1985), Gronendijk and Stokhof (1987).

has a proper embedding in a model $M = \langle D, A_1, \cdots, A_n, \cdots \rangle$ (where the A_i's represent the extensions of the non-logical predicate symbols of the DRS language) relative to some embedding g iff $\exists x \phi$ is is satisfied in M relative to some assignment to free variables. Similarly,

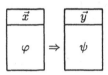

is satisfied relative to some embedding function g in M iff $\forall x \exists y (\phi \rightarrow \psi)$ is satisfied relative to some assignment.

Let us now make this more precise with a full semantic definition of proper embedding. I define simultaneously the model theoretic transition P and the satisfaction of conditions V relative to a model.

Definition 1:

$$f P_M(U, O) g \quad \textit{iff} \quad f \subseteq g \wedge dom(g) = dom(f) \cup U$$

$$f \in V_M(R(x_1, \cdots, x_n)) \quad \textit{iff} \quad R_M(f(x_1), \cdots, f(x_n))$$

$$f P_M(K^\cap \gamma) g \quad \textit{iff} \quad f P_M(K) g \wedge g \in V_M(\gamma)$$

$$f \in V_M(\neg K) \quad \textit{iff} \quad \neg \exists g \, f P_M(K) g$$

$$f \in V_M(K \Rightarrow K') \quad \textit{iff} \quad \forall g \, (f P_M(K) g \rightarrow \exists h \, g P_M(K') h)$$

$$f \in V_M(K \vee K') \quad \textit{iff} \quad \exists g \, f P_M(K) g \vee \exists h \, f P_M(K') h$$

I assume a sufficiently large set *MOD* of models, viz. those formed from maximal consistent saturated sets of first order formulas.

In order to formulate an insightful correspondence between DRT and first order logic, it is useful not only to define P but also its *lifted* counterpart $\mathcal{P}$ on sets of model sequence pairs (MSP)'s for a DRS K. $\mathcal{P}(K)$:Pow(MSP) $\rightarrow$ Pow(MSP) is defined distributively or pointwise over the set of model sequence pairs , exploiting K's effect on an input MSP (M, g) to produce a certain output MSP (M', g'), where . That is for a context X:

$$[\mathcal{P}(K)](X) = \{(M', g') : \exists(M, g) \in X(M = M' \wedge gP_M(K)g')\}$$

One can associate with each DRS K both a set of pairs $\langle M, f \rangle$ where f is a proper embedding of K in a Tarskian model M. Such sets are called information states and they correspond to a a first order formula. This connection is made precise in the following lemma (proved by induction). I use here the formulation of Fernando (1994). Let $\{\langle w, 0 \rangle : M \in MOD\} = \sigma_0$
Characterization Lemma for DRT:

- For every first order formula χ with a set of free variables U, there is a DRS (U, C) such that $\mathcal{P}(U, C)[\sigma_0] = \{(M, f) : Dom(f) = U$ and $M \models \chi[f]\}$

- Every DRS (U, C) has a characteristic formula χ where U is the set of free variables in χ and $\mathcal{P}(U, C)[\sigma_0] = \{(M, f) : Dom(f) = U$ and $M \models \chi[f]\}$

DRT is a dynamic theory of discourse interpretation. The idea is that each sentence has a *context change potential* (CCP) that changes a discourse context when the information contained within the sentence is added to the context. The background context is represented as a DRS; the contribution of the individual sentence is the context DRS extended with the conditions and discourse referents contributed by the processing of the sentence. Thus, the CCP of a sentence is a relation between DRSs. More precisely, let C_ϕ be the set of conditions of the DRS ϕ, χ_ϕ be the characteristic formula of ϕ, and let

$$\phi^\cap \psi = ((U_\psi \cup U_\phi), (C_\psi \cup C_\phi)), \text{ if } (\chi_\phi \wedge \chi_\psi) \text{ is first order consistent.}$$

The CCP of an unambiguous sentence S, which by a procedure known as the DRS construction procedure yields a DRS K_S,[3] can now be represented as the DRS transition predicate T_{drs}, between consistent DRSs. We define T_{drs} as the set

$$\{(K, K_S, K^\star) : K \text{ is a DRS}, K_S, K^\star \text{ are consistent DRSs and } K^\star = K^\cap K'\}$$

---

[3] For details see Kamp and Reyle (1993).

The CCP of a sentence may also be represented as a relation between information states (assuming once again the presence of the DRS construction procedure). A model-theoretic context is a set of model sequence pairs (MSPs), and so a natural candidate for the model-theoretic notion of the CCP of a sentence S is just the lifted function P applied to the DRS derived from S.

Using the Characterization Lemma, Fernando (1995) proves a precise equivalence between the representational and model-theoretic conceptions of CCP for the simple core fragment of DRT and he has also shown that the notion of equivalent CCP is r.e. The representational level and the model theoretic level yield bisimulation-equivalent notions of context and CCP.[4] Let $\Phi_\perp$ be the set of absurd formulas, and let MOD be defined as above). Define Acc as the smallest set of states such that, where $[\mathcal{P}(\phi)](\sigma)$ represents the application of the function $\mathcal{P}(\phi)$ to σ:

$$\sigma_0 \in Acc \wedge (\sigma \in Acc \rightarrow [\mathcal{P}(\phi)](\sigma) \in Acc)$$

Thus, we can define within Acc for any consistent ϕ, $\sigma_\phi = [\mathcal{P}(\phi)](\sigma_0)$. Suppose that $\mathcal{P}(\phi^\cap \psi) = \mathcal{P}(\phi) \circ \mathcal{P}(\psi)$ and define $\phi \leftrightarrow \psi$ iff $\forall \phi' (\phi^\cap \phi' \in \Phi_\perp \leftrightarrow \psi^\cap \phi' \in \Phi_\perp)$.

Theorem 2.1 *(Fernando 1995)*: $\phi \leftrightarrow \psi$ iff $\sigma_\phi = \sigma_\psi$; *and further,* $\leftrightarrow$ *and* $\Phi_\perp$ *are r.e.*

$\leftrightarrow$ defines a bisimulation relative to the function $\cap$, and $\mathcal{P}(K)[\sigma_0]$, in effect, exploits $\leftrightarrow$ to induce a bisimulation relative to state transitions, the model-theoretic interpretation of a DRS on a set of MSP's. Further, bisimilarity on state transitions is strongly extensional; bisimilar transitions on states have identical outputs when applied to the empty information state, which explains the first part of the theorem. For the second part of the theorem, note that $\leftrightarrow$ is defined proof theoretically, by an operation on DRSs. Given the *Characterization Lemma*, every DRS is equivalent to a first order formula. Hence, the notions $\leftrightarrow$ and $\Phi_\perp$ are first order definable and so r.e.

The effect of this theorem is to show that the model-theoretic and the representational conceptions of CCP in dynamic semantics are equivalent. Far from "eliminating representationalism," the inclusion in MSPs of functions from discourse referents to objects makes them essentially representational;

---

[4]A similar representation theorem for a slightly different conception of model theoretic context is sketched in Asher (1993). For more on the notion of bisimulation, see Aczel (1989) and Park (1988).

this information is not about the world but about how our information about the world is structured.[5] But using MSPs is useful because they indicate the minimal amount of representational information needed to treat phenomena for which DRT was designed. I apply this strategy now to pragmatic-semantic contexts.

3 DRT_1—DRT with events

Without events or temporal constants of any kind, the basic DRT fragment is unable to make sense of change. In the timeless models of DRT_0, the following simple discourse is (unintuitively) inconsistent:

(4)   John sat down. John got up.

To analyze change, DRT follows a Davidsonian approach and introduces eventualities as additional arguments of conditions derived from verbs. The only change to the definition of DRSs and DRS conditions from section 1 is to add an eventuality type discourse referents and to change the set of Predicates— those derived from verbs will have one more argument place than before. This makes possible a satisfactory analysis of change. I will here introduce ways of treating tense that make the discourse (4) consistent, though they do not give a complete temporal interpretation of tensed discourse. Much of the temporal interpretation of discourse is pragmatically determined. The notion of CCP equivalence remains r.e. for DRT_1.

DRT_0 and DRT_1 define distinct consequence relations for some natural language discourses by offering a translation into a formal language that is given a model theoretic interpretation. So for instance, a discourse like (4), when translated (via the DRS construction procedure) as a DRS ϕ in DRT_0, is inconsistent, provided we make certain assumptions about the relationship between the predicate sit down and get up (viz. one cannot simultaneously get up and sit down). However, the translation of (4) is consistent in DRT_1. We could represent this as (4) $\vdash_0 \perp$ but $\nvdash_1 \perp$.

---

[5]Many authors who have rejected the representationalism of DRT do not seem to acknowledge that the use of assignment functions in their semantic values amounts to an equal representational commitment.

4   Beyond Dynamic Semantics

DRT, and dynamic semantics in general, make an important contribution to our understanding of how the discourse context affects pronominal and temporal interpretation. Neither temporal nor pronominal anaphora is adequately accounted for in a static semantic framework like Montague Grammar or classical (Barwise and Perry 1983) situation theory, even if the static semantic theory incorporates some contextual sensitivity such as that suggested by Kaplan (1977). The key novelty in dynamic semantics that is brought out by the characterization theorem is that in dynamic semantics it matters how information is introduced into the discourse—viz. what variable or discourse referent this information is predicated of. These discourse referents are now understood to be a fundamental part of information states and discourse contexts. Is this really part of the content? Well, they might be but it seems more intuitive to think of them as part of how information is *packaged*. Dynamic semantics invites us to make a distinction between information content and information packaging.

We can represent information packaging within possible worlds semantics at a cost. We must replace the notion of information as represented by sets of worlds with a set of n-tuples consisting of a world and a sequence of objects—one object, roughly, for each indefinite introduced into the discourse (this needs to be refined when indefinites occur within the scope of other logical operators and quantifiers). But this is a technical trick without much philosophical substance and it obscures the central point of dynamic semantics: the information conveyed by a text is more than just truth conditional content (a set of possible worlds) it also involves some sort of "information packaging" which tells us how to understand a variety of linguistic anaphoric phenomena.[6] A representational structure giving us a notion of information packaging is needed here. For dynamic semantics this information packaging is quite minimal— we can make do with just variables.

Dynamic semantics and its limited view of information packaging does not do justice, however, to the complex interaction of pragmatic and semantic factors in discourse. This leads to incorrect predictions about the way anaphors, both temporal and pronominal,are treated in DRT. In effect the notion of information packaging in DRT is too weak to support its analysis of the phenomena; it tries to do too much with truth conditions and not

---

[6]Stalnaker's critical evaluation in his 1996 SALT talk of dynamic semantics attempts to defend classical possible worlds semantics. But he does so by packing the representational component, information packaging, into the possible worlds themselves. Further, because he doesn't really propose a theory of information packaging, his account isn't doesn't offer an alternative to the view developed here.

enough with information packaging.

4.1 Temporal Anaphora

A clear example of where DRT goes wrong is with its analysis of temporal anaphora. DRT is right to make the contributions of tense depend upon the discourse context. But DRT attempts to make the temporal structure of a text almost completely dependent on the tense forms used in the text. In most natural languages, however, the temporal structure of the events introduced in a text is underdetermined by the sequence of tense forms. In particular, the rule above, which is a consequence of the DRS construction procedure, is false for French—a point of which some of the earliest workers on tense in DRT were aware—or English. Consider the following examples (from Lascarides and Asher 1993):

(5)   John entered the room. Fred greeted him.

(6)   John fell. Fred pushed him.

These two discourses employ the same sequence of tense forms, yet they suggest different temporal structures. DR-theorists have been forced to revise the construction procedure and to abandon the view that the tense forms and the order of the sentences in a discourse alone determine temporal structure. This conclusion follows not only from an examination of the English simple past but also from a careful look at the data concerning the French plus que parfait (Bras and Asher 1994) and the English pluperfect (Lascarides and Asher 1993b).

One might ask, what in combination with tense sequences determines the temporal structure? One proposal is that a more developed view of discourse structure determines tense structure. Originally suggested by the computer scientist Jerry Hobbs, this thesis has been worked out in the context of a formal discourse semantics by Asher and Lascarides (Lascarides and Asher 1993a, 1993b) using a more elaborate analysis of information packaging than that found in standard DRT. Specifically, it is the rhetorical connection between the propositions that often supplies the information needed to determine temporal structure.

4.2 The Contextual Treatment of Definites

There have been two treatments of definite descriptions vying for contention since the times of Russell and Strawson. One is Russell's famous 1905 analysis of definites in terms of first order quantifiers; the ϕ ψ's is to be translated as $\exists x(\forall y(\phi(y) \leftrightarrow y = x) \wedge \psi(x))$. Russell's analysis works well enough for the example he discusses at length, *the present King of France is not bald*; but there are other uses of definite descriptions such as (7) that render his analysis very problematic:

(7)   If I invite a celebrity, the celebrity never comes.

Such uses of definites have motivated an entirely different sort of analysis from Russell's that begins with Strawson's paper of 1950 and that has proved popular amongst linguists. Most of those who have adopted this position have argued that definite descriptions generate presuppositions of familiarity; the individual that is the denotation of the definite must be familiar to the speaker's audience or already mentioned in the text. More generally, such presuppositions must be understood as holding in the context in which the sentence generating the presuppositions is to be interpreted. In the linguistic literature, this notion of holding is glossed with the help of two other notions: binding and accommodation. The presupposition of familiarity is said to be bound if the discourse referent or variable introduced by the definite is identified with a discourse referent or variable introduced by some antecedent NP. All presuppositions must either be bound or accommodated in the discourse context. When such an antecedent NP is not available, advocates of this approach say that the presupposition is accommodated by adding to the antecedent context a variable with the appropriate properties. Presupposition theorists have not said much about uniqueness— perhaps justifiably so in view of the apparent failure of any uniqueness claim or presupposition in (7). One of the success stories of dynamic semantics has been to give structure to contexts through the recursive structure of conditions so that constraints could be placed on accommodation (van der Sandt 1992); dynamic semantics made sense of the inherent dynamism in the notions of presupposition already evident in Gazdar's work in the late seventies.[7]

The presuppositional view of definites gives a nice analysis of the anaphoric behavior of the definite in (7), and it can be extended by using accommodation to handle Russell's example. However, neither the presuppositional view nor Russell's makes sense of the behavior of definites in so called "bridging" examples, where an antecedent for the definite is not explicitly given in the text but constructed from other information sources. Here are some examples.

(8)   Mary moved from Brixton to St. John's Wood. The rent was less expensive

---

[7]For details see Beaver (1994, 1996).

(9)   I met two interesting people last night at a party. The woman was a member of Clinton's Cabinet.

It is easy to see why the presupposition theory of definites (let alone the Russellian theory) fails to account for these examples. Consider for a moment (9). According to the approach to presuppositions mentioned above, the woman must be bound to some woman already introduced in the text, or its presupposition of familiarity must be accommodated. Since no woman has been explicitly mentioned prior to the discourse, one is added to the context, which gives the wrong the truth conditions.

Several proposals have been devised to supplement the presupposition theory to get the right sort of anaphoric connections for the definites in (8)-(9) — in particular the addition of lexical information or world-knowledge (Bos et al. 1995). It is possible to get the right anaphoric connection for the woman in (9) if we develop extended rules for binding that exploit the information that *woman* is a subtype of the noun *people* in the antecedent noun phrase. But this sort of approach won't make sense of (8). No lexical information or nonlinguistic bit of world knowledge about the parts of London will yield the right interpretation of the definite the rent. World knowledge about London will rather predict the opposite, since Brixton is commonly acknowleged to have apartments and houses with lower rents than St. John's Wood, an interpreter relying solely on world knoweldge or lexical information would conclude that the rent referred to the rent of a dwelling in Brixton. But this is plainly not what was intended.

What sort of information is required to get the right interpretations of these definite descriptions? An important clue to what seems to be the right answer comes from investigating parallel uses of indefinites, as this example from Charniak (1983) attests:

(10)   a.   Jack was going to commit suicide. He got a rope on Tuesday.

         b.   Jack was going to commit suicide. He got the rope on Tuesday.

An interpreter is likely to construct the same link in (10a-b) between the propositions expressed by the two clauses: Jacks gets the rope he uses to commit the suicide on Tuesday. It is the rhetorical relation between the two propositions that enables the interpreter to resolve the definite in the appropriate way.

Information about rhetorical structure also governs the interpretation of recalcitrant examples like (8). Our approach has a much more liberal notion of binding than the traditional one; identity is only one of many ways in which two discourse referents can be linked. In (8) the second sentence explains why the agent moved, and it this relation of Explanation that guides the binding of the definite *the rent*. The rent's being less expensive in St. John's would explain why the agent moved, and so the inference to an Explanatory connection between the two clauses ensures that the discourse referent introduced by *the rent* to is linked to the disocurse referent introduced by *St. John's Wood*. The discourse structure in (9) is somewhat different; there the second sentence serves as an elaboration of the event describedin the first. The fact that we have an elaboration leads us to bind the presupposition in a particular way—viz., the woman is identified as one of the people the speaker met last night. Such connections are not part of the compositional semantics of such discourses even on dynamic semantics's conception of such. They are part of information packaging. And interestingly, these rhetorical connections between propositions in a text play an important role in interpreting not only temporal anaphora but definite descriptions.[8]

4.3   Lexical Ambiguity

Few compositional theories of meaning tackle the problem of lexical ambiguity, since they are designed to articulate principles of how word meanings combine together. But lexical ambiguity is also a matter for compositional semantics and discourse interpretation, since it is usually the case that combining words with ambiguous meanings leads to a decrease in ambiguity rather than an increase in ambiguity. Sometimes syntactic factors or semantic requirements like selectional restrictions reduce ambiguity, but in many cases the resolution of semantic ambiguities depends on other pragmatic factors — in particular, ambiguity resolution often depends on the discourse structure of the discourse context, as argued in Asher and Lascarides (1995). Consider the following example from that paper:

(11)   The judge demanded to know where the defendant was. The barrister apologised and said that he was drinking across the street. The court bailiff found him asleep beneath the bar.

The word *bar* in the last sentence is ambiguous even when taken as a noun. But it is not ambiguous in this discourse. Standard disambiguation techniques in AI try to use word senses in the same clause to disambiguate other words going for a most probable interpretation (different word senses being associated with different frequencies) in a particular domain. But here this would net the wrong result, since this would predict that bar is being used

---

[8]This approach to definite descriptions is worked out in detail in Asher and Lascarides (in press).

in its courtroom sense. But if we reconstruct what is going in the discourse and exploit the way the various sentences are related to each other in disambiguation, we get a quite different story— one that is recounted in Asher and Lascarides (1995). Very roughly, the text is a narrative; the proposition expressed by each sentence is related by the discourse relation of Narration to the proposition expressed by the previous sentence. Narration imposes strong spatio-temporal constraints on the actors and events involved, so that once we expect that the defendant is drinking at a pub, we assume as a default that he stays there unless the narrator indicates otherwise. So we interpret the last sentence such that the bailiff finds the defendant in the drinking establishment not in the court slumped underneath the bar there. Once again discourse structure seems to play an important role in discourse interpretation, and it is a role that cannot be captured within dynamic semantics as it stands.

4.4 Propositional and Concept Anaphora

If pronominal anaphoric reference to indefinites identifies a need for discourse referents in the analysis of information, then anaphoric reference to chunks or segments of text suggests other representational commitments. Consider the following text:

(12)   One plaintiff was passed over for promotion three times. Another didn't get a raise for five years. A third plaintiff was given a lower wage compared to males who were doing the same work. But the jury didn't believe this (any of this).

There appear to be two possible antecedents (depending to some degree on how this is stressed) in the fourth sentence of (8)—the proposition expressed by the discourse as a whole and the proposition expressed by the last sentence. But in dynamic semantics neither proposition as it stands is a possible antecedent. It was such references to abstract objects that motivated Asher (1993) to develop a non-trivial extension to Kamp's DRT in which a richer conception of information packaging segments the information given in a discourse into bits that are supported by the data such as in (8).

Similarly, concept anaphora can be subject to the effects of discourse structure. One form of concept anaphora is verb phrase ellipsis, exemplified in the following examples.

(13)   a.   John said that Mary hit him. But Sam did.
        b.   John said that Mary hit him. Sam did too.

Aided by prosodic stress which gives certain important clues about the discourse context, speakers almost uniformly prefer the embedded verb phrase meaning to go in for the elided VP in (13a)—i.e. $\lambda x(x hit\ y)$, while in (13b) they prefer the verb phrase meaning derived from the main VP— $\lambda x(x said\ that\ Mary\ hit\ x)$. A theory that does not take account of the discourse structure generated by particles like *but* or *too* will not be able to make sense of these preferred interpretations.

5   Beyond Dynamic Semantics: $SDRT_0$

A fuller   theory of discourse interpretation and a theory of the pragmatics/semantics interface incorporates pragm-atics and other information sources besides just the compositional semantics exploited in DRT. In particular, such a theory must build, I have argued, interpretation from discourse structure. There are many ways of incorporating discourse structure within dynamic semantics, but the simplest is the one which remains squarely within the confines of the first order theory discussed above. The relevant discourse structure for temporal interpretation can be formalized by generalizing the idea of eventuality discourse referents introduced in section 2. In this extension of the original DRT fragment, I introduce speech act discourse referents, π_0, π_1, π_2, etc. The signature of the fragment adds two sorts of conditions to those defined above: if π is a speech act discourse referent and K a DRS, then $\pi : K \in Conditions$. Discourse Structure is encoded by relation symbols on speech act discourse referents; e.g., $Elaboration(\pi_1, \pi_2)$ signifies that the speech act π_2 elaborates what was said in π_1. We extend the model theory of DRT_1 to handle conditions of the form $\pi : K$ by relativizing all conditions in K to values of π in a recursive fashion. I assume below that the interpretation of a predicate of the DRS language is relativized to M and a sequence α of the values speech act discourse referents. This relativization is intended to capture, albeit imperfectly, the intuition that the information in K pertains to the speech act discourse referent that characterizes it. We must, however, reinterpret DRSs so each condition defines a transition predicate; in effect $K^\cap\gamma$ is now the sequence $K; \gamma$. We can then ensure that variables declared in K but also occurring in K' (due to anaphoric equations) are properly bound.

Definition 2:

- $fP_M^\alpha(U, O)g$ iff $f \subseteq g$ and $dom(g) = dom(f) \cup U$

- $fP_M^\alpha(R(x_1, \cdots, x_n))g$ iff $R_M^\alpha(f(x_1), \cdots, f(x_n)) \wedge f = g$

- $fP_M^\alpha(K^\cap\gamma)g$ iff $\exists h(fP_M^\alpha(K)h \wedge hP_M^\alpha(\gamma)g)$

- $fP_M^\alpha(\neg K)h$ iff $f = h \wedge \neg\exists g\, f\, P_M^\alpha(K)g$

- $fP_M^\alpha(K \Rightarrow K')k$ iff$(f = k \wedge \forall g\, (fP_M^\alpha(K)g \rightarrow \exists\, h\, gP_M^\alpha(K')h))$

- $fP_M^\alpha(K \vee K')k$ iff $(f = k \wedge \exists\, g\, fP_M^\alpha(K)g \vee \exists h\, fP_M^\alpha(K')h)$

- $fP_M^\alpha(\pi : K)g$ iff $fP_M^{\alpha\cap f(\pi)}(K)g$

The last clause of the recursive definition above shows how the relativization of the interpretation of conditions in a DRS K is built up recursively when the latter itself occurs in a condition of the form $\pi : K$. Notice also how the last clause ensures that the output assignments from labelled constituents in which variables are declared will make those assignments available to the interpretation of subsequent labelled constituents. Otherwise, the definition is similar to the one for classic DRT. One can easily define the "lifting" of P above on the informationally empty discourse context σ_0; it is simply the same (and I'll use the same notation P) as given in section 2 for DRT. When the contexts are not informationally empty, matters become more complicated, as we shall soon see.

Let us call SDRT$_0$ the theory which delivers those discourse representations, which I'll call SDRSs, whose syntax and semantics is as described above. The language of SDRT$_0$ is first order. For the Characterization Lemma for DRT applies equally well to SDRT$_0$. Recall that the characterization lemma comes in two parts; the first claims that every DRS corresponds to a first order formula and the second that every first order formula corresponds to a DRS. To show that a reformulation of the Characterization Lemma holds for SDRT$_0$, we need only to show that every SDRS in this fragment corresponds to a first order formula. To show this, we have to specify in the induction on the complexity of SDRSs what formula corresponds to the new condition of the form $\pi : K$. We will assume that the model here has parametrized interpretations corresponding to the function V^α and that each predicate is parametrized for one argument, a sequence of speech acts. By the inductive hypothesis, K has a parametrized characteristic formula $\chi_K(x)$, where x is the sequence of variables corresponding to the speech acts. So the corresponding formula for $\pi : K$ is just like $\chi_K(x^\cap u_\pi)$, where u_π is a new free variable appended to the sequence of speech acts that constitute an argument of each atomic formula in χ_K. Thus, we have established:

Characterization Lemma for SDRT$_0$:

- For every first order formula χ with a set of free variables U, there is an SDRS (U, C) such that $\mathcal{P}(U, C)[\sigma_0] = \{(M, f) : Dom(f) = U$ and $M \models \chi[f]\}$

- Every SDRS of $SDRT_0$ (U, C) has a characteristic formula χ where U is the set of free variables in χ and $\mathcal{P}(U, C)[0] = \{(M, f) : Dom(f) = U$ and $M \models \chi[f]\}$

$SDRT_0$ can be used to address many of the issues that made us dissatisfied with dynamic semantics proper—the account of definites and presupposition, the resolution of many lexical ambiguities, and some of the problems with temporal anaphora. To address these issues, however, $SDRT_0$ requires a background theory in which the semantic consequences of discourse relations are specified. These consequences include not only temporal effects, of which I will give an example below, but also spatio-temporal effects, effects on lexical choice and effects on the choice of binding relation and antecedent for presupposition.[9]

This background theory is most naturally expressed in the $SDRT_0$ language.[10] To economize on space, however, I shall use the first order translations here to give an indication of the axioms in the background theory. For example, Narration demands a consistency and coherence which many have observed and which is axiomatized in SDRT below: in words it says that if two speech acts form a narrative sequence, then the eventualities described by these speech acts must cohere together at least in the sense that the post-state of the first be consistent with the pre-state of the second (for more discussion see Asher 1996, Asher et al. 1995). I designate the main eventuality introduced in a DRS K (since we have DRSs for each clause this will be the eventuality discourse referent introduced by the main verb) by e in the characteristic formula of $\pi : K$, the prestate of an eventuality e by pre(e), the poststate of an eventuality e by post(e), the relation of temporal overlap by O_t, the relation of temporal precedence by $\prec$ and the relation of temporal inclusion by $\subseteq_t$:

$$(Narration(\pi_1, \pi_2) \wedge \chi_{\pi_1:K_1}(e_1) \wedge \chi_{\pi_1:K_2}(e_2)) \rightarrow O_t(post(e_1), pre(e_2))$$

Other uncontroversial axioms about eventualities and their pre-states and

---

[9]For details concerning these other semantic consequences, see Asher and Lascarides 1995, forthcoming, and Asher et al. 1995. This background theory elaborated in these articles defines semantic consequences of eight of the most prominent discourse relations found in expository text.

[10]In Lascarides and Asher 1993a, this background theory was made part of another component of reasoning—what I call below the "glue logic". But the consequences of discourse structure should be expressed as DRS or SDRS conditions, since the SDRS language is what gives the content of the discourse.

post-states ensure that we can conclude the following fact:

Fact 1 $(\text{Narration}(\pi_1, \pi_2) \wedge \chi_{\pi_1 : K_1}(e_1) \wedge \chi_{\pi_1 : K_2}(e_2)) \rightarrow e_1 \prec e_2$

There are other axioms capturing the temporal consequences of other discourse relations. Here are a few examples.

$Elaboration(\pi_1, \pi_2) \rightarrow e_2 \subseteq_t e_1$
$Explanation(\pi_1, \pi_2) \rightarrow \neg e_2 \prec e_1$
$(Explanation(\pi_1, \pi_2) \wedge (\neg(State(e_2) \vee Process(e_2)))) \rightarrow e_2 \prec e_1$

This is not an exhaustive list of the axioms for the temporal consequences of discourse structure, let alone a list of the axioms needed for lexical disambiguation or the analysis of presuppositions. But these axioms do give an idea of how one could use these temporal order axioms and other axioms to compute the contribution of discourse relations to the interpretation of constituents, supplementing the compositional semantics. We can derive formulas about the temporal order of eventualities mentioned in a text from the $SDRT_0$ background theory. This enables us to show that $SDRT_0$ is a non-conservative extension of DRT_1 in the following sense. Recall discourse (4) above:[11]

Fact 2 *There is a formula ϕ in the language of DRT_1 such that (4) $\nvdash_1 \phi$ but (4) $\vdash_2 \phi$.*

5.1   CCPs of new information in $SDRT_0$

There are again two conceptions of CCP for an unambiguous sentence S in $SDRT_0$. The representational conception of the CCP of S is a relation between SDRSs—in particular between an SDRS of $SDRT_0$ representing the discourse context, an SDRS derived from S, which I'll assume is just $\pi_S : K_S$ where K_S is the DRS produced from S using the DRS construction procedure, and an "output" SDRS.[12] Note that each piece of new information to be added to the contextual SDRS introduces its own speech act discourse referent. But otherwise $SDRT_0$ conceptions are similar in appearance to those in DRT. The model theoretic conception of the CCP of an unambiguous sentence too resembles its DRT counterpart; it is a relation between contexts or sets model embedding function pairs, just as was the

---

[11]Of course a more complex DRS constrution procedure might also net us the appropriate temporal information for (4) but as we saw in section 4, there is ample evidence to suggest that a purely semantic construction procedure will not produce the right results.

[12]This is a simplification, since discourse structure can also occur within a single sentence. We'll see how to rid ourselves of this assumption shortly.

case with DRT_0. What changes dramatically once we move to $SDRT_0$ is the complexity of computing the change wrought in the discourse context by "adding" the information contained in S. To compute the CCP of ϕ relative to a discourse context τ, one must compute a discourse relation between π and some available "attachment" point, which is a subset of the speech act discourse referents in the discourse context.

But in order to do this, we must change our conception of what a discourse context is. It can no longer simply be an information state, for such a state does not provide us with the information by means of which to do the computation of attachment.

One approach defines the CCP of a formula to be the relation that, given a discourse context allows *any* possible attachment by means of any discourse relation to make up an allowable CCP transition. That is, the CCP of ϕ relative to a discourse context τ is that relation that yields $\tau^{\cap}\phi^{\cap}R(\pi_1, \pi_2)$, where π_2 is the speech act discourse referent of ϕ and π_1 is some speech act discourse referent in τ and R some discourse relation symbol.

This proposal, however, would not enable us to make *any* predictions about the temporal structure of a discourse such as (6), let alone the meanings of definite descriptions such as in (8). The conception of CCP needs to be more restrictive. A natural alternative is to treat the construction of the appropriate SDRS in which the relevant discourse relation attaches the new information to the given context as a black box. That is, the CCP of ϕ relative to a discourse context τ is that relation that yields $\tau^{\cap}\phi^{\cap}R(\pi_1, \pi_2)$, assuming that $\pi_2 : \phi$ and that π_1 is some "appropriate" (in a sense explicitly defined only within the black box) speech act discourse referent in τ and R the "appropriate" discourse relation used to attach the information in ϕ. While this black box approach might seem a little bit like "cheating," it is customary in semantics to argue in this way. Such an approach has the virtue that it allows us to separate out processing questions from the logical character of the final representation. This black box is in effect an oracle—call it O.

What is the logical status of $SDRT_0$ defined transitions relative to this oracle O for appropriate SDRSs? Given that the characterization lemma still holds for the representations produced by $SDRT_0$ (they are in effect just DRS's), it is a straightforward matter to adapt theorem 1 to such a conception of CCP, as labelled by SDRSs defined within $SDRT_0$. The CCP transitions are more restrictive, allowing fewer model assignment pairs to be related in a transition. Further, the underlying logical theory of $SDRT_0$ is more than just first order logic, since it contains non-logical axioms about the semantic consequences of discourse structure—e.g., the temporal axioms discussed in the previous subsection. So to distinguish the absurd formulas relative to this underlying theory from those of section 3, I'll call the the set of absurd formulas for SDRT_0, $\Phi^0_\perp$.

Letting Acc be as in the preamble to theorem 1 of section 2, each SDRS ϕ labels a transition on states $\mathcal{P}^0(\phi)$; these transitions are distinct from the simpler transitions we appealed to for DRT, because they are engendered by SDRT_0 conceptions of transitions and discourse contexts). Clearly $\mathcal{P}^0$ transitions are distinct from those transitions induced simply by DRSs; they are more restrictive. But as before we may define:
$\sigma^0_\phi = [\mathcal{P}(\phi)](\sigma_0)$.
And again we may suppose that $\mathcal{P}^0(\phi^\cap\psi) = \mathcal{P}^0(\phi) \circ \mathcal{P}^0(\psi)$. But our notion of $\leftrightarrow^0$ is more complex because of the complexity of the update relation in SDRT_0, which outputs a set of SDRSs—one for each attachment site.
$\phi \leftrightarrow^0 \psi$ iff $\forall\phi'$:

- $\forall\chi\exists\chi'$:

    1. $(T^0(\phi, \phi', \chi)$ is defined iff $T^0(\psi, \phi', \chi')$ is defined, and
    2. $T^0(\phi, \phi', \chi) \in \Phi^0_\perp$ iff $T^0(\psi, \phi', \chi) \in \Phi^0_\perp$

.

- $\forall\chi'\exists\chi$:

    1. $(T^0(\phi, \phi', \chi)$ is defined iff $T^0(\psi, \phi', \chi')$ is defined, and
    2. $T^0(\phi, \phi', \chi) \in \Phi^0_\perp$ iff $T^0(\psi, \phi', \chi) \in \Phi^0_\perp$

Theorem 5.1 $\phi \leftrightarrow_0 \psi$ iff $\sigma^0_\phi = \sigma^0_\psi$; and further, $\leftrightarrow_0$ and $\Phi^0_\perp$ are r.e. relative to O.

Theorem 2 is a straightforward adaptation of theorem 1 to the basic SDRT fragment and its underlying logic. Note, however, that the theorem relies on essentially an oracle for giving us the "appropriate" output SDRS. That is, our transition predicate on SDRSs relies on this oracle. Can we do any better?

To do better requires a method for calculating the more restrictive and appropriate CCP notion given an input SDRS (our representation of the given discourse context) and some new information ϕ that we also represent as an SDRS of SDRT_0. To calculate this more restrictive CCP notion, we have to: (1) put constraints on which speech act discourse referents may act as attachment points in the antecedently given discourse context, and (2) articulate mechanisms for constraining what are the admissible discourse relations by means of which we can bind the new information to the given discourse structure.

Without going too much into the gory details of how SDRS's are constructed, we can still specify the CCP of a formula SDRT_0 in more detail.

I understand the constraints and mechanisms needed to compute $SDRT_0$'s restrictive CCP notion as forming a particular logic G, a "glue" logic, for deducing a new SDRS from a contextually given SDRS and an SDRS representing new information. As SDRT has pretensions to being a computational semantic theory (see Lascarides and Asher 1993a), G should be a logic that is at least decidable and preferably lower in complexity than NP-complete. So G must be distinguished from the underlying logic of $SDRT_0$, which is at least as strong as first order logic and hence undecidable.

In SDRT, the glue logic G is a nonmonotonic logic. From a computational point of view, this adds to the complexity of G deductions, but nonmonotonicity seems unavoidable when trying to compute discourse structure. While we sometimes have sufficient information in the message itself to deduce what rhetorical function the speaker intended to have this new information serve, in many cases we do not and we must make a best guess as to what the discourse relation is. Recall, for instance, the "push-fall" example (6) above on which it seems natural to conclude that the information conveyed by the second sentence serves as an explanation for what happened in the first. But now consider (14):

(14)   John fell. Fred pushed him. Unable to stop himself, John slid off the edge of the cliff to his death.

In (14) we do not infer that the proposition that Fred pushed John explains why John fell. Rather, we infer that the second sentence introduces an event that is subsequent to the first. The inference to Explanation in (6) is here defeated by further information—a trademark of nonmonotinicity.

To build SDRSs, the glue logic G must exploit a variety of information sources-lexical information and information about the structure of an SDRS principally, but it may also make use of nonlinguistic information or world knowledge. As G should be a computationally tractable logic, the language of G, L_G, should have a tractable semantics. To date it has been sufficient to make the language L_G a quantifier free fragment of a first order language augmented by a weak conditional operator $>$, which formalizes generic or defeasible rules of interpretation ($A > B$ means "if A then normally B"). This language has the following syntax with formulas defined recursively from predicates Ψ and constants α.

- α ::= Speech Act Discourse Referents of SDRT

- 1-place predicates Ψ^1 ::= $\{\phi : \phi \text{ is an SDRS condition }\} \cup \{Event\text{-}proposition, Stative\text{-}proposition\}$

- 2-place predicates Ψ^2; = $\{\langle.,.\rangle\} \cup \{ D\text{: }D \text{ is a discourse relation symbol }\}$

- $\Phi ::= \Psi^n(\alpha_1, \cdots, \alpha_n) | \neg\Phi | \Phi_1 > \Phi_2 | \Phi \wedge \Phi_2$

Predicates like *Event-proposition* and *State-proposition* are needed to capture aspectual information in SDRSs for the purpose of calculating temporal structure. But these are in effect SDRS conditions—they are just the typing conditions for certain types of discourse referents. We define the semantics of L_G relative to a world $w \in W$ and a selection function $\star : Pow(W)XW \rightarrow Pow(W)$, on which we can place certain constraints to get a particular nonmonotonic logic.[13] The modal semantics of L_G is itself rather unexceptional:

Semantics for L_G

- The usual rules for the truth of atomic and Boolean combinations of formulas apply.

- $[\![\phi > \psi]\!]_w = 1$ *iff* $\star (\| \phi \|, w) \subseteq \| \psi \|$

The monotonic notion of validity in G is easily axiomatized The definition of the notion of nonmonotonic consequence relation $\mathrel{|\!\sim}$ is more involved. Essentially, the idea is to turn $>$ into $\rightarrow$ whenever this is consistent. There are several ways of working out this intuition technically—the most elegant being that found in Morreau (1995). $\mathrel{|\!\sim}$ gives rise to a proof theoretic equivalent.[14]

To turn G into a logic for building SDRSs, we need to add axioms for inferring discourse relations. These axioms typically exploit $>$ and the presence of certain SDRS conditions in the given SDRS (representing the discourse context) and in the SDRS representing the new information to be integrated into the context; details can be found in a number of places (e.g., Lascarides and Asher 1993a, Asher et al. 1995). But roughly they are of the following form:

$$(< \pi_1, \pi_2 > \wedge\phi(\pi_1) \wedge \psi(\pi_2)) > D(\pi_1, \pi_2)$$

which is intended to say that if π_1 and π_2 are to be related to each other in a discourse context, and conditions ϕ obtain in the SDRS characterizing π_1 and conditions ψ obtain in the SDRS characterizing π_2, then normally discourse relation D holds between π_1 and π_2.[15]

---

[13] For details, see Asher and Morreau (1991) or Morreau (1995).

[14] Again for details see Asher (1995) or Morreau (1995).

[15] I have changed and simplified the syntax and semantics of L_G somewhat from that suggested but not explicitly formalized, say, in Lascarides and Asher (1993a) and other works.

What we would like to say is that axioms allow us to conclude that certain discourse relations hold by default, when the consequences of their holding are consistent with what compositional and lexical semantics yields. But one cannot do that without rendering $\vdash_G$ undecidable because this would require the testing of the consistency of arbitrary first order conditions. Instead G is allowed access to certain information about conditions–including some of the consequences of discourse structure determined in the SDRT_0 background theory. This information is encoded in G as *coherence constraints*. These mimic the information encoded in the background theory of SDRT to a limited extent; for instance, we have:

- $Narration(\pi_1, \pi_2) \to [e_1 \prec e_2](\pi_2)$
- $[e_1 \prec e_2](\pi_2) \to \neg[O_t(e_2, e_1)](\pi_2)$

In words the first axiom says that if Narration holds then a certain condition must be true of π_2—namely that the condition $e_1 \prec e_2$ holds of π_2 or more particularly is a condition in the SDRS that π_2 characterizes. The second axiom is an example of how to encode information about the incompatibilities between various DRS conditions in L_G.

For a glue logic capable of generating appropriate SDRSs to analyze temporal anaphora and lexical disambiguation, we are able keep G at a manageable level of complexity, as was shown in Lascarides and Asher (1993a).

Fact 3 $\vdash_G$ *in SDRT is decidable.*

As the gloss on the axioms for discourse relations suggests, we work within a very particular set of models for G in using G to build SDRSs. Each world of these models is an SDRS or discourse structure, and the satisfaction of atomic L_G formulae in such models is just the presence of the appropriate condition in the SDRS. For instance:

- For individual constants π, $[\![\pi]\!] = \pi$.
- If Ψ is a discourse relation symbol, then $[\![\Psi]\!]_w = \{\langle\pi, \pi'\rangle : \Psi(\pi, \pi')$ *is a condition in w*$\}$.
- If Ψ is not a discourse relation symbol and not of the form , then: $[\![\Psi]\!]_w = \{\pi' : \exists K' \Psi$ is a condition or a generalization of a condition in K' and $\pi' : K'$ is a condition in $w\}$
- $[\![<\cdot, \cdot>]\!]_w = \{< \pi_1, \pi_2 >:$ For some discourse relation symbol D, $D(\pi_1, \pi_2)$ is a condition in $w\}$

We can't capture these particular models with the expressive power of L_G. But we can codify this connection between G and SDRT by means of a function μ that "transfers" information from SDRSs to formulas of L_G. This function isn't part of L_G; it's the link between G and the SDRSs G is supposed to reason about. It's the link between SDRT's logic of information packaging and its logic of information content.

Now we can express the notion of an $SDRT_0$ transition defined relative to some unambiguous sentence S. I define the T_0 transition relation for $SDRT_0$, which relates K, K' and a new SDRS $K^\dagger$. Intuitively, $K^\dagger$ is an SDRS where the old and new information have been merged together. More specifically, $K^\dagger$ includes (a) the old information K, (b) the new information K' derived from S, and (c) an attachment of K' with a rhetorical relation to an available attachment point in K. The relation $T_0(K, K', K^\dagger)$ is constrained so that it can hold only if part (c) of $K^\dagger$ is computed via G. More formally, let $Avl(K)$ be the set of available attachment sites in K, and let K_π stand for the SDRS ϕ such that $\pi : \phi$ occurs in K. Further, let $Pred_\pi$ be the label of the SDRS constituent in which π is declared or (equivalently) in which a condition of the form $\pi : K$ occurs, let K_π stand for the SDRS constituent labelled by π, and finally let $\alpha[\beta/\gamma]$ be the result of replacing γ in α with β. Then the T_0 predicate for SDRT is defined as:

- **The Update Relation**
  $T_0(K, K', K^\dagger)$ *iff* $\exists \pi \in Avl(K)$ such that:

  1. $(\mu(K_\pi), \mu(K')) \hspace{-2pt}\mid\hspace{-4pt}\sim (R(\pi, \pi') \wedge \varphi)$; and

  2. $K^\dagger = K[K^+/K_{Pred(K_\pi)}]$, where:

  3. $K^+ = Update_{drt}(K_{Pred_\pi}, \langle\{\pi'\}\{\pi' : K'(\varphi), R(\pi, \pi')\}\rangle)$, where $K'(\varphi) = K'$ together with those conditions specified in φ, where φ is that information needed to satisfy the coherence constraints on R.

The resulting SDRS for the discourse incorporates the new information and a rhetorical relation R that is computed on the basis of the axioms of G together with information about the semantic content of the old and new information (i.e., $\mu(K_\pi)$ and $\mu(K')$) where the SDRS K_π is the one characterized by π. The background SDRT theory ensures that the appropriate semantic consequences of the discourse links are included in the content of the discourse. Clearly, if the set of theorems in G is decidable or even r.e., then we can eliminate the reference to the oracle O in theorem 2 and show that $SDRT_0$ transitions are r.e. *tout court*. This constitutes a decided improvement over theorem 2.

6 Reasons to be dissatisfied with $SDRT_0$

$SDRT_0$ allows us to accomplish a great deal with a very little extra bit of information packaging— a vanishingly small amount in fact. But the theory is unsatisfying because of its extensional character.

First, the extensional character of the state transitions defined in $SDRT_0$ won't do to handle anaphoric references to abstract entities. We have imposed few constraints on the interpretation of the speech acts, and importantly none that links the speech act with the information content of what was said in the speech act. Because of this and because we have not introduced propositions as objects whose content and behavior is linked to the SDRSs that describe them, the interpretation of propositional anaphoric constructions such as

(15)   John got an A on his test, but Sam doesn't believe it.

is hopeless. In principle, we could identify the discourse referent introduced by *it* in the second clause with the speech act discourse referent introduced by the first clause, but there would be nothing in the semantics that determined that what Sam didn't believe was that John got an A on his test. Further, examples like (12) can't be analyzed within the theory.

A related difficulty with $SDRT_0$ is that we cannot really capture the meanings of conditions of the form $R(\pi_1, \pi_2)$, when the semantics of the discourse relation R involves an appeal to the content of what was said in π_1 or π_2. We can specify the first order consequences of such conditions by means of axioms—e.g., the spatial and temporal effects of discourse relations. On the other hand, an SDRS condition like $Elaboration(\pi_1, \pi_2)$ should entail that the content or proposition labelled by π_2 entails that labelled by π_2. A condition like $Explanation(\pi_1, \pi_2)$ also involves a relation between the propositions labelled by π_1 and π_2. But these relations cannot be expressed in $SDRT_0$. It is difficult to see exactly how these relations can be treated in a purely extensional first order framework for familiar reasons.

A third reason for abandoning the extensional framework of $SDRT_0$ for a different theory comes with dialogue. In dialogue, a speech act may have a rhetorical or discourse function that it simply cannot have in monologue or that is very rare in monologue. SDRT represents such different functions as different discourse relations; but we can also think of these as different types of speech acts. In dialogue, there are a host of discourse relations in dialogue that cannot commit one— as one is forced to do in $SDRT_0$—to the truth of what the participants said. For example consider a correction like the following:

(16)   a.   A: John distributed the copies.

      b.   B: No, it was Sue who distributed the copies.

Corrections are a typical type of speech act (or discourse relation) in dialogue. In corrections, it is clear that the content of the dialogue is not the relational composition of the model assignment pairs verifying each of A's and B's assertions; for that would net us an empty truth conditional content. Rather there is a complex relation between the contents of what is said in the two speech acts; the second speech act is an attempt to correct what the second speaker sees as deficiencies in the content of the first speech act.[16]

We can just represent such disagreements as given by (16) in $SDRT_0$, because the interpretations of conditions is relativized to a sequence of speech acts. If one speech act asserts p and the another not p, $SDRT_0$ represents this disagreement without inconsistency. But it also does not represent these speech acts as *disagreeing* with each other; their contents are simply incomparable. This isn't right either. To make sense of corrections and other such speech acts in dialogue, we have to be able to talk about the contents of the two speech acts. And that we cannot really do in $SDRT_0$.

One further complication is that these different contents may involve anaphoric links such as (19):

(19)   a.   A: John shot a man during the robbery

      b.   B: No he didn't shoot him.

In order to make sense of the anaphoric connections, the contents associated with speech acts that are linked by discourse relations must have linked interpretations.

---

[16]Mark Danburg Wyld has made a careful study of these nonveridical or "divergent" discourse relations and speech acts for his dissertation (Danburg-Wyld forthcoming). Another kind of divergent discourse relation is *Counterevidence* of which (17) is an example.

(17)   a.   A: Smith shot the guard at the bank.

      b.   B: He has witnesses that say he was out of town at the time of the robbery.

Another type of speech act that Danburg-Wyld isolates is one in which the speaker denies a discourse relation implicated to hold between two propositions by some other speaker. Here is an example:

(18)   a.   A: John went to jail. He embezzled pension funds.

      b.   B: That's not why he went to jail. He was convicted of tax fraud.

We have uncovered three reasons for going beyond an extensional theory of discourse interpretation: a theory of abstract entity anaphora, an account of the semantics of discourse relations, and a treatment of nonextensional discourse relations in dialogue. The theory in the next section addresses these concerns.

7 An Intensional Theory of Discourse Interpretation: SDRT$_1$

To handle propositional anaphora and VP ellipsis, we must change the signature of SDRT$_0$ to include conditions that represent the identification of a discourse referent with some DR-theoretic structure that represents a proposition or a property. A first try would be to introduce conditions of the form $z = K$, where K is a DRS or a lambda abstracted DRS (representing a property or verb phrase denotation). But this isn't quite right, because K is not a singular term in the syntax of SDRT$_0$. We need a theory internal representative of K, $\sharp K$ for each DRS. $\sharp K$ is a singular term and denotes an object in a model of SDRT$_1$. So we extend the signature of SDRT$_0$ to include such singular terms. This marks a departure from standard DRT, in which there are no singular terms at all with a constant interpretation. We also suppose discourse referents ranging over the objects denoting by singular terms. Finally, to handle dialogue or multilogue, we will replace conditions of the form $\pi : K$ with $\pi : (x, \sharp K)$, where x is a discourse referent representing the speaker in the dialogue whose speech act π is.

As the purpose of introducing singular terms of the form $\sharp K$ was to have a way of referring to contents in SDRT, so this must be reflected in the interpretation of these terms. The idea that first comes to mind (examined in some detail for DRT in Frank (1997) and in recent unpublished work by Kamp) is to exploit the correctness definition and to assign these terms sets of world embedding function pairs, having fixed a model. This amounts to making *dynamic propositions* entities in the models. Because these terms may share discourse referents with the context in which they occur and these shared discourse referents must be assigned the same value, the interpretation of such terms will be sensitive to the discourse context in which they are to be interpreted. So we will assign to $\sharp K$ a discourse context—i.e., a set of world assignment pairs relative to a given model and assignment (elements of a discourse context). The set of world assignment pairs denoted by $\sharp K$ is determined relative in distributive fashion, in keeping with the tradition of dynamic semantics.[17] Since assignments are always finite

---

[17]The interpretation of conditions involving $\sharp K$ will also change so as to exploit the

and we can assign an order to the discourse referents in their domain, we can think of the assignments in these pairs as finite sequences of objects in the domain of the model. Nevertheless, this requires some minimal amount of set theory in our model and thus marks a complication in the logical foundations of discourse interpretation. On the other hand, we no longer need to relativize the interpretation of conditions to labels for SDRS constituents to interpret conditions of the form $\pi : \natural K$, since we will have an interpretation for $\natural K$ as a whole. So our satisfaction definition below will have one fewer parameter than the previous one. Further, we can return to treating conditions as "tests" using the interpretation function V.

To carry out these changes, we will pass from an extensional semantics to an intensional one. Our models will contain not only a domain of individuals and interpretations of the primitive nonlogical relation symbols but also a set of worlds. The interpretation of primitive relation symbols will be a function from worlds to extensions. The interpretation of $\natural K$, however, will also necessitate the presence of certain sets in the models—sets of pairs of worlds and and assignment functions.

A potential source of problems is that we want the interpretation of $\natural K$ to be sensitive to the assignments made to discourse referents in the SDRS in which $\natural K$ occurs as a term, as well as to the assignments made to discourse referents in other terms $\natural K'$ which are discourse related to $\natural K$ and which hence may support anaphoric connections to discourse referents in $\natural K$ as in 19. This generates two problems. The first is that the interpretation of $\natural K$, because it relies on the state transition $P_M(K)$, must be defined relative to an input world assignment pair where the assignment does not assign objects to all the discourse referents that occur as terms in conditions of K *but* are not declared in K either (since they are declared in the universes of other K' where $\natural K$ and $\natural K'$ are discourse related). So we must modify the interpretation of $\natural K$ and K to deal with such "improper" DRSs or SDRSs. The interpretation of $\natural K$ will in fact be an interpretation of $\natural K$ paired with the background assignment.

The second problem is that, on pain of violating well-foundedness, we don't want the assignments that are part of the interpretation of $\natural K$ to include the assignment made to $\natural K$. To this end, I define for any discourse referent x occuring as a term in the conditions of K but not in the universe of K:

- $f_K^- = \{< x, f(x) >: x \text{occurs in K but not in} U_K\}$

---

meaning of $\natural K$, as will the conditions involving speech act discourse referents if these are definable in terms of the contents associated with the speech acts. I won't go into these definitions here, but for an attempt in this area see Asher 1993. Many but not all the discourse relations can be given definitions in terms of the contents associated with their arguments.

Because every SDRS is well-founded, by exploiting f^- rather than f, we will ensure that our semantics is well-founded too:

Definition 3:

- $(w, f)P_M(U, O)(w', g)$ *iff* $w = w' \wedge f \subseteq g$
  $$\wedge \operatorname{dom}(g) = \operatorname{dom}(f) \cup U.$$

- $(w, f) \in V_M(R(x_1, \cdots, x_n))$ *iff* $R_{w,M}(f(x_1), \cdots, f(x_n))$

- $(w, f) \in V_M(x = \sharp K)$ *iff* $f(x) := V_M(\sharp K, f)$

- $(w, f)P_M(K^\cap \gamma)(w', g)$ *iff* $(w, f)P_M(K)(w', g) \wedge (w', g) \subset V_M(\gamma)$

- $(w, f) \in V_M(\neg K)$ *iff* $\neg \exists h \, (w, f)P_M(K)(w, h)$

- $(w, f) \in V_M(K \Rightarrow K')$ *iff* $\forall k \, ((w, f)P_M(K)(w, k) \rightarrow$
  $$\exists h \, (w, k)P_M(K')(w, h))$$

- $(w, f) \in V_M(K \vee K')$ *iff* $\exists g \, (w, f)P_M(K)(w, g) \vee$
  $$\exists h(w, f)P_M(K')(w, h)$$

- $(w, f) \in V_M(\pi : x, \sharp K)$ *iff* $\langle f(\pi), f(x), V_M(\sharp K) \rangle \in Say_{w,M}$

- $V_M(\sharp K, f) = \{(w', h) : \exists g(f_K^- \subseteq g \wedge (w', g)P_M(K)(w', h) \wedge \operatorname{dom}(g) = \operatorname{dom}(f_K^-) \cup \{x : x \text{ occurs in a condition of } K \text{ but not in } U_K\})\}$

As the satisfaction definition stands, the interpretation of terms of the form $\sharp K$ is still not sufficiently constrained. As I mentioned above, anaphoric relations may obtain between components of an SDRS that are discourse related to each other. This means that two complex singular terms $\sharp K$ and $\sharp K'$ may share discourse referents. To have the anaphoric links make sense, we must give the same assignments to the shared discourse referents. The constraint (on models) that the contents associated with any two discourse linked speech acts be *equipollent* accomplishes this. What the two subclauses in the definition of equipollence do is to set up a bisimulation between the contents of the two related speech act with respect to the assignments to the common discourse referents that appear in their respective characterizing SDRSs.[18]

Definition of Equipollence: $\natural K_1$ and $\natural K_2$ are equipollent relative to M, written $\natural K_1 \sim_M \natural K_2$, iff:

- $\forall w', h(w', h) \in V_M(\natural K_1) \rightarrow \exists g \exists w''((w'', g) \in V_M(\natural K_2)) \wedge$
  $\forall x \in Dom(h) \cap Dom(g)h(x) = g(x)))$

- $\forall w', h((w', h) \in V_M(\natural K_2) \rightarrow \exists g \exists w''((w'', g) \in V_M(\natural K_1) \wedge$
  $\forall x \in Dom(h) \cap Dom(g)h(x) = g(x))$

Equipollence Constraint

- $\forall w', g((w', g) \in V_M(\pi_1 : x_1, \natural K_1) \cap V_M(\pi_2 : x_2, \natural K_2)$
  $\wedge R_{w', M}(g(\pi_1), g(\pi_2))) \rightarrow \natural K_1 \sim_M \natural K_2$

$\sim_M$ is an equivalence relation. So we can show that if $\natural K \sim_M \natural K'$ and $\natural K' \sim_M \natural K''$ that we can show that there is a bisimulation on the values of $\natural K$ and $\natural K''$ with respect to the discourse referents common to $\natural K$, $\natural K'$ and $\natural K''$, which is what we really need to assure semantic coreference in the anaphoric equations.

The syntax and semantics of SDRT$_1$ allows us to identify pronouns referring to propositions, facts or properties with the values of $\natural K$ or with the value of a lambda abstracted SDRS.[19] A discourse context is on this view also a dynamic proposition, consisting of a relational structure of dynamic propositions. In SDRT$_0$ we had no propositional constituents in the discourse context— only speech acts. In SDRT$_1$ dynamic propositions are full citizens.

Every SDRS of SDRT$_1$ has a translation into a higher order intensional logic as, say, developed in Gallin (1975). We have an additional type in our logic besides that of worlds, entities and truth values: the type of an assignment. To ensure that this type is well-founded, we will have to build it up inductively in the following fashion using functional types and Currying cartesian products. Let π be the type of discourse referents or stores in our type theory (as in Muskens 1996):

- $g_0 : \pi \rightarrow e$

---

[18]You might think that equipollence is too strong and all that is needed is some conditional dependence–like that given by the first clause of equipollence, but this would allow us to have some assi gnments in the meaning of $\natural K_2$ that did whatever they wished to the drefs or variables declared in K_1. To rule out this unwanted possibility, the second clause needs to be added.

[19]To do the latter, it would appear attractive to have a third store of assignments for VPs that we can pick up and use in new contexts, but I won't work out the details here.

- $g_{n+1} : \pi \rightarrow (w \rightarrow (\bigcup_{i \sqsubset n} g_n \rightarrow t)$

- $g = \bigcup_{n \in \omega} g_n$

Let's call the higher order logic with this set of types TY_3*. Besides the basic DRT translation into first order logic we now have the translation of dynamic propositional variables as variables of type $w \rightarrow (g \rightarrow t)$, while the translation of $\natural K$: $tr(\natural K) = {}^\wedge tr(K)$. We then exploit this translation in the following characterization lemma, together with the "lifting" of P defined above to the empty discourse context: $\sigma_0^1 = \{(M, w, 0) : M \text{ is an intensional model}, w \in W_M\}$.

Characterization Lemma for SDRT$_1$:

- For every higher order intensional formula χ with a set of free variables U, there is an SDRS of SDRT$_1$ (U, C) such that $\mathcal{P}^1(U, C)[\sigma_0^1] = \{(M, f) : Dom(f) = U \text{ and } M \models \chi[f]\}$

- Every SDRS of SDRT$_1(U, C)$ has a characteristic formula χ of higher order intensional logic where U is the set of free variables in χ and $\mathcal{P}^1(U, C)[\sigma_0^1] = \{(M, f) : Dom(f) = U \text{ and } M \models \chi[f]\}$

As in Muskens's (1996) system, we have lost in SDRT$_1$ the SDRT$_0$ equivalence between a logic with purely objectual variables and the target dynamic theory. For we now have to introduce discourse referents as a primitive type into the intensional logic to get the equivalence with SDRT.

The work of Cocchiarella (1989) has shown how to build axiomatizations of theories with stratified comprehension, and our "stratification" of propositional types in TY_3* can borrow these ideas to get an axiomatization, which together with the techniques of Fernando leads to a notion of bisimulation over transitions defined by SDRS's intensionally construed and simultaneously over the state transitions defined by their model theoretic interpretations in the same general vein as theorems 1 and 2. All the same definitions pertinent to the definition of $\leftrightarrow^1$, the notion of transition equivalence for SDRT$_1$, and for $\Phi_\perp^1$ and accessible states carry over here. It is a simple exercise to modify the definition of the update relation in SDRT$_0$ to fit SDRT$_1$. The update relation exploits only the representational structures and the glue logic G. The syntax of the representations in SDRT$_1$ has changed from SDRT$_0$ but only in ways that are inessential to the definition of the transition relation.

A natural query is to ask whether one could eliminate the complexity of the well-founded type of embedding functions in the models. In fact it appears that one can and thereby get a much simpler intensional version of SDRT, which I'll call SDRT$_2$.[20]

The idea here is to make the assignments relevant to interpreting the intensional contexts part of the overall context and not part of the semantic value of the intensional terms. Again, we want the interpretation of $\natural K$ to be sensitive to the assignments made to discourse referents in the SDRS in which $\natural K$ occurs as a term, as well as to the assignments made to discourse referents in other terms $\natural K'$ which are discourse related to $\natural K$ and which hence may support anaphoric connections to discourse referents in $\natural K$ as in 19. The interpretation of $\natural K$ relies on the state transition $P_M(K)$; but it must be defined relative to an input world assignment pair where the assignment assigns objects to all the discourse referents that occur as terms in conditions of K *but* are not declared in K either (since they are declared in the universes of other K' where $\natural K$ and $\natural K'$ are discourse related). Further of course, these input world assignment pairs must respect the assignments made to discourse referents by the "outside" assignment. To do this, I will need to keep track of not only the outside assignment but also of previous assignments in previous intensional contexts in order to interpret conditions of the form $\natural K$. And to do this I will redefine the notion of a context to have two assignments—one for the extensional discourse referents and one for the "intensional" discourse referents (the discourse referents that occur within terms of the form $\natural K$). The conditions of the form $\pi : x, K$ will not then function simply as tests as in definition 3 but will actually change the assignments to the intensional contexts, in a way similar to definition 2.

Definition 4:

- $(w,f,k)P_M(U,O)(w',g,k')$ $\quad$ *iff* $\quad$ $w = w' \wedge k = k' \wedge f_w \subseteq g_w$ $\wedge \operatorname{dom}(g,w) = \operatorname{dom}(f,w) \cup U$

- $(w,f,k)P_M(R(x_1,\cdots,x_n))(w',f',k')$ *iff* $w = w' \wedge f = f' \wedge$ $k = k' \wedge R_{w,M}(f_w(x_1),\cdots,f_w(x_n))$

- $(w,f,k)P_M(x = \natural K)(w',f',k')$ *iff* $w = w' \wedge f = f' \wedge$ $k = k' \wedge f(x) := V_M(\natural K,k,w)$

- $(w,f,k)P_M(K^\cap\gamma)(w',f',k')$ *iff* $\exists w'',f'',k''$ $((w,f,k)P_M(K)(w'',f'',k'') \wedge (w'',f'',k'')P_M(\gamma)(w',f',k'))$

---

[20]Something like this approach has been suggested for the treatment of conditionals by Matthew Stone (1997) and Robert van Rooy (1997).

- $(w, f, k)P_M(\neg K)(w', f', k')$ *iff* $w = w' \wedge f = f' \wedge k = k'$
$\neg \exists h \, (w, f, k)P_M(K)(w, h, k)$

- $(w, f, k)P_M(K \Rightarrow K')(w', f', k')$ *iff* $w = w' \wedge f = f' \wedge k = k'$
$\forall g \, ((w, f, k)P_M(K)(w, g, k) \rightarrow \exists \, h \, (w, g, k)P_M(K')(w, h, k))$

- $(w, f, k)P_M(K \vee K')(w', f', k')$ *iff* $w = w' \wedge f = f' \wedge k = k'$
$\exists \, g \, (w, f, k)P_M(K)(w, g, k) \vee \exists h(w, f, k)P_M(K')(w, h, k)$

- $(w, f, k)P_M(\pi : x, \sharp K)(w', f', k')$ *iff* $w = w' \wedge f = f' \wedge$
$\langle f(\pi), f(x), V_M(\sharp K, f, k, w) \rangle \in Say_{w,M}$
$\wedge \, \forall w' \in V_M(\sharp K, f, k, w)(w', f \cup k, f \cup k)P_K(w', f \cup k, k')$

- $V_M(\sharp K, f, g, w) \;=\; \{w' : \exists k \, (w', f \cup g, f \cup g)P_M(K)(w', k, k))\}$

This way of proceeding is in some ways much closer to the original formulation of SDRT$_0$. We don't need a notion of equipollence because the intensional assignments are updated each time a conditon of the form $\pi : K$ is interpreted and so variables within the constituent K that are declared in other portions are interpreted appropriately. The propositions that we refer to in SDRT$_2$ aren't *dynamic* but rather static; all dynamic elements are located in the assignment functions, which are not part of semantic values in the interpretation of SDRT$_2$.

The real payoff of SDRT$_2$ is a correspondence with a standard intensional logic with only objectual variables. Every SDRS of SDRT$_2$ has a translation into higher order intensional logic as, say, developed in Gallin (1975). Besides the basic DRT translation into first order logic we now have the translation of $\sharp K$: $tr(\sharp K) = {}^\wedge tr(K)$. We then exploit this translation in the following characterization lemma, together with the "lifting" of P defined above to the empty discourse context: $\sigma_0^2 = \{(M, w, 0, 0) : M$ *is an intensional model*, $w \in W_M\}$.

Characterization Lemma for SDRT$_2$:

- For every higher order intensional formula χ with a set of free variables U, there is an SDRS of SDRT$_2(U, C)$ such that $[\mathcal{P}^1(U, C)](\sigma_0^2) = \{(M, w, f, g) : M \models \chi[f \cup g]$, where $Dom(f) = U \wedge Dom(g) = U*$, and where $U*$ is the set of free variables x_i such that for some formula ϕ, ${}^\wedge \phi$ occurs in χ and x_i occurs in $\phi\}$

- Every SDRS of SDRT$_2(U, C)$ has a characteristic formula χ of higher order intensional logic where U is the set of free variables in χ, $U*$

is the set of free variables x_i such that for some formula ϕ, $^{\wedge}\phi$ occurs in χ, x_i occurs in ϕ and $\mathcal{P}^1(U, C)_{\sigma_0^2} = \{(M, f, g) : Dom(f) = U, Dom(g) = U* \text{ and } M \models \chi[f \cup g]\}$

This characterization lemma, in conjunction with the use of generalized Henkin models for higher order modal logic, leads to a notion of bisimulation over transitions defined by SDRS's intensionally construed and simultaneously over the state transitions defined by their model theoretic interpretations in the same general vein as theorems 1 and 2. Once again the definitions pertinent to the definition of $\leftrightarrow^2$, the notion of transition equivalence for SDRT$_2$, and for $\Phi_\perp^1$ and accessible states carry over from our earlier theories or are easily modified as is the definition of the update relation.

Given that $\vdash_G$ (including the coherence constraints) is decidable, we have the following:

Theorem 7.1 $\phi\leftrightarrow^1\psi$ iff $\sigma_\phi^2 = \sigma_\psi^2$; and further, $\leftrightarrow^2$ and $\Phi_\perp^2$ are r.e.

8   More Information Packaging: SDRT$_3$

SDRT$_1$ and SDRT$_2$, like SDRT$_0$, have a small amount of information packaging. The advantage of this is that it is easy to extend the correspondence between model-theoretic and syntactically or representationally defined state transitions that is found in dynamic semantics. We can in fact do alot with these theories. We can represent various forms of ellipsis and anaphora having to do with reference to abstract objects; we can study dialogue; and we can analyze the other phenomena mentioned in connection with SDRT$_0$.

But a proper theory of anaphora and ellipsis still remains outside the purview of SDRT$_{0,1,2}$. These theories of discourse interpretation do not reflect, for instance, important pragmatic constraints on anaphora. A widely noticed fact is that discourse structure governs which antecedents to pronouns and which temporal discourse referents are permissible antecedents. It is very difficult to refer with a pronoun to an object mentioned in some proposition that is not discourse related to the proposition expressed by the sentence containing the pronoun or on the "right frontier" of the discourse structure. But as a study of the satisfaction definitions 2 and 3 shows, a pronoun can refer to the value of any discourse referent introduced in the discourse, no matter how far back.

A plausible-sounding solution to our first difficulty is to make use of some sort of "down-date" of the assignment functions. But already this makes much more complex the problem of getting a correlation between representationally defined and model-theoretically defined transitions. Second, while pronominal anaphora and temporal anaphora obey an "availability" constraint imposed by discourse structure, studies of texts have shown that definite descriptions may pick up antecedents that do not meet this constraint (Asher 1993). So it looks as if downdating the assignments (removing some variables from their domains) will not work for definite descriptions. We need a richer notion of information packing within which to represent the different status of discourse referents. In particular, this richer notion must reflect aspects of the structure of an SDRS.

An additional advantage of adding structure is that we can internalize within the notion of a discourse context the G governed notion of CCP and make sense of the information transfer function as being part of our notion of a discourse context. If we want to have a clear conception of the state transition engendered by new information that comes from a variety of sources after the processing of the verbal message, then it behooves us to incorporate the additional information needed to compute the transition into our representation of the discourse context. So far we have not constructed a completely model theoretic notion of CCP for SDRT; we have merely used the relational composition of DRT and exploited the link between the model theory and the representational formulations of dynamic discourse interpretation. But this doesn't really give us a theory in which the model-theoretic notions of context are incrementally built up as new information is added. By adding to information packaging information about the structure of the SDRS along the lines of Asher (1996), we will be able to do so.

A related issue is that even in $SDRT_{1,2}$ we have a lousy semantics for attitudinal constructions, a semantics which does not distinguish between logically equivalent beliefs. One solution to the problem of logical equivalence of beliefs is to exploit the structure of the object of belief in a context sensitive way (see Asher 1986, Kamp 1990, or Perry and Crimmins, for example. If these objects of belief are just the sort of propositions that are involved in our representation of discourse (and why not in view of the intimate link between saying and believing), then this also suggests that our view of information structure should at least include the logical structure of the SDRS and maybe even something about the concepts that figure within the conditions of the SDRS. Once again the whole SDRS seems to be relevant to information packaging—and not only for pragmatic processing issues but for the semantics of certain constructions as well.

A sufficient enrichment of information packaging to address the concerns raised in the previous paragraphs requires several modifications to our conception of a discourse context. First we will want to represent explicitly the structure of an SDRS— we can do this by encoding this structure on the set of speech act discourse referents. Second, we need to say something about the structure of the contents—the SDRSs—associated with each speech act, at least enough to be able to use the "glue" logic to compute the appropiate discourse relation for attachment, given a DRS—i.e. a compositional semantic representation of the new information—together with lexical information and pragmatic principles. We now make these changes more precise on the conceptually simpler version $SDRT_2$.

To this end let a discourse context (relative to a chosen model) be a tuple consisting of a world, as before, and, in addition, a triple consisting of the set of speech act discourse referents X, a strict partial ordering R on X, a function μ from X into formulas of the "glue" logic encoding information about the conditions associated with that element of X, and an element $x^0 \in X$ designating the current speech act. μ is our information transfer function; if x is not a speech act discourse referent $\mu(x)$ is the conjunction of all the conditions in which x figures as an argument or if x is a speech act discourse referent then $\mu(x)$ is the conjunction of all the conditions in the DRS representing the content of the speech act x; in other words μ keeps a record of the actual conditions used in the discourse and files them with each associated discourse referent.

- (w, X, R, μ, x^0)

x^0 and R together tell us which speech act discourse referents may act as suitable attachment points among the set X of all speech acts in the antecedent context. Most researchers in discourse theory have made the "right frontier" of the discourse structure the area in which possible attachments of new information may be made (though for some complicating details see Asher 1993), a constraint easily representable with our new notion of discourse context.

We then define that two such contexts τ and σ stand in the CCP relation inductively similarly to the way we defined CCP transitions between simpler contexts in definitions 2 -4. As in definition 4, the world and assignment components of the discourse contexts do all the work in the definition of satisfaction of conditions. The principal change is the way information packaging gets treated— viz. in the first clause in the recursion. The input and output discourse contexts' information packaging is now much more finely structured, and so this requires several clauses to make sure that each part of the structure is modified in the appropriate fashion. Further, as the inference of the appropriate discourse connection between the new information and the context makes use of conditions in SDRSs, we must define the transition P in a somewhat different way; this recursive definition defines P for an SDRS with an arbitrary set of conditions. As in definitions 2, 3 and 4, we will suppose that $M_\tau = M_\sigma$ in any CCP transition defined by an SDRT$_1$ representation, and we will make the valuation V of conditions sensitive to contexts, as in definition 3. σ_0^3 represents the empty discourse context in SDRT$_3$, and DR is the set of discourse relations.

Definition 5:

- $(\sigma_1, f, k)P_M(U, Con)(\sigma_2, g, h)$ *iff* :

    1. $w_{\sigma_1} = w_{\sigma_2} = w$
    2. $f_w \subseteq g_w \wedge dom(g, w) = dom(f, w) \cup U$
    3. $\forall \gamma \in Con\ (\sigma_2, g, k)P_M(\gamma)(\sigma_2, g, h)$
    4. $X_{\sigma_2} = X_{\sigma_1} \cup \{\pi : \pi \in U\}$
    5. $x_{\sigma_2}^0 \in \{\pi : \pi \in U\}$, if $\{\pi : \pi \in U\}$ is non-empty;
       $x_{\sigma_2}^0 = x_{\sigma_1}^0$, otherwise; and $x_\emptyset^0 = \emptyset$.
    6. $\mu_{\sigma_2} = \mu_{\sigma_1} \cup \{(\phi, x_2^0) : \phi \in Con\}$
    7. *either* $(\sigma_1 = \emptyset \wedge R_{\sigma_2} = R_{\sigma_1} = 0)$ *or* $\exists \pi \in X_{\sigma_1} \exists D \in$ DR
       $(R_{\sigma_1}(\pi, x_{\sigma_1}^0) \wedge (\sigma_2, g, h)P_M(D(\pi, x_{\sigma_2}^0))(\sigma_2, g, h)$
       $\wedge(\langle \pi, x_{\sigma_2}^0 \rangle \wedge \mu_{\sigma_1}(\pi) \wedge \mu_{\sigma_2}(x_{\sigma_2}^0)\mathrel{\vert\!\sim}_G D(\pi, x_{\sigma_2}^0) \wedge$
       $R_{\sigma_2} = $ *Transitive Closure*$(R_{\sigma_1} \cup (\pi, x_{\sigma_2}^0))$

- $(\sigma, f, g)P_M(R(x_1, \cdots, x_n)(\sigma, f', g')$ *iff* $\wedge f = f' \wedge g = g'$
  $(R_{w_\sigma, M}(f_{w_\sigma}(x_1), \cdots, f_{w_\sigma}(x_n))$

- $(\sigma, f, g)P_M(x = \sharp K)(\sigma, f', g')$ *iff* $f(x) := V_M(\sharp K, f, \sigma) \wedge$
  $f = f' \wedge g = g'$

- $(\sigma, f, g)P_M(\neg K)(\sigma, f', g')$ *iff* $\neg \exists h \exists \sigma'\ (\sigma, f)P_M(K)(\sigma', h)$
  $\wedge f = f' \wedge g = g'$

- $(\sigma, f, g)P_M(K \Rightarrow K')(\sigma, f', g')$ *iff* $\forall k \forall \sigma'\ ((\sigma, f)P_M(K)(\sigma', k)$
  $\rightarrow \exists h \exists \sigma''\ (\sigma', k)P_M(K')(\sigma'', h)) \wedge f = f' \wedge g = g'$

- $(\sigma, f, g)P_M(K \vee K')(\sigma, f', g')$ *iff* $\exists g \exists \sigma'\ (\sigma, f)P_M(K)(\sigma', g)$
  $\vee \exists h \exists \sigma''(\sigma, f)P_M(K')(\sigma'', h) \wedge f = f' \wedge g = g'$

- $(\sigma, f, g)P_M(\pi : x, \sharp K)(\sigma, f', k)$ *iff* $f = f' \wedge$
  $\langle f(\pi)f(x), V_M(\sharp K, f, g, \sigma)\rangle \in Say_{w_\sigma, M} \wedge$
  $\forall \sigma' \in V_M(\sharp K, f, g, \sigma)(f_w \subseteq g'_w \wedge (\sigma', f \cup g, f \cup g)P_K(\sigma', k, k)$

- $V_M(\sharp K, f, g, \sigma) = \{\sigma' : \exists k(\sigma', f \cup g, f \cup g)P_M(K)(\sigma', k, k))\}$

In definition 5, we make use of the glue logic G by means of which we compute the discourse relation used to attach the new information to an available attachment point. This logic exploits the information transferfunction μ from speech acts to information about those speech acts (expressed in the language L_G). μ is not part of G but rather part of our conception of discourse context.

The question as to whether $SDRT_3$ engenders r.e. transitions between states is an important but delicate one. Because we have complicated the notion of contexts, we no longer have a simple correspondence between a first order or higher order formula and a state transition labelled by a DRS (or SDRS). But our contexts can be broken into an intensional part and a part governed by a formula that captures the relation between the information packaging of the two contexts. The language of the relation governing the information packaging contains names of formulas and functions from formulas to the set of free variables in it, together with the function μ that takes us from a formula to lexical information associated with its non-logical constants. Suppose that the function μ gives information from lexical look-up; then μ is clearly computable.

Given that G is decidable and that μ involves only lexical look-up, the relational expression between the information packaging of the input and output contexts (viz. the relation between input set X of speech act discourse referents, current discourse referent and strict partial order on X and the corresponding output elements) is also *decidable*.

The information packaging language L_1 is L_G together with the relation symbol T. L_1 contains names of L formulas as well as names of variables, and variables for sets of variables. It also contains a symbol representing the strict partial ordering on sets of variables. T encodes the transitions concerning the structure of SDRSs as they are modified by new information. To turn the definition of L_G satisfaction into a definition of L_1 *satisfaction*, we need only add a clause for T. Given that every SDRS has a translation into a formula of higher order intensional logic, we can also use T to characterize the structure of such a formula. Suppose that T encodes that transition between structures relevant to determining the sort of information packaging we have supposed in the contexts defined in $SDRT_3$. Calculating this transition involves G, but since G is decidable and determining the extension of T just involves the manipulation of finite structures, L_1 validity is also decidable.

We now translate an $SDRS_3$ context into a pair of formulas—one a formula of intensional higher order logic as before and the other a formula of L_1 that encodes the structure of that context—including of course a set of speech act discourse referents X.

Definition

(χ, δ) is a *characterizing pair of formulas* iff χ is a formula of higher order intensional logic and δ is a formula of L_1 such that $\forall x \in X_\delta \mu(x)$ is L_1 satisfied by the discourse context $\mathcal{P}(K_\chi)[\sigma_0^3]$, where K_χ is the SDRS that is the translation of χ.

Characterization Lemma for SDRT$_3$:

- For every pair of characterizing formulas (χ, δ) such that χ has free variables U, there is an SDRS (U, C) such that $\mathcal{P}(U, C)[\sigma_0^3] = \{(M, f) : Dom(f) = U \text{ and } M \models \chi[f]\}$ and $(U, C)L_1$ satisfies δ.

- Every SDRS$_3(U, C)$ has a pair of characterizing formulas (χ, T) such that: χ has a set of free variables U and $\mathcal{P}(U, C)[\sigma_0^3] = \{(M, f) : Dom(f) = U \text{ and } M \models \chi[f]\}$ and δ is L_1 satisfied by (U, C)

We use the characterization lemma to define state transitions labelled by SDRSs by keeping the two parts apart. If we can keep L_1 validity decidable (which means keeping G decidable), then we can show that the transitions of SDRT$_3$ also remain r.e. for the same reasons as in fact 3. But how do these state transitions compare to our previously defined state transitions? Because of the information packaging formula, the transitions are much more restrictively defined than in SDRT$_{0,1}$; e.g., two logically equivalent SDRSs may not induce the same state transition in SDRT$_3$.

This sensitivity of the transitions to packaging has advantages I have already mentioned. There are natural language constructions like attitude reports whose semantics is plausibly sensitive not only to truth conditional content but to information packaging as well—thus blurring the distinction with which we set out in the paper. This leads to the thought that the *entire* structure of the SDRS is relevant to information packaging and even occasionally to content. And this suggests one other view of discourse structure—the one originally proposed in Asher (1993).

So far we have been interested in constructing the minimal notions of context needed to attack certain problems and phenomena associated with a theory of the pragmatic-semantic interface. I haven't said anything about the interpretation of certain intensional predicates—predicates of propositions. Candidates for trouble are the predicates for truth, belief and other propositional attitudes. The paradoxes can easily be reintroduced within this framework by predicates that exploit both information packaging and model theoretic content in their semantics. Self-referential paradoxes as of now do not arise, because we have split the information packaging off from the model-theoretic content. The original SDRT of Asher (1993) grew out of a concern with a semantics for abstract objects of the sort needed for an adequate theory of belief reports. This interest in propositions yields a collection of objects in which information packaging and information content are intertwined. Consequently, that theory did not make make the split between information packaging and contents (at least not explicitly) but took the values of $\sharp K$ to be SDRSs themselves. We could call that theory SDRT$_4$. Models for SDRT$_4$ involved techniques familiar to those analyzing the paradoxes—e.g., Frege structure. SDRT$_4$ can also be understood as defining a CCP, but unlike that of our earlier theories it can be shown to be nonaxiomatizable (Asher 1993).

9 Conclusion

In the approach to a variety of systems for discourse interpretation vand problems in discourse interpretation that I have sketched here, I understand the distinction between information packaging and information content to be an essential one. Both content and packaging are needed to understand discourse interpretation, but they should be kept separate if we hope to have a computationally tractable approach to the construction of discourse representations. The distinction between information packaging and content is, however, also a logical distinction that gives us, as the last paragraphs suggest, a slightly different way of thinking about the self-referential paradoxes and about certain problematic constructions in semantics. But both packaging and content are parts of discourse meaning, and this strongly suggests that there are more levels to discourse meaning than just syntax and just model-theoretic semantics.

10 References

Peter Aczel [1988] *Non-well-founded sets.* Center for the Study of Lnaguage and Information. Lecture Notes, no. 14, Stanford California.

Nicholas Asher [1986] Belief in Discourse Representation Theory, *Journal of Philosophical Logic*, **15**, pp.127–189.

Nicholas Asher [1987] A Typology for Attitude Verbs, *Linguistics and Philosophy*, **10**, pp.125–197.

Nicholas Asher [1993] *Reference to Abstract Objects in Discourse*, Kluwer Academic Publishers.

Nicholas Asher [1995] Commonsense Entailment: A Logic for Some Conditionals, in G. Crocco, L. Farinas del Cerro, A. Herzig (eds.), *Conditionals and Artificial Intelligence*, Oxford: Oxford University Press, 1995, pp.103-147.

Nicholas Asher [1996] The Mathematical Foundations of Discourse Interpretation, in P. Dekker, M. Stokhof (eds.), *Proceedings of the Tenth Amsterdam Colloquium in Formal semantics*, ITLI lecture notes series, University of Amsterdam.

Nicholas Asher, Michel Aurnague, Myriam Bras, Pierre Sablayrolles, and Laure Vieu [1995] De l'espace-temps dans l'analyse du discours, *Semiotiques* vol. 9, pp. 11-62.

Nicholas Asher and Tim Fernando [1997] Effective Labelling for Disambiguation, *Proceedings of the Second International Workshop in Computational Linguistics*, Tilburg, the Netherlands.   ·

Nicholas Asher and Hans Kamp [1989] Self-Reference, Attitudes and Paradox: Type-Free Logic and Semantics for the Attitudes, in G. Chierchia, B. Partee and R. Turner, (eds.) *Properties Types and Meanings*, Kluwer Academic Publishers, 1989, pp. 85-159.

Nicholas Asher and Alex Lascarides [1994] Intentions and Information in Discourse, in *Proceedings of the 32nd Annual Meeting of the Association of Computational Linguistics (ACL94)*, pp.35-41, Las Cruces.

Nicholas Asher and Alex Lascarides [1995] Lexical Disambiguation in a Discourse Context, *Journal of Semantics*, **12**, pp.69-108.

Nicholas Asher and Alex Lascarides [in press] Bridging, *Journal of Semantics*,

Nicholas Asher and Michael Morreau [1991] Commonsense Entailment: A Modal, Nonmonotonic Theory of Reasoning, *Proceedings of IJCAI 91*, San Mateo, CA: Morgan Kaufman Press, pp.387-392.

Jon Barwise and John Perry [1983] *Situations and Attitudes*, Cambridge, MA: MIT Press.

David Beaver [1994], An Infinite Number of Monkeys, Technical Report, ILLC, Universiteit van Amsterdam.

David Beaver [1996], Presupposition, in J. van Bentham A. ter Meulen (eds.), *Handbook of Logic and Linguistics*, Amsterdam: Elsevier Publishers, pp. 939-1008.

Johann Bos, J., A-M. Mineur and P. Buitelaar [1995] Bridging as coercive accommodation, technical report number 52, Department of Computational Linguistics, Universität Saarbruücken.

Myriam Bras and Nicholas Asher [1994] Le raisonnement non monotone dans la construction de la structure temporelle de textes en franais," *Reconnaissance des Formes et Intelligence Artificielle* (RFIA), Paris, 1994.

Nino Cocchiarella [1989] Philosohpical Perspectives on Formal Theories of Predication, *Handbook of Philosophical Logic*, eds. D. Gabbay and F. Guenthner, Dordrecht: Reidel, pp.253-326.

Mark Danburg-Wyld [forthcoming] *Speech Act Theory and Discourse Representation Theory*, Ph.D. thesis, University of Texas at Austin.

David Dowty [1979] *Word Meaning and Montague Grammar*, Dordrecht: Reidel.

Tim Fernando [1994] What is a DRS?, *Proceedings of the First International Workshop on Computational Semantics*, Tilburg.

Anette Frank [1997] *Context Dependence in Modal Constructions*, Ph.D. Thesis, University of Stuttgart.

Daniel Gallin [1975]: *Intensional and Higher-Order Modal Logic*, Amsterdam: North Holland Publishing Co.

Irene Heim [1982] *The Semantics of Definite and Indefinite Noun Phrases*, Ph.D. Dissertation, University of Massachusetts, Amherst.

Jerry Hobbs, Mark Stickel, Doug Appelt, Paul Martin [1993] Interpretation as Abduction, *Artificial Intelligence*, **63**, pp.69-142.

Hans Kamp [1981] A Theory of Truth and Semantic Representation, in J. Groenendijk, T. Janssen and M. Stokhof (eds.), *Formal Methods in the Study of Language*, Mathematics Center Tracts, Amsterdam.

Hans Kamp [1990] Propositional Attitudes, in C. Anthony Anderson, Joseph Owens (eds.), *The Role of Content in Logic, Language, and Mind*, CSLI Lecture Notes **20**, University of Chicago Press, pp.27-90.

Hans Kamp and Uwe Reyle [1993] *From Discourse to Logic: Introduction to Modeltheoretic Semantics of Natural Language, Formal Logic and Discourse Representation Theory*, Kluwer Academic Publishers.

David Kaplan [1977] *Demonstratives*, John Locke Lectures, Oxford: Oxford University Press.

Alex Lascarides and Nicholas Asher [1993a] Temporal Interpretation, Discourse Relations and Commonsense Entailment, in *Linguistics and Philosophy*, **16**, pp.437-493.

Alex Lascarides and Nicholas Asher [1993b] Semantics and Pragmatics for the Pluperfect, *Proceedings of the European Chapter of the Association for Computational Linguistics* (EACL 93), pp. 250-259, Utrecht, the Netherlands.

William C Mann and Sandra Thompson [1986] Rhetorical Structure Theory: Description and Construction of Text, ISI technical report number RS-86-174.

Richard Montague [1974] *Formal Philosophy*, New Haven: Yale University Press.

Michael Morreau [1995] Allowed Arguments *Proceedings of IJCAI 95*, San Mateo, CA: Morgan Kaufman Press, pp. 1466-1472.

Reinhard Muskens [1996] Combining Montague semantics and discourse representation. *Linguistics and Philosophy*, 19(2).

David Park [1988] Conurrency and automata on infinite sequences, in P. Deussen (ed.), *Proceedings of the 5th GI Conference*, Lecture Notes in Computer Science, vol. 104, Berlin: Springer Verlag, pp. 167–183.

John Perry and Mark Crimmins [1993] *Journal of Philosophy*.

Barbara Partee [1973] Some Structural Analogies Between Tenses and Pronouns in English, *Journal of Philosophy* **70**, pp. 601-609.

Livia Polanyi [1985] A Theory of Discourse Structure and Discourse Coherence. In *Papers from the General Session at the Twenty-First Regional Meeting of the Chicago Linguistics Society*, pp. 25–27.

Robert van Rooy [1997] *Attitudes and Changing Contexts*, Ph.D. Thesis, University of Stuttgart.

Rob van der Sandt [1992] Presupposition Projection as Anaphora Resolution, *Journal of Semantics*, **19**(4).

Robert Stalnaker, R. [1978] Assertion, in Cole, Peter (ed.) *Syntax and Semantics volume 9: Pragmatics*, Academic Press.

Vallduví, E. [1990] *The Informational Component*, PhD. thesis, University of Pennsylvania, Philadelphia.

Author address

Department of Philosophy
The University of Texas
Austin TX 78712-188
USA
email: nasher@mail.la.utexas.edu

Complete Sets and Structure in Subrecursive Classes

Harry Buhrman[2] & Leen Torenvliet

ABSTRACT In this expository paper, we investigate the structure of
complexity classes and the structure of complete sets therein. We give an
overview of recent results on both set structure and class structure induced
by various notions of reductions.

1 Introduction

After the demonstration of the completeness of several problems for **NP**
by Cook [Coo71] and Levin [Lev73] and for many other problems by Karp
[Kar72], the interest in completeness notions in complexity classes has
tremendously increased. Virtually every form of reduction known in computability theory has found its way to complexity theory. This is usually
done by imposing time and/or space bounds on the computational power
of the device representing the reduction.

Early on, Ladner et al. [LLS75] categorized the then known types of
reductions and made a comparison between these by constructing sets that
are reducible to each other via one type of reduction and not reducible
via the other. They however were interested just in the relative strength
of the reductions and not in comparing the different degrees of complete
sets that are induced by these reducibilities. This question was picked up
much later by Watanabe [Wat87] for deterministic exponential time and
others following him for other classes. A recent survey of this can be found
in [Buh93]

Complete sets under some type of reduction form an interesting, rich,
and hence much studied subject in complexity theory. A complete set can
be viewed as representing an entire complexity class. Through the translation of the reduction one can with the help of the complete set, decide
all questions of membership for any set in the class. The resource bound
on the reduction is (to be of interest) always much less than the resource

---

[1] Received October 1996; revised version February 25, 1997.
[2] Partially supported by NWO

bound that defines the complexity class. A complete set is, as such, a "most difficult object" in the complexity class.

On the other hand by that very same representation property, we observe that the complete set codes all the information for all the sets present in the complexity class. A complete set therefore necessarily has to have large parts that are computationally easily recognizable. Thus, if a complete set is "most difficult" this usually is not pertinent to all of the set, but rather to "significant parts." It then becomes an interesting question, which parts of the complete sets are difficult, and, more pressing, which are not.

Complete sets in complexity classes are also usually the only sets that arise naturally. Once a set has been identified as being of a certain complexity, it is by experience soon after exposed as being complete in the corresponding class. This phenomenon was also noted in computability theory where every naturally arising non-recursive r.e. set turned out to be complete. This initiated E. Post [Pos44] to formulate his problem and program.

The different types of reductions impose structure on both complexity class and complete sets. Different types of reductions induce different notions of completeness which may give rise to a different degree structure. Stronger forms of reduction give more restrictions on how the information is stored in the complete sets than do weaker forms. Hence stronger reduction types impose more structure on the sets that are complete.

The most interesting complexity class to study complete sets on is the class **NP**. However this is also by far the most reluctant class to reveal its secrets. Exponential time (and larger) classes have the tremendous advantage that they allow for diagonalization against polynomial time bounded reductions and therefore for the construction of sets and degrees that have the desired properties. The bulk of the results obtained (and therefore the bulk of the results reviewed in this paper) pertains to exponential time (and larger) complexity classes.

In the different sections of this paper we pay attention to the structure of complexity classes as well as the structure of complete sets in these classes under various types of reductions. We treat several subjects that have received much attention in recent years. We must perforce limit our attention to some selected topics. The selection of the topics was almost always inspired by the fact that these topics were our own subject of (a) research (paper) at some point in time. We do not wish to implicitly or explicitly valuate other related and unrelated topics not mentioned in this paper, nor do we wish to make any claims about giving a complete survey on the topics that *do* appear in this paper. Having stated this disclaimer, we can now come to a list of topics that we will treat.

Class Structure: Degrees of Complete Sets

Here we investigate the question which reduction types give rise to different degrees of complete sets (collapse or separation). We survey the work of Watanabe [Wat87] and Buhrman et al. [BHT91, BST93]

Class Structure: Isomorphism

This is actually a refinement of the degree structure question. Of importance is which sets in (the complete) many-one degrees are isomorphic. This question was also inspired by the fact that in recursive set theory the many-one degree collapses to a single isomorphism degree [Myh55].

Class Structure: Measure Theory

Until recently, measure theory was a subject that was not applicable to complexity classes (or more general to effective classes) because of their inherent countability. Lutz [Lut90] however, introduced "resource bounded measure" with which many interesting properties about "abundance" of sets in complexity classes could be derived.

Set Structure: Sparse Complete Sets

Here we discuss some new developments on the most apparent internal structural properties of a complete set—the number of elements it has per length.

Set Structure: Redundant Information

Some strings in a complete set are essential to its completeness. E.g., for many-one reductions some strings are the image of a string under this reduction and some are not. Removing the former may destroy completeness, removing the latter has no consequence. Which (subsets of) strings are crucial?

Set Structure: Instance Complexity

As we already noted much of a complete set is of trivial complexity. Which is the part that makes it difficult and how dense is this part? To measure the complexity of single instances we use the measure introduced by Orponen et al. [OKSW94].

Set Structure: Post's Program Revisited

Autoreducibility is a special form of reducibility introduced by Trakhtenbrot [Tra70]. This notion has received considerable attention recently [BFT95] because of its potential to discover answers to the fundamental questions. Autoreducibility is a structural notion that complete sets in some complexity classes do and complete sets in other complexity classes do not have and is therefore a potential separator of complexity classes. As such this can be viewed as a new instantiation of Post's Program.

Several expository papers were written on the structure of complete sets in complexity classes. The first were published in 1990 [Hom90, KMR90]. On complete sets with special structure (sparse sets) a survey was published in 1992 ([HOW92]). The present authors presented a survey in 1994 ([BT94]) in a paper that has roughly the same structure as this paper. The field is however rapidly expanding and needs surveys such as this and as [Hom97] for constant update.

2   Preliminaries

All sets (and languages) in this paper are subsets of Γ^*, where $\Gamma = \{0,1\}$, and are denoted by capital letters A, B, C etc. Strings are elements of Γ^* and are denoted by small letters x, y etc. The characteristic function of a language A is denoted by $\chi_A : \Gamma^* \mapsto \Gamma$. The characteristic sequence of a language A is the infinite string

$$\chi_A(\lambda)\chi_A(0)\chi_A(1)\chi_A(00)\chi_A(01)\chi_A(10)\chi_A(11)\dots$$

and is also denoted by χ_A. The complement of a set A, $\Gamma^* - A$, is denoted by $\overline{A}$. The complement of a class of languages $\mathcal{X}$, $2^{\Gamma^*} - X$, is denoted by $\overline{\mathcal{X}}$. The *class complement* of a class $\mathcal{X}$, $\{A \mid \Gamma^* - A \in \mathcal{X}\}$, is denoted by $\text{Co} - \mathcal{X}$. The length of a string x is denoted by $|x|$ and the cardinality of a set A is denoted by $\|A\|$. We assume (standard) enumerations of resource bounded Turing Machines $M_1, M_2, \dots$ where the resource bounds on the machines vary to satisfy our needs. The set accepted by a Turing machine M is called its *language* and is denoted by $L(M)$. The main complexity classes considered in this paper are **LOG**, **NLOG P**, **NP**, **PSPACE**, **E**, **NE**, **EXP** and **NEXP**, which are the classes of languages recognized by:
deterministic logarithmic space,
nondeterministic logarithmic space,
deterministic polynomial time,
nondeterministic polynomial time,
deterministic polynomial space,
deterministic linear exponential time,
nondeterministic linear exponential time,
deterministic exponential time, and
nondeterministic exponential time bounded Turing machines respectively.

2.1   Reductions

Complete sets are always complete under some form of reduction. Many different models for this concept are found in the literature. As the concept of complete sets and therefore the concept of reductions is central to this paper, we take some space here to present a uniform machine based approach in which all reductions are modeled by Oracle Turing Machines.

2.1.1   Oracle Turing Machines

An oracle Turing machine is a standard multi-tape Turing machine with two extra tapes :

1. a write only tape called the QUERY-tape.

2. a read only tape called the ANSWER-tape.

These tapes will be called the oracle tapes. Furthermore we add an extra state: the QUERY-state. We use the following convention for access to the oracle tapes: M is allowed to write on the QUERY-tape a string q, called *query*, then at some point it decides to go into the QUERY-state. Subsequently the QUERY-tape is cleared[3], and depending on the oracle, something is written on the ANSWER-tape. Now M is allowed to read the ANSWER-tape, until a next QUERY-state is reached. A Turing machine equipped with the above described extra tapes and state is called an *oracle Turing machine*. In the above discussion it was not clear what the role of the oracle is. An oracle is just a set, say A. When an oracle Turing machine M writes a string q on the QUERY-tape and enters the QUERY-state, the oracle writes down – in one step – on the ANSWER-tape the value of the characteristic function of A on q, that is $\chi_A(q)$. Informally, M asks oracle A whether y is a member of A and finds the answer on its ANSWER-tape. We note here that the role of the oracle can be more complex in the sense that it could write down not only one character but a whole string of characters. Examples of this can be found in [ABJ91, FHOS93]. Let A be a set and M be an oracle Turing machine We say that M accepts x *relative to A* if M has an accepting computation on input x, with A as oracle. We say that $L(M, A)$ is the set of strings accepted by M relative to A. As usual, we can talk about polynomial time oracle machines and computations.

2.2   Adaptive and Non-Adaptive

Essentially there are two ways an oracle machine can compute its queries:

1. adaptive: M is allowed to read the ANSWER-tape at *any* time during the computation and may compute the next query depending on the contents of the ANSWER-tape. In this case the queries are *dependent* on the oracle.

---

[3]With clearing a tape we mean that after clearing, the only symbols on the tape are blanks.

2. non-adaptive: M is not allowed to read the ANSWER-tape *before* it enters the last QUERY state. In this case the computation of all the queries depends solely on the input and the program, and is *independent* of the oracle.

Sometimes we want to to talk about the set of (possible) queries that M could ask on input x.

Definition 2.1 *Let M be an oracle Turing machine.*

- $Q(M, x, A)$ *is the set of all queries M wrote on its QUERY-tape with x as input and A as oracle.*

- $Q(M, x) = \bigcup_{A \subseteq \Sigma^*} Q(M, x, A)$. *This denotes the set of all possible queries M could ask.*

2.2.1   Reductions

Oracle machines are used to model (almost) all different types of reductions. To achieve this we add restrictions to the oracle machine M. More formally a *restriction* is a 4 tuple:
$r = <$N, COMP, ACCEPT-RESTR, QUERY-TYPE$>$. Where,

1. N is a function from $I\!N \times \Sigma^* \to I\!N$. This function depends on the index of M and the input. The function is the number of queries M is allowed to make during the computation. With "number of queries" we mean the number of times that M entered the QUERY-state.

2. COMP can be adaptive or non-adaptive.

3. ACCEPT-RESTR is a function from $\Sigma^* \times \Sigma^*$ to $\mathcal{P}(\{0,1\})$, where $\mathcal{P}(\{0,1\})$ denotes the *power* set of $\{0,1\}$. This function depends on the INPUT-tape and the ANSWER-tape. When this function takes on the value $\{0\}$ or $\{1\}$, the machine is forced to reject or accept respectively.

4. QUERY-TYPE is a set of additional constraints on the type of queries M is allowed to make. E.g., all the queries should start with a 0 or should be smaller in length than the input.

We say that an oracle Turing machine M_i (we assume an effective enumeration of oracle Turing machines) *obeys* restriction r, if for all input strings x:

- N $= \emptyset$ or $M_i(x)$ does not make more than N(i, x) queries, and

- COMP $= \emptyset$ or $M_i(x)$ generates it's queries in a COMP (i.e., either adaptive or non-adaptive) fashion, and

- ACCEPT-RESTR $= \emptyset$ or if $M_i(x)$ halts then
  $M_i(x) \in$ ACCEPT-RESTR(x, y), for y the string on the ANSWER-tape
  when $M_i(x)$ halts, and

- QUERY-TYPE $= \emptyset$ or $M_i(x)$ only wrote down queries q that satisfy the
  constraints in QUERY-TYPE.

We say that M is an r-restricted oracle machine, if r is a restriction and
M is an oracle Turing machine, that obeys restriction r.

Definition 2.2 *A r reduces to B (A $\leq_r^{rec}$ B) iff there exists a recursive
r-restricted oracle Turing machine M, such that $A = L(M, B)$.*

In this paper, we will mainly talk about polynomial time oracle machines. Note that this does not necessarily means that the ACCEPT-RESTR
function is computable in polynomial time. This notion will be called polynomial time reduction. Several forms of polynomial time reduction were
first defined and compared by Ladner, Lynch and Selman [LLS75]. In the
following we will not redefine the existing notions of reducibility. We will
capture them in a machine based framework. We think that the most natural way to think about a reduction is as an oracle Turing machine with
several restrictions on the access it has to the oracle. The most general
one is the Turing reduction which has no restrictions at all. The definitions
found in the literature are by no means uniform in this sense. Sometimes
they define reductions as functions. Other times the machine based point
of view is used.

The approach we take has also the advantage that it gives a taxonomy
of the reductions in four natural groups. Several new reductions emerge
from this taxonomy by varying the 4 different aspects of the reductions.
Sometimes already existing reductions come out. For example *adaptive* conjunctive reductions are the same as non-adaptive conjunctive reductions,
but it is probably not true that adaptive parity (or majority) reductions
are the same as their non-adaptive counterparts.

Definition 2.3 *A reduces r to B in polynomial time (A $\leq_r^p$ B) iff there
exists an r-restricted polynomial time oracle machine M such that $A =
L(M, B)$.*

We will now show that some of the standard reductions found in the literature are easily captured by our formalism. To start with the most general
restriction:

1. T $= <\emptyset, \emptyset, \emptyset, \emptyset>$. ($\leq_T^p$)
   This restriction does not restrict the class of oracle machines. This
   reduction is called Turing reduction.

2. tt$= <\emptyset,$ non-adaptive$, \emptyset, \emptyset>$. ($\leq_{tt}^p$)
   The oracle machines are restricted in the way they generate their
   queries. This reduction is called truth-table reduction.

3. btt $= <n_b,$ non-adaptive, $\emptyset, \emptyset>$. $(\leq_{btt}^p)$
   $n_b(i, x) = i.$
   This reduction is called bounded truth-table reduction.

4. k-tt $= <n_k,$ non-adaptive, $\emptyset, \emptyset>$. $(\leq_{k-tt}^p)$
   $n_k(i, x) = k$, k a constant.
   This reduction is called k-truth-table reduction. Actually this defines a whole class of reductions. One for every constant k.

5. k-T $= <n_k, \emptyset, \emptyset, \emptyset>$. $(\leq_{k-T}^p)$
   $n_k(i, x) = k$, k a constant.
   This reduction is called k-Turing. The bounded Turing reduction $(\leq_{b-T}^p)$ is defined by replacing n_k by $n_b(i, x) = i$.

6. ctt $= <\emptyset,$ non-adaptive, $f_c, \emptyset>$. $(\leq_c^p)$
   $$f_c(x, y) = \left\{ \begin{array}{ll} \{1\} & \text{if } \forall i, y_i = 1. \quad (y = y_1 \ldots y_n) \\ \{0\} & \text{otherwise.} \end{array} \right.$$

   This reduction is called conjunctive truth-table reduction.

7. dtt $= <\emptyset,$ non-adaptive, $f_d, \emptyset>$. $(\leq_d^p)$
   $$f_d(x, y) = \left\{ \begin{array}{ll} \{1\} & \text{if } \exists i, y_i = 1. \quad (y = y_1 \ldots y_n) \\ \{0\} & \text{otherwise.} \end{array} \right.$$

   This reduction is called disjunctive truth-table reduction.

8. m $= <n_m, \emptyset, f_m, \emptyset>$. $(\leq_m^p)$
   $n_m(i, x) = 1.$
   $$f_m(x, y) = \left\{ \begin{array}{ll} \{y\} & \text{if } y = 0 \text{ or } y = 1. \\ \emptyset & \text{otherwise.} \end{array} \right.$$

   This reduction is called a many-one reduction. Sometimes it will be more elegant to use the following equivalent definition: $A \leq_m^p B$ iff there exists a total polynomial time computable function f such that $x \in A$ iff $f(x) \in B$. Obviously f can be constructed from an oracle machine that obeys the restriction m and vice versa.

9. m̂ $= <n_{\hat{m}}, \emptyset, f_{\hat{m}}, \emptyset>$. $(\leq_{\hat{m}}^p \; p)$
   $n_{\hat{m}}(i, x) = 1.$
   $$f_{\hat{m}}(x, y) = \left\{ \begin{array}{ll} \{y\} & \text{if } y = 0 \text{ or } y = 1. \\ \{0, 1\} & \text{otherwise.} \end{array} \right.$$

   This reduction will be called extended many-one.

10. m,li $= <n_m, \emptyset, f_m, \text{LI}>$ $(\leq_{m,li}^p \; p)$
    n_m and f_m as above.
    LI $= \forall y \in Q(M_i, x) : |y| > |x|.$
    This constraint says that the queries have to be bigger (in length) than the input.

This reduction is called many-one length increasing. As in the case of the many-one reduction we sometimes use the equivalent functional definition in terms of total polynomial time computable functions that are length increasing.

11. m,1-1 = $<n_m, \emptyset, f_m, \text{ONE-ONE}>$ $(\leq^p_{m,1\text{-}1} p)$
n_m and f_m as above.
ONE-ONE = $\forall y \in Q(M_i, x) : \forall x' < x : y \notin Q(M_i, x')$. This constraint says that each query is asked once. Clearly this implies injectivity. This reduction is called many-one and one to one reduction. We will use sometimes the existence of a total polynomial time computable function that is one-one.

12. m,1-1,li $=<n_m, \emptyset, f_m, \text{ONE-ONE-LI}>$ $(\leq^p_{m,1\text{-}1,li})$
n_m and f_m as above.
ONE-ONE-LI = ONE-ONE and LI. This means that both the constraints (ONE-ONE and LI) have to be satisfied in order to satisfy ONE-ONE-LI. This reduction is called many-one, length increasing and one to one. Again the functional equivalent way is sometimes chosen: there exists a total polynomial time computable function that is one-one and length increasing.

13. m,1-1,eh $=<n_m, \emptyset, f_m, \text{ONE-ONE-EH}>$ $(\leq^p_{m,1\text{-}1,eh})$
n_m and f_m as above.
EH = $\forall y \in Q(M_i, x) : 2^{|y|} > |x|$.
This constraint says that the queries do not decrease more than exponentially in length. ONE-ONE-EH = ONE-ONE and EH. This reduction is called many-one, one to one and exponentially honest. The same comment applies here: the functional variant must be exponentially honest, i.e. not decrease more than an exponential in the length of the argument.

14. pos = $<\emptyset, \emptyset, f_{pos}, \emptyset>$ $(\leq^p_{pos})$
Let **POS** be the class of all positive boolean formulas. These are formulas, that can be represented using only disjunctions and conjunctions as connectives. For x a boolean variable, $x := 1(0)$ means $x := \top(\bot)$. $\phi = 1(0)$ if it evaluates to true (false).
$f_{pos}(x, y) = \{\phi(x_1 := y1, \ldots, x_i := y_i)\}$ $(y = y_1 \ldots, y_i)$.
For $\phi \in$ **POS**.
This reduction is called positive Turing reduction. The positive truth-table, positive bounded-truth-table and the positive k-truth-table reductions are defined as truth-table, bounded truth-table or k-truth-table reduction with f_{pos} as ACCEPT-RESTR.

Another way of looking at this reduction is as follows: M_i is a positive Turing reduction if for all oracles A and B it holds that if $A \subseteq B$ then $L(M, A) \subseteq L(M, B)$.

Reductions stronger than $\leq_m^p$-reductions can be modeled by circuits that have oracle gates, or circuits that compute functions. In particular so-called $\mathbf{AC}^0$, $\mathbf{NC}^0$ and $\mathbf{NC}^1$ reductions have recently received some attention. Here $\mathbf{AC}^k$ and $\mathbf{NC}^k$ are the classes of languages recognized families of circuits of polynomial size and depth $\log^k$ of unbounded and bounded fan in respectively.

2.3 Resource Bounded Measure

Classical Lebesque measure is an unusable tool in complexity classes. As these classes are all countable, everything we define in such a class has measure 0. Yet, we might wish to have a notion of "abundance" and "randomness" in complexity classes. Lutz [Lut87, Lut90] introduced the notion of *resource bounded measure*, and gave a tool to talk about these notions inside complexity classes.

Definition 2.4 *A* martingale *d is a function from Γ^* to $\mathcal{R}^+$ with the property that $d(w0) + d(w1) \leq 2d(w)$ for every $w \in \Gamma^*$.*

Definition 2.5 *A p-martingale is a martingale $d : \Gamma^* \mapsto \mathcal{Q}^+$ that is polynomial time computable.*

Definition 2.6 *A martingale d* succeeds *on a language A if*

$$\limsup_{n \mapsto \infty} d(\chi_A[0 \ldots n-1]) = +\infty$$

We write $S^\infty[d] = \{A \mid d \text{ succeeds on } A\}$

Definition 2.7 *Let $\mathcal{X}$ be a class of languages.*

- *$\mathcal{X}$ has p-measure 0 ($\mu_p(\mathcal{X}) = 0$) iff there exists a p-martingale d such that $\mathcal{X} \subseteq S^\infty[d]$.*
- *$\mathcal{X}$ has p-measure 1 ($\mu_p(\mathcal{X}) = 1$) iff $\mu_p(\overline{\mathcal{X}}) = 0$*
- *$\mathcal{X}$ has p-measure 0 in $\mathbf{E}$ ($\mu_p(\mathcal{X}|\mathbf{E}) = 0$) iff $\mu_p(\mathcal{X} \cap \mathbf{E}) = 0$*
- *$\mathcal{X}$ has p-measure 1 in $\mathbf{E}$ ($\mu_p(\mathcal{X}|\mathbf{E}) = 1$) iff $\mu_p(\overline{\mathcal{X}} \cap \mathbf{E}) = 0$*

2.4 Completeness and Degrees

Definition 2.8 *Let $\leq_r$ be a reduction and $\mathcal{C}$ be a complexity class. A set C is r-hard for $\mathcal{C}$ iff $\forall A \in \mathcal{C}, A \leq_r C$. If, moreover, $C \in \mathcal{C}$ then C is r-complete for $\mathcal{C}$.*

Definition 2.9 *Let $\leq_r$ be a reduction and A be a set. The r-degree of A, denoted $\mathbf{a}$ is the class $\{B \mid A \leq_r B \wedge B \leq_r A\}$*

Definition 2.10 *A set C is weakly-r-hard for $\mathcal{C}$ if $\mu_p\{A \mid A \in \mathcal{C} \wedge A \leq_r C\} \neq 0$. If, moreover $C \in \mathcal{C}$ then C is weakly-r-complete for $\mathcal{C}$.*

2.5 Complexity of Instances

The computational complexity of a single string is always a constant. Indeed, for each string we can define a Turing machine that recognizes just that string and nothing else. This machine works in constant time (where the constant depends on the size of the single string recognized). Yet, we think that some instances of a set may be computationally harder than others. Kolmogorov complexity and generalized Kolmogorov complexity provide means to talk about (descriptional) complexity of individual strings. Instance complexity is a notion closely related to Kolmogorov complexity but of a more computational nature.

Consider the class of Turing machines that on each input always output 1 (accept) or 0 (reject) or ? (don't know). Such a Turing machine is said to be *consistent* with a set A iff, for all inputs x such that $M(x) \neq ?$ it holds that $M(x) = \chi_A(x)$. The t-bounded instance complexity of an instance x with respect to a set A, $\mathrm{ic}^t(x : A)$ is the size of the smallest–in length–Turing machine M such that

1. M is consistent with A

2. M runs in time bound t for all inputs

3. $M(x) \neq ?$

For a precise definition see [OKSW94].

3 Degrees of Complete Sets

Complete sets under some kind of reduction in a complexity class have (by definition) the property that they all reduce to each other under that reduction. I.e., they form a degree. The different reductions that are available thus form different complete degrees. Most of these degrees are ordered by inclusion by the comparative strength of the reductions. One very natural question to ask is whether two consecutive complete degrees in this ordering differ. It is trivially true that all the degrees of complete sets under polynomial time reductions on **P** are the same (namely **P** itself). On **P** all polynomial time degrees are known to *collapse* to the many-one degree (but known *not* to collapse to the even smaller isomorphism degree.) Therefore showing two degrees of complete sets under polynomial time different on any complexity class $\mathcal{C}$, implies that this class is not equal to **P**. Unfortunately, until now the only complexity classes on which separation of degrees has been a successful undertaking are classes which encompass exponential time.

Theorem 1 ([Wat87, BHT91, HKR93, BST93])
*For $\mathcal{C} \in \{$**E**, **EXP**, **NE**, **NEXP**$\}$*

1. *The degree of the $\leq^p_{1-tt}$-complete sets collapses to the degree of the $\leq^p_m$-complete sets.*

2. *The degrees of the $\leq^p_m$-, $\leq^p_{btt}$-, $\leq^p_c$-, $\leq^p_d$-, $\leq^p_{tt}$- and $\leq^p_T$-complete sets are all different and, moreover, when not obviously ordered by inclusion, incomparable.*

In [BST93] also the degrees of query-bounded reductions are compared and it turns out that

Theorem 2 ([BST93]) *For $C \in \{$**E**, **EXP**, **NE**, **NEXP**$\}$*

1. *For any $k \geq 2$, $\leq^p_{k-c}$-, and $\leq^p_{k-d}$-completeness are incomparable on C*

2. *For any k and l, with $k < l \leq 2^k - 2$, $\leq^p_{k-T}$- and $\leq^p_{l-tt}$-completeness are incomparable on C.*

These two theorems reveal all relations between degrees of complete sets on exponential time and beyond under all reductions known from [LLS75]. Recently, Lutz [Lut94] introduced a new notion of complete set on exponential time, the *weakly* complete set. A set A is weakly hard under reduction $\leq_r$, if a non-significant part (i.e., a class with non-zero p-measure) of **E** reduces to A. It was not known until [ASMZ96] whether the degrees of weakly complete sets under various reduction types are different. As it turns out, these degrees behave the same as the classical complete degrees. All are different except for the many-one and 1-tt, which coincide. (See Theorem 10)

In connection with the isomorphism problem (See Theorem 6), degrees defined by reducibilities even stronger than $\leq^p_m$ have been studied. In particular in [AAR96] the $\mathbf{AC^0}$ degree is shown to collapse to the $\mathbf{NC^0}$ degree for every complexity class C that is closed under $\mathbf{NC^1}$ reductions.

On the class **NP** all relations of polynomial time complete degrees necessarily remain open questions (without a proof that $\mathbf{P} \neq \mathbf{NP}$). Even from the assumption $\mathbf{P} \neq \mathbf{NP}$ it is still open whether any two reductions induce different degrees on **NP**. From the stronger assumption that $\mathbf{E} \neq \mathbf{NE} \cap \mathrm{Co} - \mathbf{NE}$, Selman [Sel82] showed that the reductions many-one and Turing differ, but not the completeness notions. A first result in the direction of separation of completeness notions on **NP** was achieved by Longpré and Young [LY90] who showed that for each k there exist many-one **NP**-complete sets that A and B such that $A \leq^p_{2-d} B$, where the reduction needs linear time, but A does not reduce many-one to B using less than n^k time. The most recent development on **NP** is a theorem by Lutz and Mayordomo [LM94] who prove that complete degrees differ on **NP** from the very strong assumption that **NP** has nonzero measure in **EXP**.

Theorem 3 *If $\mu(\mathbf{NP} \mid \mathbf{EXP}) \neq 0$ then*

1. *The $\leq_m^p$ -complete degree is different from the $\leq_{2-T}^p$ -complete degree for* **NP** *[LM94].*

2. *The $\leq_{2-tt}^p$ -complete degree is different from the $\leq_{2-T}^p$ -complete degree for* **NP** *[May94b].*

A last result we would like to mention in this section is about the many-one and Turing degrees on **PSPACE**. Here also, it is of course not known whether this class differs from **P** and therefore an assumption is necessary. Assuming that either randomized completeness notions differ or that **PSPACE** has a set with a dense subset of high generalized Kolmogorov complexity, Watanabe and Tang [WT89] show that the many-one and Turing complete degrees differ on **PSPACE**. From recent work of Buhrman and Fortnow [BF96] it follows that there exists a relativized world in which the $\leq_m^p$ and the $\leq_{1-tt}^p$ complete set on **PSPACE** differ.

On **NP** most problems remain open. With respect to $\leq_m^p$- and $\leq_{1-tt}^p$- complete degree on **NP**, Fortnow [For96] informed us of the existence of an oracle where these degrees are different.

Even under assumptions that imply **P** $\neq$ **NP** these problems remain hard. In particular the following seem most urgent

- Can the complete degrees for **NP**-under various reduction types be separated under the assumption **P** $\neq$ **NP**?

- Assume $\mu(\textbf{NP}|\textbf{EXP}) \neq 0$. Can the $\leq_{btt}^p$-complete degree for **NP** be separated from the $\leq_{tt}^p$-complete degree, and can the $\leq_{tt}^p$-complete degree be separated from the $\leq_T^p$-complete degree?

- Assume $\mu(\textbf{NP}|\textbf{EXP}) \neq 0$. Does the $\leq_{1-tt}^p$-complete degree on **NP**-coincide with the $\leq_m^p$-complete degree?

4   Isomorphism

Special among the complete degrees are the degrees of isomorphism. Berman and Hartmanis [BH77] proved that all natural **NP**-complete sets are inter-reducible via length-increasing 1-1 reductions and could therefore, via a polynomial time analog of the Cantor-Bernstein-Myhill theorem show that these sets are all isomorphic under polynomial time computable isomorphisms.

After a long period of unsuccessful work in trying to prove isomorphism of all $\leq_m^p$-complete sets in **NP**, and oracle proof that the conjecture might not hold (most prominent, the conjecture fails relative to a random oracle [KMR89]) opinion shifted against the idea that the isomorphism conjecture might be true. It was not until 1992 that Fenner, Fortnow and Kurtz [FFK92] proved the existence of an oracle relative to which the isomorphism conjecture holds. In 1985, Joseph and Young [JY85] constructed unnatural sets, the so-called k-creative sets for which every 1-1 polynomial time computable honest function is a productive function. Hence these sets are very unlikely to be isomorphic to SAT. Also, there is a strong belief in the existence of so-called one-way functions. These are functions that are honest and polynomial time computable, but not polynomial time invertible. The image of SAT under such a function may be a complete set that is not isomorphic to SAT. This conjecture (in some sense opposing the isomorphism conjecture) became known as the "Encrypted Complete Set Conjecture" (ECC). It has also met with relativized counter examples [Rog95]. Modern versions of the isomorphism conjecture are more in the direction of stronger reductions. Most of the sets known to be $\leq_m^p$-complete in **NP** are also complete under much stronger reductions. Allender [All88] was the first to show that sets in **PSPACE**, complete under 1-L reductions (which is a function computable by a logspace bounded Turing machine that has a one-way input head) are polynomial time isomorphic. Burtschick and Hoene [BH92b] showed on the other hand that these sets are *not* necessarily isomorphic under 1-L computable isomorphisms.

The following theorem by Agrawal and Biswas, is the most general theorem known for the 1-L reductions.

Theorem 4 ([AB93]) *Let C be a complexity class that is closed under lin-log reductions, e.g.,* **P, NP, PSPACE**. *The sets complete for C under 1-L reductions are all p-isomorphic.*

The first order projection (defined by Valiant in [Val82]) is another example of a very strong form of reduction. Allender, Balcázar and Immerman showed in [ABI93] that for the first order projections an isomorphism theorem holds

Theorem 5 ([ABI93]) *Let C be a nice complexity class, e.g.* **P, NP, PSPACE**. *All sets complete for C under first order projections are isomorphic under first order isomorphisms.*

It is natural to try to weaken the reduction types in order to get stronger and stronger versions of isomorphism theorems for **NP**. One way to weaken the first order projections is to consider reductions that can be defined by circuits. A circuit can be equipped with oracle gates that can compute answers to queries. The class $\mathbf{NC^0}$ is the class of circuits of polynomial size and constant depth with bounded fan-in. The class $\mathbf{AC^0}$ is the class of circuits of polynomial size and constant depth with unbounded fan-in. There has recently be some good progress with respect to $\mathbf{AC^0}$ reductions by Agrawal, Allender and Rudich.

Theorem 6 ([AAR96]) *All $\mathbf{AC^0}$ complete sets for* **NP** *are $\mathbf{AC^0}$ isomorphic*

The proof of the above theorem relies heavily on the nonuniformity of the reduction. It is open whether an isomorphism theorem is true for **NP** when uniform reductions are used. On the other hand it is true that the existence of one-way polynomial time functions implies the existence of one-way AC^0 functions [AAR96]. Hence the above theorem heavily rocks the belief in the ECC.

5   Measure Theory

Having established a difference between the degrees of complete sets under various reductions, one might continue to investigate the part(s) of the complexity class that consist(s) of complete sets. It is easily seen that a single complete set, by padding, gives a countable class of complete sets. But "countable class in countable class" gives no structural information about the class. A more informative concept is the resource bounded measure of a degree. For instance if a complete degree would have measure 1 in a complexity class, then every non-negligible part of that complexity class would have a complete set. That is, complete sets are all over the place. Even if one can just show that a complete degree has non-zero measure, (note that this can either mean that it has measure one or that it is not measurable) then the complete degree has a non-empty intersection with any measure one class in the complexity class. A complete set can then be found in any rich enough substructure. Until now results obtained seem to point in another direction. It seems that indeed chaos (as classes of random sets do not have measure 0) is more abundant than structure, even in the small universe of complexity classes.

Theorem 7

1. The $\leq^p_m$ -complete degree for **E** has measure 0 in **E** [JL93, May94a].

2. For all k, the $\leq^p_{k-tt}$ -complete degree for **E** has measure 0 in **E** [BM95].

3. The $\leq^p_{btt}$ -complete degree for **E** has measure 0 in **E** [ASNT94].

The question remains open for $\leq^p_T$-complete sets for **E**. Some progress has been made however. Allender and Straus showed the following.

Theorem 8 ([AS94]) *For almost every set A in* **EXP**, **BPP**^{A} = **P**^{A}.

This theorem shows that if **BPP** would be equal to **EXP** then the Turing complete sets for **EXP** would not have measure 0. Hence a proof that the Turing complete sets for **EXP** have measure 0 would separate **BPP** from **EXP**. It is well known that there exist relativized worlds where **EXP** = **BPP**, and since Theorem 8 relativizes it follows that there exist relativized worlds where the Turing complete sets for **EXP** do not have measure 0, but in fact have measure 1. On the other hand Ambos-Spies [AS96] has informed us of the construction of an oracle where the Turing complete degree for **EXP** has measure 0.

The class of weakly complete sets, a generalization of the classical notion of completeness, would not even have been studied without the introduction of resource bounded measure. A comparison between weakly complete sets and other complete sets is of course the first goal. It turns out that weakly complete sets differ from classical complete sets, both in the sense that their degrees are different and in the sense that they behave differently.

Theorem 9 ([JL94])

1. *Every language that is weakly $\leq_m^p$-complete for **E** is weakly $\leq_m^p$-complete for **EXP**.*

2. *There is a language in **E** that is weakly $\leq_m^p$-complete for **EXP**, but not for **E**.*

Which is rather surprising if one takes into account that, by padding, a set is $\leq_m^p$-complete for **E** if and only if it is $\leq_m^p$-complete for **EXP**.

As was shown recently by Ambos-Spies, Mayordomo and Zheng [ASMZ96], the weakly complete degrees and other complete degrees, both in **E** and in **EXP** form an interleaved structure. In particular.

Theorem 10 *[ASMZ96]*

1. *For any set A, A is weakly-$\leq_m^p$-complete for **E** if and only if A is weakly-$\leq_{1-tt}^p$-complete for **E**.*

2. *Let $k \geq 1$ there is a set A such that A is $\leq_{k+1-d}^p$-complete for **E** (and hence weakly-complete for both **E** and **EXP**), but neither weakly-$\leq_{k-T}^p$-complete for **E** nor $\leq_{k-T}^p$-complete for **EXP***

3. *There is a set A such that A is $\leq_d^p$-complete for **E** (and hence weakly dttp-complete for **E** and **EXP**), but A is neither weakly $\leq_{btt}^p$-complete for **E** nor weakly-$\leq_{btt}^p$-complete for **EXP**.*

4. *There is a set A such that A is $\leq_T^p$-complete for **E** (hence weakly-$\leq_T^p$-complete for **E** and **EXP**), but A is neither weakly-$\leq_{tt}^p$-complete for **E** nor weakly $\leq_{tt}^p$-complete for **EXP**.*

6   Sparse Complete Sets

One of the first structural questions to ask of a *set* is how many of the 2^n strings of each length are in or out of the set. I.e., one of the first apparent questions about the structure of a set is the question of its density. Also other notions of structure seem very closely related to this basic question. For instance it is known that the class of sets (Turing) reducible to sparse sets coincides with the class of sets reducible to P-selective sets and coincides with the class of sets that are recognizable by families of small (polynomial size) circuits and coincides with the class of sets that are recognizable in polynomial time given a polynomial amount of advice. In other words, the amount of information that can be stored in a set seems closely related to the number of strings that a set has per length. It therefore comes as no big surprise that a sparse set cannot be $\leq_m^p$-complete for **EXP**. This follows from an old theorem by Berman [Ber77]. For **NP** this question is a lot harder (if $\mathbf{P} = \mathbf{NP}$ then any set in **P** is $\leq_m^p$-hard for **NP** so also the sparse ones). It was answered by Mahaney [Mah82], building on earlier work of Fortune [For79], who showed that sparse sets could not be $\leq_m^p$-complete unless $\mathbf{NP} = \mathbf{P}$. For **P** the question needs reformulation. Under $\leq_m^p$-reductions any set is complete in **P**, but under $\leq_m^{logspace}$-reductions this is only true if $\mathbf{P} = \mathbf{LOG}$. An analogous question about the density of $\leq_m^{logspace}$-complete sets in **P** can thus be posed. Indeed, it was conjectured by Hartmanis in [Har78] that no sparse set can be complete for **P** under logspace reductions unless $\mathbf{P} = \mathbf{LOG}$. It was not until recently that this conjecture was proven true.

Theorem 11 ([Ogi95, CS95]) *If a sparse set S is hard for* **P** *under many-one reductions computable in logspace, then* $\mathbf{LOG} = \mathbf{P}$.

This Theorem was extended by Van Melkebeek to bounded truth-table reductions.

Theorem 12 ([vM96]) *If a sparse set S is hard for* **P** *under bounded truth-table reductions computable in logspace, then* $\mathbf{LOG} = \mathbf{P}$.

An extensive survey of the (im)possibilities of sparse complete sets, which for chronological reasons lacks the above results, but not many others was given by Hemachandra, Ogiwara and Watanabe [HOW92]. For exponential time classes it seems that the most pressing open question is the existence of Turing hard sparse sets. As mentioned above the question of a sparse Turing hard set coincides with the existence of polynomial sized circuits for the class and this question seems particularly interesting for **EXP**. (Until now the smallest class known not to be computable by polynomial size circuits is **MA(exp)**.) This result is due to Buhrman and Thierauf. (See [KW95].)

There are some steps set along this path however.

Theorem 13 ([Fu93])

1. *For $\alpha < 1$, all $\leq_{n^\alpha-T}^p$ -hard sets for* **EXP** *are exponentially dense.*

2. *For $\alpha < \frac{1}{4}$, all $\leq_{n^\alpha-T}^p$ -hard sets for* **E** *are exponentially dense.*

An incomparable theorem dealing with the part of **E** that may have small circuits (is in **P/poly**) is the following.

Theorem 14 ([LM93]) *For $\alpha < 1$, all $\leq^p_{n^\alpha - tt}$ -weakly-hard sets for* **E** *and* **EXP** *are exponentially dense.*

Finally, Homer and Mocas studied the possibility of exponential time computable sets being decidable with a fixed polynomial amount of advice. (Recall that if **EXP** has sparse complete sets then **EXP** $\subseteq$ **P/poly**.) They prove the following.

Theorem 15 ([HM93]) *for every k there exists a set A in* **EXP** *such that A is not in* $\mathbf{DTIME}(2^{n^k})/\mathbf{ADVICE}(n^k)$.

Improving upon these results seems very interesting, but also seems very hard since it would require non relativizing techniques. Wilson [Wil85] shows the existence of an oracle relative to which **EXP** has polynomial size circuits.

Specifically we note the following questions.

- For any k: are the $\leq^p_{n^k - T}$ complete sets for **EXP** exponentially dense?

- For any k: are the $\leq^p_{n^k - T}$ weakly complete sets for **EXP** exponentially dense?

- Related to [BH92a], are the $\leq^p_T$-complete sets for **NEXP** not sparse unless **NEXP** $= \Sigma^P_2$?

7 Redundant Information

As noted in the previous section, some complexity classes do not allow for sparse complete sets under some reductions. For such complexity classes in particular the question can be asked: "How dense must these sets be?" By the result of Berman [Ber77] the many-one complete sets for **EXP** for instance cannot be sparse. Schöning [Sch86], building upon the work of Yesha [Yes83], showed that for complete sets A in **EXP** and every set D in **P**, the set $A \triangle D$ is of exponential density. If sets must be really dense to be complete, we can also ask the question: "Can a small set perhaps be taken out of the set so that the remaining set is still complete?"

This question was first taken up by Tang, Fu and Liu [TFL93], who showed, inspired by Schönings theorem, that the set D in this theorem can be taken subexponential time computable. They go on to show that for arbitrary sparseness condition, there exists a single subexponential time computable sets S, such that for *any* exponential time complete sets A, the set $A - S$ is no longer exponential time complete.

The natural question to ask next after the result is obtained for many-one reductions is: "How do complete sets under weaker types of reductions behave?" The answer to this question was given in [BHT93]. They show that the observation on this structural aspect of complete sets is not limited to many one completeness *or* to deterministic exponential time.

Theorem 16
Given a recursive non-decreasing function $g(n)$ with $\lim_{n\to\infty} g(n) = \infty$. There exists $g(n)$-sparse subexponential time computable sets S_1, S_2, S_3 and S_4 such that: [TFL93] For any $\leq_m^p$-hard set A, the set $A - S_1$ is not $\leq_m^p$-hard, and [BHT93]

1. *For any $\leq_{btt}^p$-complete set A for* **EXP** *the set $A - S_2$ is not $\leq_{btt}^p$-hard.*

2. *For any $\leq_c^p$-hard set A for* **EXP** *the set $A - S_3$ is not $\leq_c^p$-hard.*

3. *For any $\leq_d^p$-hard set A for* **EXP** *the set $A \cup S_4$ is not $\leq_d^p$-hard.*

Not only the sparseness of the sets S_i is controllable, but also their "subexponentiality." The construction can be slowed down to bring the set arbitrarily close to being polynomial time computable, but not quite. Tang, Fu and Liu [TFL93] already noted that for many-one reductions it holds that a $\leq_m^p$-complete set remains $\leq_m^p$-complete if an arbitrary sparse, polynomial time computable set is taken out. This however seems to have less to do with its polynomial time computability than with its structural simplicity. In [BHT93] this question was re-addressed for P-selective sets i.s.o. P-sets (which is of course stronger). The following was shown.

Theorem 17 ([BHT93]) *For any set A that is $\leq_m^p$-, $\leq_c^p$-, $\leq_d^p$-, or $\leq_{2-tt}^p$-hard for* **EXP** *and any $p(n)$ sparse P-selective set S, the set $A - S$ remains hard w.r.t. the same reduction.*

It may seem strange that this theorem can only be proven for $\leq_{2-tt}^p$-reductions. It follows however directly from a recent result of Buhrman, Fortnow and Torenvliet [BFT95] that this result is optimal. They show the existence of a $\leq_{3-tt}^p$-complete set in **EXP** that is not $\leq_{btt}^p$-autoreducible. (We will meet this result again in Section 9). Inspection of the proof learns that the set is constructed by diagonalizing against autoreductions on inputs in the set $\{0^{b(n)} : n \in \omega\}$ where $b(n)$ is some suitably chosen gap function. They prove that a 3-tt complete set A can be constructed such that every btt-reduction (from A to A) that does not query its input must, for some n, incorrectly compute membership of $0^{b(n)}$ in A. Without essentially changing the proof, $b(n)$ can be chosen such that $\{0^{b(n)} : n \in \omega\}$ is a polynomial time computable sparse set. The following corollary then follows immediately.

Corollary 18 ([BFT95]) *There exists a 3-tt complete set A in* **EXP** *and a sparse set S in* **P** *such that $A - S$ is not btt-hard for* **EXP***.*

As a $\leq_{tt}^p$-complete set may not be $\leq_{btt}^p$-complete this last corollary does not answer all remaining questions. The following questions still remain open. (Wilson's [Wil85] oracle forces non-relativizable proofs on answers.)

- Let A be $\leq_{tt}^p$ or $\leq_T^p$-complete for **EXP**. Does there exist a sparse subexponential time computable set S such that $A - S$ is not complete?

- Let A be $\leq_{btt}^p$-complete for **EXP**. Is for every sparse $S \in \mathbf{P}$, $A - S$ still $\leq_{btt}^p$-complete.

Related to the question whether removal of sparse polynomial time computable subsets destroys completenes is the question whether exponential time complete sets *have* polynomial time computable subsets. It follows easily from the theorem of Berman [Ber77] already cited in Section 6 that all $\leq_m^p$-complete sets in **EXP** have infinite polynomial time computable subsets. This question has however remained open for **NEXP** for a long time. It was Tran [Tra95] who proved the following.

Theorem 19 ([Tra95])   *All $\leq_m^p$-complete sets for **NEXP** are not **P**-immune.*

For weaker reductions, **P**-immune complete sets can be shown to exist by straightforward diagonalization.

8   Instance Complexity

The hardness of individual instances of a complete set may not seem to be a structural question about complete sets and therefore beyond the scope of this paper. One can however ask questions about the distribution of hard and easy instances, which is certainly a question about the structure of the set. Questions like: "Are there an infinite number of hard instances in the set?" and: "Is the subset of hard instances dense or sparse?" certainly fall within the category of structural questions about sets and therefore these questions will be surveyed here with respect to complete sets. Complexity of individual strings was first studied by Kolmogorov and Chaitin (See [Har83, LV93]). Instance complexity is closely related to Kolmogorov complexity. Recall that the t-bounded Kolmogorov complexity of a string x is the size of the smallest Turing machine M that on input λ outputs x and takes no more than $t(|x|)$ steps. (Here also "size" is defined relative to a fixed universal machine.)

The following simple relationship holds between Kolmogorov complexity and instance complexity.

Proposition 20 ([OKSW94])   *For any time constructible function t, there exists a constant c such that for any set A and string x,*

$$\mathrm{ic}^{t'}(x : A) \leq K^t(x) + c,$$

where $t'(n) = ct(n)\log(t(n)) + c$.

The proposition states that the Kolmogorov complexity always is an upper bound on the instance complexity. On the other hand, we are interested in saying that a set A has instances that are hard or difficult. A natural way of expressing this is as follows:

Definition 8.1 *A set A has t-hard instances iff there exists a constant c such that for infinitely many instances x, $\mathrm{ic}^t(x : A) \geq K^t(x) - c$.*

In this definition we did not consider the $\log(n)$ factor that comes out of Proposition 20. Only considering polynomial instance complexity and admitting this $(\log(n))$ factor we also have the following notion of hard instances:

Definition 8.2 *A set A has P-hard instances if for any polynomial t there exist a polynomial t' and a constant c such that for infinitely many x, $\mathrm{ic}^t(x : A) \geq K^{t'}(x) - c$.*

In [OKSW94] it was conjectured that any set not recognizable in a certain time bound t will have t-hard instances.

Conjecture 21 ([OKSW94]) *Let A be a set not in $\mathbf{DTIME}(t)$. Then there exist infinitely many x and a constant c such that $\mathrm{ic}^t(x : A) \geq K^t(x) - c$.*

As evidence for their conjecture it is proved for any set A in $\mathbf{E} \setminus \mathbf{P}$:

Theorem 22 ([OKSW94]) *Let A be a set in $\mathbf{E} \setminus \mathbf{P}$. There exists a constant c such that for any polynomial t there exists a constant c and infinitely many x, such that, $\mathrm{ic}^t(x : A) \geq K^{2^{cn}}(x) - c$.*

More recently, Fortnow and Kummer gave more evidence for this conjecture.

Theorem 23 ([FK95]) *Let A be a set not in $\mathbf{E}$. Then for any c there exists a c' and d such that for infinitely many x, $\mathrm{ic}^{2^{cn}}(x : A) \geq K^{2^{c'n}}(x) - d$.*

We will now shift our attention to complete sets.

8.1 Complete sets and instance complexity

The natural question addressed here is: "Do complete sets have hard instances?" Of course for classes like $\mathbf{NP}$ it is important to fix the time bound of the instance complexity to polynomial. In this section we will therefore only consider $p(n)$-bounded instance complexity for $p(n)$ some polynomial.

The first partial results along these lines were obtained by Orponen in [Orp90]. In this paper it was shown that it cannot be the case that the instance complexity of $\leq_{btt}^p$-hard sets for $\mathbf{NP}$ is low for all the instances unless $\mathbf{P} = \mathbf{NP}$.

Theorem 24 ([Orp90]) *If A is self-reducible and there exist a constant c and a polynomial t such that for all x, $\mathrm{ic}^t(x : A) \leq c\log(|x|) + c$ then A is in* **P**.

It now follows, using the fact that the instance complexity can not decrease much through a $\leq^p_{btt}$-reduction, that all classes that possess self-reducible Turing complete sets—such as **PSPACE** and **NP**—cannot have $\leq^p_{btt}$-hard sets with low instance complexity everywhere unless they are equal to **P**. The above theorem yields as an easy corollary an absolute result about **EXP**. Later on we will see that this can be improved.

Corollary 25 *There does not exists a $\leq^p_{btt}$-hard set A for **EXP** such that for all x, $\mathrm{ic}^t(x : A) \leq c\log(|x|) + c$ for some constant c and polynomial t.*

Proof. Note that sets with low instance complexity everywhere, as in the corollary, are in **P/poly**. Assume for a contradiction that the corollary is not true, it now follows by a well-known theorem of Karp and Lipton [KL80] that if **EXP** has a $\leq^p_T$-hard set that is sparse, that then **EXP** $= \Sigma^P_2$. Furthermore, Σ^P_2 contains self-reducible Turing complete sets and hence by Theorem 24 above it follows that $\Sigma^P_2 = $ **P**, but now we have a contradiction: **EXP** $= \Sigma^P_2 = $ **P**.   □

The above theorem and corollary are still far away from the instance complexity conjecture. Recently Fortnow and Kummer [FK95] showed the instance complexity conjecture in a special case

Theorem 26 ([FK95]) *Every tally set not in **P** has **P**-hard instances.*

Shifting to complete sets it is shown in [BO94] that the conjecture is correct for $\leq^p_m$-complete sets for **E** and polynomial time bounds t.

Theorem 27 ([BO94]) *Let A be a $\leq^p_m$-complete set for **E**. Then there exists an exponentially dense set $C \subseteq A$, such that for all $x \in C$: $\mathrm{ic}^t(x : A) \geq K^t(x) - c$, for c some constant and $t \in \omega(n\log(n))$.*

Note that the theorem is also true for **EXP** and $\leq^p_{1-tt}$-reductions. The theorem states that there is an exponentially dense subset of any complete set for **E** that has hard instances. This is in contrast with the fact that there also exists a subset of A that is in **P** [Ber77] [4] and hence has constant instance complexity for all x. Note also that Theorem 27 implies that any set B that is $\leq^p_{btt}$-hard for **E** contains infinitely many instances x such that $\mathrm{ic}^t(x : B) \geq cK^t(x) - d$ for some constants c and d. It remains open however whether Theorem 27 can be proven for complete sets with respect to other reductions. It even remains open if we drop the density requirement.

---

[4]See [HW94] for an extensive study on immunity and $\leq^p_m$-complete sets for **EXP** and **NEXP**.

It is not known whether Theorem 27 is true for $\leq_m^p$-complete sets for **NP**. However, Orponen showed in [Orp90] that under the assumption that $\mathbf{E} \neq \mathbf{NE}$ the **NP**-hard sets have **P**-hard instances.

Theorem 28 ([Orp90]) *If* $\mathbf{NE} \neq \mathbf{E}$ *then all* $\leq_{1-tt}^p$*-hard sets for* **NP** *have* **P***-hard instances.*

A special type of complete sets in **EXP** does have p-hard instances as is shown in [FK95]

Theorem 29 ([FK95]) *If* A *is complete in* **EXP** *under honest* $\leq_T^p$*-reductions, then* A *has* **P***-hard instances.*

Specific open problems remain:

1. If A is complete for **EXP** under $\leq_{btt}^p$, $\leq_{tt}^p$ or $\leq_T^p$, does A have t-hard instances, for t a polynomial?

2. If A is complete for **EXP** under $\leq_{tt}^p$ or $\leq_T^p$, does A have **P**-hard instances?

3. Is the set of hard instances in 1 or 2 dense?

4. Do many-one complete sets for **NP** have t-hard instances?

9 Post's Program Revisited

The final structural property that we wish to address in this paper is that of auto-reducibility and the closely related notion of mitoticity.

Trakhtenbrot [Tra70] introduced the notion of autoreducibility on recursive sets. Informally a set is auto-reducible, if can be recognized by an oracle machine that on input x never queries x and uses the set itself as an oracle. As such, a set that is auto-reducible can be viewed as having a redundancy of information. The information that a string x is a member of A is also present in other strings in A and this is true *for every member of A.* The notion of autoreducibility is closely related to the notion of mitoticity introduced by Ladner in [Lad73]. Informally a set is mitotic if it can be split into two disjoint subsets, such that both parts are reducible to each other and moreover the original set is reducible to both parts (and vice versa.) As such, mitoticity is a form of ordered auto-reducibility. The parts of the set that contain information about the other parts are neatly ordered in disjoint subsets.

The term mitotic stems from biology, where the mitosis indicates the splitting of a cell into two cells that both contain the same information stored in the DNA of the original cell. As such this term is very appropriately chosen for the subsets that contain precisely the information of the original set. Ladner showed that the two seemingly different, but apparently related notions of auto-reducibility and mitoticity coincide for r.e. sets.

Ambos-Spies [AS84] was the first to carry over the notions of auto-reducibility and mitoticity to the realm of complexity theory. The auto-reducibility notion translates into:

Definition 9.1 *A set A is polynomially autoreducible (autoreducible for short) if there exists a polynomial time oracle Turing machine M such that:*

1. *$A = L(M, A)$.*

2. *for all x: $x \notin Q(M, x, A)$.*

Where $Q(M, x, A)$ is the set of queries M generates on input x with A as an oracle.

There also exists a randomized version of auto-reducibility called *coherence* [Yao90, BF92].

Translating mitoticity [AS84] to the polynomial time setting gives a less clean transition.

Definition 9.2

1. *A recursive set A is polynomial time m(T)-mitotic (m(T)-mitotic for short) if there exists a set $B \in \mathbf{P}$ such that:*
   $A \equiv^p_{m(T)} A \bigcap B \equiv^p_{m(T)} A \bigcap \overline{B}$.

2. *A recursive set A is polynomial time weakly m(T)-mitotic (weakly m(T)-mitotic for short) if there exists disjoint sets A_0 and A_1 such that:*
   $A = A_0 \bigcup A_1$ and $A \equiv^p_{m(T)} A_0 \equiv^p_{m(T)} A_1$.

Although mitoticity implies weak mitoticity, the two notions are quite different in the polynomial time setting. From [AS84] we learn the following about their relation.

Theorem 30 ([AS84])

1. *For any A, if A is m(T)-mitotic then A is m(T)-autoreducible.*

2. *On recursive tally sets m-mitoticity and m-autoreducibility are the same.*

3. *There exists a weakly m-mitotic set that is not m-autoreducible.*

4. *There exists a tally set which is T-autoreducible but not T-mitotic.*

5. *There exists a weakly T-mitotic set that is not T-autoreducible.*

The questions whether

- m-autoreducibility implies m-mitoticity, or

- m-autoreducibility implies weakly m-mitoticity, or

- T-autoreducibility implies weakly T-mitoticity,

remain open. In the following we will turn back to the complete sets.

9.1 Completeness

In this subsection we will address the question of mitoticity and auto-reducibility for complete sets. We will treat these questions separately.

9.1.1 Mitoticity

On the class **NP** we note that all natural **NP** complete sets are m-mitotic. These sets are all p-isomorphic to SAT [BH77] and SAT can easily be shown to be mitotic. Moreover all the notions of mitoticity and autoreducibility are invariant under p-isomorphisms. Thus it follows that if the isomorphism conjecture is true, then all $\leq_m^p$ -complete sets for **NP** are m-mitotic (and hence weakly m-mitotic). Similar observations are valid for all levels of the Polynomial Hierarchy and **PSPACE**. However it is open whether any complete degree (under some reduction) for these classes is completely (weakly) m(T)-mitotic. For **EXP** a little bit more can be said:

Theorem 31 ([BHT93]) *All $\leq_m^p$ -complete sets for* **EXP** *are weakly m-mitotic.*

This theorem parallels nicely the situation for r.e. sets. On the other hand it is shown also that not all $\leq_{tt}^p$ -complete sets are weakly m-mitotic.

Theorem 32 ([BHT93]) *There exists a $\leq_{3-tt}^p$ -complete set for* **EXP** *that is not weakly m-mitotic.*

For **NEXP** the situation is even less clear. It can be shown that every $\leq_m^p$ -complete set can be split into *infinitely* many disjoint subsets, such that each of these subsets is $\equiv_m^p$ to A [BHT93], but nothing is known for the mitotic case. Specific open problems are:

- Are all $\leq_m^p$-complete sets for **EXP** m-mitotic?

- Are all $\leq_m^p$-complete sets for **NP, PH, PSPACE**, or **NEXP** (weakly) m(T)-mitotic?

- Are all $\leq_T^p$-complete sets for **NP, PH, PSPACE, EXP** or **NEXP** (weakly) $m(T)$-mitotic?

9.1.2   Autoreducibility

The situation with respect to autoreducibility is somewhat better understood. Surprisingly, the parallel with recursion theory disappears with respect to complete sets. Of course the same remarks about $\leq^p_m$-complete sets that were made for mitoticity are true for autoreducibility but the $\leq^p_T$-complete sets seem to behave differently.

Theorem 33 ([BF92, BT96]) *Every $\leq^p_T$-complete set for* **NP** *is autoreducible.*

In fact in [BF92] it is shown that all $\leq^p_T$-degrees that contain a self-reducible set are completely autoreducible hence:

Theorem 34 ([BF92]) *All $\leq^p_T$ -complete sets for all levels of the Polynomial Hierarchy and* **PSPACE** *are autoreducible.*

Unfortunately the techniques in [BF92] only apply to sets within **PSPACE**. Extending the techniques in [BT96] it can be shown that also the complete sets for **EXP** are autoreducible.

Theorem 35 ([BFT95]) *All $\leq^p_T$ -complete sets for* **PSPACE** *and* **EXP** *are autoreducible.*

It can also be shown that all $\leq^p_{2-tt}$ -complete sets for **EXP** are 2-tt-autoreducible. Recently Buhrman, Fortnow and Torenvliet [BFT95] established that this is not true for $\leq^p_{3-tt}$-reductions. They show the existence of a set that is 3-tt complete for **EXP** but not $\leq^p_{btt}$-auto-reducible. Translation of this theorem to a larger complexity class gives a 3-tt complete set in **EEXPSPACE** that is not (Turing) auto-reducible. It follows that it is most urgent to determine the autoreducibility of sets complete in the class **EXPSPACE**. If all these sets are auto-reducible then **NL** $\neq$ **NP**, and if there is one that is not auto-reducible then **P** $\neq$ **PSPACE** (See also [BFvMT98].)

Other questions that remain:

- Are all $\leq^p_m$ -complete sets for **NP** or **PSPACE** m-autoreducible?

- Are all $\leq^p_{tt}$ -complete sets for **NP**, **PSPACE** or **EXP** tt-autoreducible?

- Are all $\leq^p_T$ -complete sets for **NEXP** autoreducible?

Acknowledgements

We would like to thank Dieter van Melkebeek, Lance Fortnow, Steve Homer and Peter van Emde Boas for their help and comments on earlier versions of this paper.

10 REFERENCES

[AAR96]   M. Agrawal, E. Allender, and S. Rudich. Reductions in circuit complexity: An isomorphism theorem and a gap theorem. Submitted for publication to JCSS Special Issue on the Eleventh Annual IEEE Conference on Computational Complexity. A preliminary version appeared in the proceedings of that conference, 1996, pp. 2-11., 1996.

[AB93]   M. Agrawal and S. Biswas. Polynomial isomorphism of 1-L complete sets. In *Proc. Structure in Complexity Theory 7th annual conference*, pages 75–80, San Diego, California, 1993. IEEE Computer Society Press.

[ABI93]   E. Allender, J.L. Balcázar, and N. Immerman. A first-order isomorphism theorem. In *STACS 93, Lecture Notes in Computer Science 665*, pages 163–174, 1993.

[ABJ91]   C. Àlvarez, J. Balcázar, and B. Jenner. Functional oracle queries as a measure of parallel time. In C. Choffrut and M.Jantzen, editors, *STACS 91, Lecture Notes in Computer Science 480*, pages 422–433. Springer-Verlag, 1991.

[All88]   E. Allender. Isomorphisms and 1-L reductions. *J. Computer and System Sciences*, 36(6):336–350, 1988.

[AS84]   K. Ambos-Spies. p-mitotic sets. In E. Börger, G. Hasenjäger, and D. Roding, editors, *Logic and Machines, Lecture Notes in Computer Science 177*, pages 1–23. Springer-Verlag, 1984.

[AS94]   E. Allender and M. Strauss. Measure on small complexity classes, with applications for BPP. In *Proc. 35th IEEE Symposium Foundations of Computer Science*, pages 817–818, 1994.

[AS96]   K. Ambos-Spies. Personal communication. 1996.

[ASMZ96] K. Ambos-Spies, E. Mayordomo, and X Zheng. A comparison of weak completeness notions. In *Proceedings 11th annual IEEE conference on Computational Complexity*, pages 171–178, Philadelphia, 1996. IEEE Computer Society Press.

[ASNT94] K. Ambos-Spies, H-C. Neis, and S.A. Terwijn. Genericity and measure for exponential time. In *Proceedings MFCS94, Lecture Notes in Computer Science*, volume 841, pages 221–232. Springer Verlag, 1994.

[Ber77]   L. Berman. *Polynomial Reducibilities and Complete Sets.* PhD thesis, Cornell University, 1977.

[BF92]   R. Beigel and J. Fcigenbaum. On being incoherent without
            being very hard. *Computational Complexity*, 2(1):1–17, Octo-
            ber 1992.

[BF96]   H. Buhrman and L. Fortnow. Two queries. Technical Report
            CS 96-20, University of Chicago Department of Computer Sci-
            ence, 1996.

[BFT95]   H. Buhrman, L. Fortnow, and L. Torenvliet. Using autore-
            ducibility in separating complexity classes. In *Proc. 36d IEEE
            Symposium on Foundations of Computer Science*, pages 520–
            527, Milwaukee, Wis, 1995.

[BFvMT98]   H. Buhrman, L. Fortnow, D. van Melkebeek, and L. Toren-
            vliet. Separating complexity classes using autoreducibilty.
            Technical Report FI-XCT1998-002, Fields Institute, 1998.

[BH77]   L. Berman and H. Hartmanis. On isomorphisms and density
            of NP and other complete sets. *SIAM J. Comput.*, 6:305–322,
            1977.

[BH92a]   H. Buhrman and S. Homer. Superpolynomial circuits, almost
            sparse oracles and the exponential hierarchy. In R. Shyama-
            sundar, editor, *Proc. 12th Conference on the Foundations of
            Software Technology & Theoretical Computerscience, Lecture
            Notes in Computer Science*, pages 116–127. Springer Verlag,
            1992.

[BH92b]   H. Burtschick and A. Hoene. The degree structure of 1-L
            reductions. In *STACS 92, Lecture Notes in Computer Science
            629*, pages 175–184, 1992.

[BHT91]   H. Buhrman, S. Homer, and L. Torenvliet. On complete sets
            for nondeterministic classes. *Math. Systems Theory*, 24:179–
            200, 1991.

[BHT93]   H. Buhrman, A. Hoene, and L. Torenvliet. Splittings, robust-
            ness and structure of complete sets. In *STACS 93, Lecture
            Notes in Computer Science 665*, pages 175–184, 1993.

[BM95]   H. Buhrman and E. Mayordomo. An excursion to the Kol-
            mogorov random strings. In *Proc. Structure in Complexity
            Theory 10th annual conference*, pages 197–203, Minneapolis,
            Minnesota, 1995. IEEE computer society press.

[BO94]   H. Buhrman and P. Orponen. Random strings make hard
            instances. In *Proc. Structure in Complexity Theory 9th an-
            nual conference*, Amsterdam, Holland, 1994. IEEE computer
            society press.

[BST93]   H. Buhrman, E. Spaan, and L. Torenvliet. Bounded reductions. In K. Ambos-Spies, S. Homer, and U. Schöning, editors, *Complexity Theory*, pages 83–99. Cambridge University Press, December 1993.

[BT94]   H. Buhrman and L. Torenvliet. On the structure of complete sets. In *Proc. Structure in Complexity Theory 9th annual conference*, pages 118–133, Amsterdam, Holland, 1994. IEEE computer society press.

[BT96]   H. Buhrman and L. Torenvliet. P-selective self-reducible sets: A new characterization of *P*. *J. Computer and System Sciences*, 53(2):210–217, 1996.

[Buh93]   H.M. Buhrman. *Resource Bounded Reductions*. PhD thesis, University of Amsterdam, June 1993.

[Coo71]   S. Cook. The complexity of theorem-proving procedures. In *Proc. 3rd ACM Symposium Theory of Computing*, pages 151–158, Shaker Heights, Ohio, 1971.

[CS95]   J.-Y. Cai and D. Sivakumar. The resolution of a Hartmanis conjecture. In *Proceedings 36th annual Symposium on Foundations of Computer Science*, pages 362–373, Milwaukee, Wis, 1995. IEEE Computer Societey Press.

[FFK92]   S. Fenner, L. Fortnow, and S.A. Kurtz. The isomorphism conjecture holds relative to an oracle. In *Proc. 33rd IEEE Symposium Foundations of Computer Science*, pages 30–39, 1992.

[FHOS93]   S. Fenner, S. Homer, M. Ogiwara, and A. Selman. On using oracles that compute values. In *STACS 1993, Lecture Notes in Computer Science 665*, 1993.

[FK95]   L. Fortnow and M. Kummer. Resource bounded instance complexity. In *Proceedings STACS, Springer Lecture Notes in Computer Science*, volume 900, pages 597–608, 1995.

[For79]   S. Fortune. A note on sparse complete sets. *SIAM J. Comput.*, 8:431–433, 1979.

[For96]   L. Fortnow. Personal communication. 1996.

[Fu93]   B. Fu. With quasi-linear queries, EXP is not polynomial time Turing reducible to sparse sets. In *Proc. Structure in Complexity Theory 8th annual conference*, pages 185–193, San Diego, California, 1993. IEEE computer society press.

[Har78]   J. Hartmanis. On the logtape isomorphism of complete sets. *Theoretical Computer Science*, 7:273–286, 1978.

[Har83]   J. Hartmanis. On sparse sets in NP-P. *Info. Proc. Let.*, 16:53–60, 1983.

[HKR93]   S. Homer, S. Kurtz, and J. Royer. A note on many-one and 1-truth table complete sets. *Theoretical Computer Science*, 115(2):383–389, July 1993.

[HM93]   S. Homer and S. Mocas. Nonuniform lower bounds for exponential time classes. manuscript, 1993.

[Hom90]   S. Homer. Structural properties of nondeterministic complete sets. In *Proc. Structure in Complexity Theory 5th annual conference*, pages 3–10, Barcelona, Spain, 1990. IEEE Computer Society Press.

[Hom97]   S. Homer. Structural properties of complete problems for exponential time. In L.A. Hemaspaandra and A.L. Selman, editors, *Complexity Theory Retrospective*, volume II, chapter 6, pages 135–150. Springer-Verlag, New York, 1997.

[HOW92]   L. Hemachandra, M. Ogiwara, and O. Watanabe. How hard are sparse sets? In *Proc. Structure in Complexity Theory 7th annual conference*, pages 222–238, Boston, Mass., 1992. IEEE Computer Society Press.

[HW94]   S. Homer and J. Wang. Immunity of complete problems. *Theoretical Computer Science*, 110(1):119–130, April 1994.

[JL93]   D.W. Juedes and J.H. Lutz. The complexity and distribution of hard problems. In *Proc. 34th IEEE Symposium on Foundations of Computer Science*, pages 177–185, 1993.

[JL94]   D.W. Juedes and J.H. Lutz. Weak completeness in E and E_2. manuscript, 1994.

[JY85]   D. Joseph and P. Young. Some remarks on witness functions for non-polynomial and non-complete sets in NP. *Theoretical Computer Science*, 39:225–237, 1985.

[Kar72]   R. Karp. Reducibility among combinatorial problems. In R. Miller and J. Thatcher, editors, *Complexity of Computer Computations*, pages 85–103. Plenum Press, New York, 1972.

[KL80]   R. Karp and R. Lipton. Some connections between nonuniform and uniform complexity classes. In *Proc. 12th ACM Symposium on Theory of Computing*, pages 302–309, 1980.

[KMR89]   S. Kurtz, S. Mahaney, and J. Royer. The isomorphism conjecture fails relative to a random oracle. In *Proc. 21nd Annual ACM Symposium on Theory of Computing*, pages 157–166, 1989.

[KMR90]   S. Kurtz, S. Mahaney, and J. Royer. The structure of complete degrees. In A.L. Selman, editor, *Complexity Theory Retrospective*, pages 108–146. Springer-Verlag, 1990.

[KW95]   J. Koebler and O. Watanabe. New collapse consequences of np having small circuits. In *Proceedings ICALP, Springer Lecture Notes in Computer Science*, volume 944, pages 196–207, 1995.

[Lad73]   R. Ladner. Mitotic recursively enumerable sets. *J. Symbolic Logic*, 38(2):199–211, 1973.

[Lev73]   L. Levin. Universal sorting problems. *Problemy Peredaci Informacii*, 9:115–116, 1973. in Russian.

[LLS75]   R. Ladner, N. Lynch, and A. Selman. A comparison of polynomial time reducibilities. *Theoretical Computer Science*, 1:103–123, 1975.

[LM93]   J.H. Lutz and E. Mayordomo. Measure, stochasticy, and the density of hard languages. In *STACS 93, Lecture Notes in Computer Science 665*, pages 38–47, 1993.

[LM94]   J.H. Lutz and E. Mayordomo. Cook versus Karp-Levin: Separating reducibilities if NP is not small. In *STACS 1994, Lecture Notes in Computer Science*. Springer-Verlag, 1994.

[Lut87]   J.H. Lutz. *Resource-Bounded Category and Measure in Exponential Complexity Classes*. PhD thesis, Department of Mathematics, California Institute of Technology, 1987.

[Lut90]   J. Lutz. Category and measure in complexity classes. *SIAM J. Comput.*, 19(6):1100–1131, December 1990.

[Lut94]   J.H. Lutz. Weakly hard problems. In *Proc. Structure in Complexity Theory 9th annual conference*, Amsterdam, Holland, 1994. IEEE computer society press.

[LV93]   Ming Li and P.M.B. Vitányi. *An Introduction to Kolmogorov Complexity and Its Applications*. Springer-Verlag, 1993.

[LY90]   L. Longpré and P. Young. Cook reducibility is faster than Karp reducibility in NP. *J. Computer and System Sciences*, 41:389–401, 1990.

[Mah82]   S. Mahaney. Sparse complete sets for NP: solution of a conjecture of Berman and Hartmanis. *J. Comput. System Sci.*, 25:130–143, 1982.

[May94a]   E. Mayordomo. Almost every set in exponential time is p-bi-immune. *Theoretical Computer Science*, 1994. to appear.

[May94b]   E. Mayordomo. *Contributions to the study of resource-bounded measure.* PhD thesis, Universitat Politècnica de Catalunya, 1994.

[Myh55]   J. Myhill. Creative sets. *Zeit. Math. Log. Grund. Math.*, 1:97–108, 1955.

[Ogi95]   M. Ogihara. Sparse P-hard sets yield space-efficient algorithms. In *Proceedings 36th annual Symposium on Foundations of Computer Science*, pages 354–361, Milwaukee, Wis, 1995. IEEE Computer Societey Press.

[OKSW94]   P. Orponen, K-I Ko, U. Schöning, and O. Watanabe. Instance complexity. *J. Assoc. Comput. Mach*, 41(1):96–121, 1994.

[Orp90]   P. Orponen. On the instance complexity of NP-hard problems. In *Proc. Structure in Complexity Theory 5th annual conference*, pages 20–27, Barcelona, Spain, 1990. IEEE Computer Society Press.

[Pos44]   E. Post. Recursively enumerable sets of integers and their decision problems. *Bull. Amer. Math. Soc.*, 50:284–316, 1944.

[Rog95]   J. Rogers. The isomorphism conjecture holds and one-way functions exist relative to an oracle. In *Proc. Structure in Complexity Theory 10th annual conference*, pages 90–101, Minneapolis, Minnesota, 1995. IEEE computer society press.

[Sch86]   U. Schöning. Complete sets and closeness to complexity classes. *Math. Systems Theory*, 19:29–41, 1986.

[Sel82]   A. L. Selman. Analogues of semirecursive sets and effective reducibilities to the study of NP complexity. *Information and Control*, 52(1):36–51, January 1982.

[TFL93]   S. Tang, B. Fu, and T. Liu. Exponential time and subexponential time sets. *Theoretical Computer Science*, 115(2):371–381, July 1993.

[Tra70]   B. Trakhtenbrot. On autoreducibility. *Doklady Akad. Nauk. SSSR*, 192:1224–1227, 1970. In Russian, translation in Soviet Math. Dokl. 11 (1970), 814–817.

[Tra95]   N. Tran. On P-immunitiy of nondeterministic complete sets. In *Proc. Structure in Complexity Theory 10th annual conference*, pages 262–263, Minneapolis, Minnesota, 1995. IEEE computer society press.

[Val82]   L. Valiant.   Reducibility   by   algebraic   projections. *L'Enseignement mathématique*, 28:3–4, 1982.

[vM96]   D. van Melkebeek. Reducing P to a sparse set using a constant number of queries collapses P to L. In *Proc. Structure in Complexity Theory 11th annual conference*, pages 88–96, Philadelphia, Pennsylvania, 1996. IEEE computer society press.

[Wat87]   O. Watanabe. A comparison of polynomial time completeness notions. *Theoretical Computer Science*, 54:249–265, 1987.

[Wil85]   C.B. Wilson. Relativized circuit complexity. *J. Comput. System Sci.*, 31:169–181, 1985.

[WT89]   O. Watanabe and S. Tang. On polynomial time Turing and many-one completenes in PSPACE. In *Proceedings of the 4th Structure in Complexity Theory Conference*, pages 15–23, Eugene, Oregon, June 1989. IEEE Computer Society Press.

[Yao90]   A.C. Yao. Coherent functions and program checkers. In *Proc 22nd ACM Symp. Theory of Computing*, pages 84–94, 1990.

[Yes83]   Y. Yesha. On certain polynomial-time truth table reducibilities of complete sets to sparse sets. *SIAM J. Comput.*, 12:411–425, 1983.

Authors addresses

Harry Buhrman
Centrum voor Wiskunde en Informatica,
Kruislaan 413, 1098 SJ Amsterdam
email: buhrman@cwi.nl

Leen Torenvliet
University of Amsterdam, Department of Computer Science
Plantage Muidergracht 24, 1018 TV Amsterdam
email: leen@wins.uva.nl

Kernels and cohomology groups for some finite covers

David M. Evans & Darren G. D. Gray[2]

ABSTRACT We extend work of G. Ahlbrandt and M. Ziegler to give a classification of the finite covers with fibre group of prime order p for the projective space over the field with p elements, and for the Grassmannian of k-sets from a disintegrated set (for $k \in \mathbb{N}$).
AMS classification: 03C35 and 20B27.

This paper is a contribution to the study of the fine detail of the class of (countable) totally categorical structures, in particular the almost strongly minimal ones. The approach we adopt is the one initiated by G. Ahlbrandt and M. Ziegler in [1] and [2] and is purely algebraic. The results we obtain are explicit classification results (under restrictive hypotheses) and are phrased in the terminology of finite covers. It may be helpful if we give a brief impression of them without using this terminology.

Corollary 2.13 represents a classification of certain strongly minimal $\aleph_0$-categorical structures where the associated strictly minimal set is a projective geometry over a prime field. Theorems 3.6 and 3.12 classify certain almost strongly minimal structures in which the associated strictly minimal set is disintegrated. In all these cases it is assumed that the relative automorphism group of the structure over the strictly minimal set is abelian of prime exponent. A key question is whether there exists an expansion of the structure which is biinterpretable with the strictly minimal set (*splitting*). One corollary of our results is that this happens in all cases we consider.

---

[1] Received on the August 6, 1996; revised version on the May 6, 1997.
[2] Supported by an EPSRC Research Studentship.

1   Introduction

1.1   Finite covers

If W is any set then the symmetric group $\mathrm{Sym}(W)$ on W can be considered as a topological group by taking as open sets arbitrary unions of cosets of pointwise stabilisers of finite subsets of W. In this topology, closed subgroups are precisely automorphism groups of first-order structures with domain W. In fact, if H is a subgroup of $\mathrm{Sym}(W)$ then the closure of H in $\mathrm{Sym}(W)$ is the set of elements of $\mathrm{Sym}(W)$ which, for each $n \in \mathbb{N}$, preserve each H-orbit on W^n. Thus we employ the following notation and terminology.

Definition 1.1 A *permutation structure* is a pair $\langle W; G \rangle$ where W is a non-empty set (the *domain*), and G is a closed subgroup of $\mathrm{Sym}(W)$ (the group of *automorphisms*). We shall usually write $G = \mathrm{Aut}(W)$ and refer simply to 'the permutation structure W.' If A is a subset of W and B a subset of W (or more generally of some set on which $\mathrm{Aut}(W)$ is acting in an obvious way), then $\mathrm{Aut}(A/B)$ denotes the permutations of A which extend to elements of $\mathrm{Aut}(W)$ fixing every element of B. We regard $\mathrm{Aut}(W)$ as a topological group with the subspace topology from $\mathrm{Sym}(W)$: a base of open neighbourhoods of the identity consists of subgroups $\mathrm{Aut}(W/X)$ for finite $X \subseteq W$. We shall write permutations on the left of the elements of W.

Permutation structures are obtained by taking automorphism groups of first-order structures on W, and we often regard a first-order structure as a permutation structure without explicitly saying so (by taking for the group of automorphisms of the permutation structure the automorphism group of the first-order structure). In this paper we will be primarily be concerned with the following permutation structures.

Definition 1.2 Let $\mathbb{F}$ be a finite field and V a vector space over $\mathbb{F}$. So V is a permutation structure with automorphism group $\mathrm{GL}(V)$, the group of invertible linear transformations of V. Let $k \in \mathbb{N}$, and let $[[V]]^k$ denote the set of k-dimensional subspaces of V. The group of permutations induced on $[[V]]^k$ by $\mathrm{GL}(V)$ is closed and the kernel of this action consists of the scalar linear transformations. Thus we may regard $[[V]]^k$ as a permutation structure with automorphism group $\mathrm{PGL}(V)$ (the quotient of $\mathrm{GL}(V)$ by the scalar transformations). This is the *Grassmannian* of k-subspaces of V. In the case $k = 1$, we also refer to this as the *projective space* of V.

Definition 1.3 Let D be any set and $k \in \mathbb{N}$. Let $[D]^k$ denote the set of subsets from D of size k. Then the group of permutations induced on $[D]^k$ by $\mathrm{Sym}(D)$ is closed and we refer to the permutation structure $\langle [D]^k; \mathrm{Sym}(D) \rangle$ as the *Grassmannian* of k-sets from (the disintegrated set) D.

We now give the group-theoretic definition of *finite cover*.

Definition 1.4 If C, W are permutation structures, then a finite-to-one surjection $\pi : C \to W$ is a *finite cover* if its fibres form an $\mathrm{Aut}(C)$-invariant partition of C, and the induced map $\mu : \mathrm{Aut}(C) \to \mathrm{Sym}(W)$ given by $\mu(g)w = \pi(g\pi^{-1}(w))$ for $g \in \mathrm{Aut}(C)$ and $w \in W$ has image $\mathrm{Aut}(W)$. We refer to μ as the *restriction* map. The *kernel* of the finite cover is $\ker \mu = \mathrm{Aut}(C/W)$. If this is the trivial group we say that the cover is *trivial*. We say that the cover is *split* if there is a closed complement to $\mathrm{Aut}(C/W)$ in $\mathrm{Aut}(C)$.

Remark. To provide a reference-point for model theorists, we give a model-theoretic version.

Definition 1.5 Let C and W be first-order structures. A finite-to-one surjection $\pi : C \to W$ is a *finite cover* of W if there is a 0-definable equivalence relation E on C whose classes are the fibres of π, and any relation on W^n (respectively, C^n) which is 0-definable in the 2-sorted structure (C, W, π) is already 0-definable in W (respectively, C).

A finite cover (in the sense of 1.5) $\pi : C \to W$ induces a homomorphism

$$\mu : \mathrm{Aut}(C) \to \mathrm{Aut}(W),$$

given by putting $\mu(g)(w) = \pi(g\pi^{-1}(w))$ for all $g \in \mathrm{Aut}(C)$ and $w \in W$. In fact, if W is countable $\aleph_0$-categorical, then 1.5 is equivalent to saying that the fibres of π are the classes of an $\mathrm{Aut}(C)$-invariant equivalence relation on C, and the map $\mathrm{Aut}(C) \to \mathrm{Aut}(W)$ induced by π has image $\mathrm{Aut}(W)$ (Lemma 1.1 of [7] ensures that Definition 1.5 implies the surjectivity). The cover is split if there is an expansion of (C, W, π) which is a trivial cover.

If $\pi : C \to W$ is a finite cover than the associated restriction map $\mu : \mathrm{Aut}(C) \to \mathrm{Aut}(W)$ is a continuous homomorphism and so the kernel of the cover $K = \mathrm{Aut}(C/W)$ is a closed normal subgroup of $\mathrm{Aut}(C)$. As all K-orbits on C are finite, it follows that K is compact (and in fact, profinite). By Lemma 1.1 of [7], μ maps open sets to open sets and closed subgroups to closed subgroups. In particular, the induced isomorphism $\mathrm{Aut}(C)/K \to \mathrm{Aut}(W)$ is a homeomorphism.

Definition 1.6 If C, C' are permutation structures with the same domain and $\pi : C \to W$ and $\pi' : C' \to W$ are finite covers with $\pi(c) = \pi'(c)$ for all $c \in C = C'$ then we say that π' is a *covering expansion* of π if $\mathrm{Aut}(C') \le \mathrm{Aut}(C)$.

Suppose $\pi : C \to W$ is a finite cover. For each $a \in W$ let $C(a)$ denote the fibre above a, that is $\{x \in C : \pi(x) = a\}$. We also define, for any $a \in W$, the *fibre group* $F(a)$ of the cover at a as the permutation group induced by $\mathrm{Aut}(C)$ on $C(a)$. The *binding group* at a is a normal subgroup of the fibre group, and is the permutation group induced on the fibre $C(a)$ by the kernel K. Clearly, if $\mathrm{Aut}(W)$ acts transitively on W then all of the fibre groups are isomorphic as permutation groups, as are the binding groups.

A finite cover is *principal* if its kernel is the product of the fibre groups. We can regard any finite cover $\pi : C \to W$ as a covering expansion of a principal finite cover $\pi_0 : C_0 \to W$ in a canonical way: we take as $\mathrm{Aut}(C_0)$ the group $K_0 \mathrm{Aut}(C)$, where K_0 is $\prod_{w \in W} F(w)$, the product of the fibre groups of π. It is easy to see that a principal finite cover is split.

In this paper we shall be concerned with determining, for primes p, the finite covers of the projective space over the field with p elements, and the Grassmannian of k-sets from a disintegrated set, where the fibre groups are of order p (acting transitively on a set with p elements). So these can be considered as covering expansions of an appropriate principal cover $\pi_0 : C_0 \to W$, and our classification is up to conjugacy of the automorphism groups in $\mathrm{Aut}(C_0)$. The strategy we use is that of G. Ahlbrandt and M. Ziegler from [2]: we first determine the possible kernels of covering expansions of π_0 and then we use cohomological methods to determine the actual covering expansions which can give rise to each kernel. The results are summarised in Corollary 2.13 (for the projective space case) and Theorems 3.6 and 3.12 (for the case of Grassmannians of disintegrated sets), but in particular, we note that all these covers split. In the rest of this section, we summarise the machinery we use. This is mostly taken from [2] and [10].

1.2 Kernels

Suppose $\pi_0 : C_0 \to W$ is a finite cover with abelian kernel K_0. So K_0 is a closed normal subgroup of $\Gamma_0 = \mathrm{Aut}(C_0)$ and we have the short exact sequence

$$1 \to K_0 \to \Gamma_0 \overset{\mu}{\to} G \to 1$$

where μ is restriction to W, and $G = \mathrm{Aut}(W)$. Recall that $\Gamma_0/K_0 \cong G$ as topological groups. Now consider Γ_0 acting on K_0 by conjugation. As K_0 is abelian, K_0 is in the kernel of this action, and so we get an action of $G = \Gamma_0/K_0$ on K_0. From now on we shall write K_0 additively, with the G-action on the left. Thus $gk = hkh^{-1}$, for $g \in G$, $k \in K_0$ and any $h \in \mu^{-1}(g)$. We have the following basic fact (see Lemma 6.2.1 of [8] for a proof).

Lemma 1.7 *With this notation K_0 is a topological G-module.* $\square$

If π_0 is a principal finite cover (so $K_0 = \prod_{w \in W} \mathrm{Aut}(C_0(w))$) then π_0 is split. Let T be a closed complement to K_0 in $\mathrm{Aut}(C_0)$ and suppose K is a closed submodule of K_0. Then KT is a closed subgroup of $\mathrm{Aut}(C_0)$ and so can be thought of as the automorphism group of a split covering expansion of π_0 with kernel K. This gives part of the following, which is a result of Ahlbrandt and Ziegler (Lemma 2.1 of [2]):

Theorem 1.8 *Suppose W is a permutation structure with automorphism group G and $\pi_0 : C_0 \to W$ is a principal finite cover of W with abelian kernel K_0. Regard K_0 as a topological G-module. Then a subgroup K of K_0 is the kernel of some covering expansion of π_0 if and only if it is a closed G-submodule of K_0.* $\square$

We shall mainly use this in the case where the fibre and binding groups are cyclic of order p, for some prime p. In this case, we can identify K_0 with the G-module $\mathbb{F}_p{}^W$ of functions from W into $\mathbb{F}_p$, the field of integers modulo p (the G-action is given by $(gf)(w) = f(g^{-1}w)$, for $f \in \mathbb{F}_p{}^W$, $g \in G$, and $w \in W$). So we are interested in the closed G-invariant subspaces of $\mathbb{F}_p{}^W$. These can sometimes more easily be described by making use of a simple instance of Pontriagin duality (for full details see [8]).

Definition 1.9 Let $\mathbb{F}_p W$ be the vector space of formal linear combinations of elements of W, and regard this as a G-module in the natural way. Let X be a subspace of $\mathbb{F}_p W$ and define its *annihilator* in $\mathbb{F}_p{}^W$ to be

$$X^{\perp} = \{f \in \mathbb{F}_p{}^W : \Sigma_{w \in W} a_w f(w) = 0 \text{ for all } \Sigma_{w \in W} a_w w \in X\}.$$

Note that $X^{\perp} \leq Y^{\perp}$ if and only if $Y \leq X$.

Theorem 1.10 *The closed G-invariant subspaces of $\mathbb{F}_p{}^W$ are precisely the annihilators $X^{\perp}$ of G-invariant subspaces X of $\mathbb{F}_p W$.* $\square$

In summary, to determine kernels of finite covers of W where the fibre and binding groups are of prime order p, it is enough to determine the G-submodules of the permutation module $\mathbb{F}_p W$.

1.3 Derivations

Here we follow rather closely the approach of [2] as modified by Hodges and Pillay in [10].

Recall that if G is a group and M is a G-module, then a *derivation* from G to M is a map $d : G \longrightarrow M$ which satisfies $d(gh) = d(g) + gd(h)$ for all $g, h \in G$. An *inner* derivation is a derivation of the form d_a (for $a \in M$) where $d_a(g) = ga - a$ for all $g \in G$. The set of all such derivations forms an abelian group (with pointwise addition of functions), and the inner derivations form a subgroup. The quotient group is denoted by $H^1(G, M)$, and is referred to as the first cohomology group of G on M. If M is a topological G-module then the continuous derivations form a subgroup of the group of all derivations, and this clearly contains all the inner derivations. We denote the quotient group of continuous derivations modulo inner derivations by $H^1_c(G, M)$.

Suppose that $\pi_0 : C_0 \to W$ is a finite cover of the countably infinite permutation structure W, and suppose from now on that the kernel $K_0 = \mathrm{Aut}(C_0/W)$ is abelian. Then conjugation in $\mathrm{Aut}(C)$ gives K_0 the structure of a topological $\mathrm{Aut}(W)$-module. Let $\mu : \mathrm{Aut}(C_0) \to \mathrm{Aut}(W)$ be the restriction homomorphism. Suppose K is a (closed) G-invariant subgroup of K_0 such that there exists a closed subgroup H_0 of $\mathrm{Aut}(C_0)$ with $H_0 \cap K_0 = K$ and $\mu(H_0) = G$. The following is from ([10], Corollary 18).

Corollary 1.11 *There is a one-to-one correspondence between the set of conjugacy classes of closed subgroups H of $\mathrm{Aut}(C_0)$ which satisfy $\mu(H) = G$ and $H \cap K_0 = K$, and $H_c^1(G, K_0/K)$.* $\square$

In applications here, $\pi_0 : C_0 \to W$ will be a principal finite cover and we will be interested in classifying covering expansions of this which have as kernel some particular G-invariant closed subgroup K of K_0. Corollary 1.11 indicates that to do this we should compute the cohomology group $H_c^1(G, K_0/K)$. If this is trivial, then we can use the following.

Corollary 1.12 *Suppose $\pi_0 : C_0 \to W$ is a principal finite cover with abelian kernel K_0 and K is an $\mathrm{Aut}(W)$-invariant closed subgroup of K_0. If $H_c^1(\mathrm{Aut}(W), K_0/K) = \{0\}$, then there is a covering expansion of π_0 with kernel K. It is unique (up to conjugacy in $\mathrm{Aut}(C_0)$) and split.*

Proof. Existence of a split covering expansion follows from Theorem 1.8 and the remarks preceding it. The uniqueness follows from (1.11): the automorphism groups of any two covering expansions of π_0 with kernel K are conjugate in $\mathrm{Aut}(C_0)$. $\square$

The following curious lemma will replace the use of envelopes in [2]. It allows us to deduce triviality of the cohomology groups we are concerned with from known results about 1-cohomology of finite general linear and symmetric groups.

Lemma 1.13 *Let Γ be a Hausdorff topological group and M a compact topological Γ-module. Suppose there exists $(G_i : i < \omega)$, an increasing chain of subgroups of Γ such that $G = \bigcup_{i<\omega} G_i$ is dense in Γ. Suppose also that for each i we have an open, G_i-invariant subgroup M_i of M, and that $M_{i+1} \leq M_i$ for all $i < \omega$ and $\bigcap_{i<\omega} M_i = \{0\}$. Suppose further that for all i, any continuous derivation from G_i to M/M_i is inner. Then any continuous derivation $d : \Gamma \longrightarrow M$ is inner.*

Proof. Note first that if two continuous derivations $\Gamma \longrightarrow M$ agree on a dense subgroup, then they must be equal. So (as inner derivations are continuous) it will suffice to prove that $\delta = d|G$ is inner. The hypotheses imply that M is metrizable, with a metric θ such that the diameters of the M_i tend to zero (in fact, M as a topological group is isomorphic to the inverse limit of the finite groups M/M_i).

For every $i < \omega$ there exists $a_i \in M$ such that for all $g \in G_i$ we have

$$\delta(g) + M_i = ga_i - a_i + M_i.$$

By compactness of M we may assume that the a_i converge to some $a \in M$. Let d_a denote the inner derivation obtained from a. Thus, for $g \in G_i$, for every $j > i$ there exists $m_j \in M_j$ such that

$$\theta(\delta(g), d_a(g)) = \theta(ga_j - a_j + m_j, ga - a).$$

Now, the m_j tend to 0 as j tends to infinity, and so (by continuity of the Γ-action) $\theta(\delta(g), d_a(g))$ can be arbitrarily small. So $\delta(g) = d_a(g)$. But this holds for all i, and so we conclude that $d = d_a$, as required. $\square$

The following is easy, but useful.

Lemma 1.14 *Let Γ be a topological group and M a continuous Γ-module. Let N be a closed submodule of M and suppose that $H_c^1(\Gamma, M/N)$ and $H_c^1(\Gamma, N)$ are trivial. Then $H_c^1(\Gamma, M)$ is trivial.* $\square$

We shall also require the following version of the 'long exact sequence of cohomology'. A proof can be found in [8] (or see ([5], III.6.1) for the discrete case). If M is a G-module, then $H^0(G, M)$ is the submodule of G-fixed elements of M.

Lemma 1.15 *Suppose G is a group and*

$$0 \to K \to M \to N \to 0$$

is an exact sequence of G-modules. Then there is an exact sequence of abelian groups:

$$0 \to H^0(G, K) \to H^0(G, M) \to H^0(G, N) \to$$

$$\to H^1(G, K) \to H^1(G, M) \to H^1(G, N).$$

If, moreover, G is a topological group and the the short exact sequence is a sequence of compact topological G-modules in which the homomorphisms are continuous, then there is a long exact sequence as above in which the H^1 terms are replaced by H_c^1. $\square$

2   Projective spaces

Throughout this section p will be a prime and $\mathbb{F}_p$ will denote the field with p elements. If n is a cardinal then $V(n, p)$ denotes the vector space of dimension n over $\mathbb{F}_p$.

2.1 Kernels

Let $V = V(\aleph_0, p)$, $W = [[V]]^1$ and $G = \mathrm{GL}(V)$. Let $\pi_0 : C_0 \longrightarrow W$ be a principal finite cover with fibre groups cyclic of order p. In this subsection we determine all possible kernels of covering expansions of π_0. According to Theorem 1.8 and the remarks following it we may identify the kernel K_0 of π_0 with the G-module of functions $\mathbb{F}_p{}^W$, and we want to know the closed G-submodules of K_0. By Theorem 1.10 this problem is equivalent to determining all the G-submodules of the permutation module $\mathbb{F}_p W$. To do this we first investigate the finite case. So let $V_n = V(n, \mathbb{F}_p)$ and $G_n = \mathrm{GL}(V_n)$. We will determine precisely the G_n-submodule structure of the G_n-module $\mathbb{F}_p[[V_n]]^1$. Crucial to this study are the following natural 'incidence maps.'

Definition 2.1 Let k and l be integers satisfying $0 \leq l < k \leq n$. Then we define the map $\beta_{k,l}^n : \mathbb{F}_p[[V_n]]^k \longrightarrow \mathbb{F}_p[[V_n]]^l$ by $\beta_{k,l}^n(w) = \sum\{w' : w' \in [[w]]^l\}$ for $w \in [[V_n]]^k$, and extend linearly to the whole of $\mathbb{F}_p[[V_n]]^k$. So $\beta_{k,l}^n$ maps a k-dimensional subspace w of V_n to a formal sum of all the l-dimensional subspaces of w. This map is clearly a homomorphism of G_n-modules.

The images of these maps (with $l = 1$) provide us with a stock of submodules of $\mathbb{F}_p[[V_n]]^1$. Also consider the map $\beta_{1,0}^n : \mathbb{F}_p[[V_n]]^1 \longrightarrow \mathbb{F}_p$, the so-called *augmentation map*. Obviously, the kernel of this is an $\mathbb{F}_p G_n$-submodule of $\mathbb{F}_p[[V_n]]^1$ (known as the *augmentation submodule*) and it is easy to see that $\ker \beta_{1,0}^n$ is of codimension one in $\mathbb{F}_p[[V_n]]^1$.

Suppose $0 \leq s \leq t$. Denote the number of s-dimensional subspaces of $V(t, p)$ by $\begin{bmatrix} t \\ s \end{bmatrix}_p$ (a *Gaussian coefficient*). It is not difficult to compute this in terms of s, t and p and show that it is coprime to p.

Lemma 2.2 *Let l and k be integers satisfying $0 \leq k < l < r \leq n$. Then we have that $\beta_{l,k}^n \beta_{r,l}^n = \begin{bmatrix} r-k \\ l-k \end{bmatrix}_p \beta_{r,k}^n$.*

Proof. The homomorphism $\beta_{r,l}^n$ maps an r-dimensional subspace x of V_n to the formal sum of the l-dimensional subspaces of x. Let $a = \beta_{r,l}^n(x)$. Now the coeffcient of a k-dimensional subspace y of V_n in $\beta_{l,k}^n(a)$ is equal to the number of l-dimensional subspaces of x which contain y, which is $\begin{bmatrix} r-k \\ l-k \end{bmatrix}_p$. $\square$

Lemma 2.3 *Let r, l and k be integers satisfying $0 \leq k < l < r \leq n$. The we have that $\mathrm{im}\, \beta_{r,k}^n \leq \mathrm{im}\, \beta_{l,k}^n$.*

Proof. This follows from the above because $\begin{bmatrix} r-k \\ l-k \end{bmatrix}_p$, considered as an element of $\mathbb{F}_p$, is non-zero. $\square$

So we have the following chain of submodules:

$$0 \leq \operatorname{im} \beta_{n,1}^n \leq \operatorname{im} \beta_{n-1,1}^n \leq \cdots \leq \operatorname{im} \beta_{3,1}^n \leq \operatorname{im} \beta_{2,1}^n \leq \mathbb{F}_p[[V_n]]^1.$$

Futher, we have:

Lemma 2.4 *The two submodules* $\operatorname{im} \beta_{n,1}^n$ *and* $\ker \beta_{1,0}^n$ *of* $\mathbb{F}_p[[V_n]]^1$ *are incomparable.*

Proof. This follows from the fact that $\operatorname{im} \beta_{n,1}^n$ is one-dimensional, and $\begin{bmatrix} n \\ 1 \end{bmatrix}_p$ is coprime to p. $\square$

We now give a result due to P. Delsarte which is the the most important (and difficult) ingredient needed for calculating the submodule structure of $\mathbb{F}_p[[V_n]]^1$. In its original form, the result involves so-called (non-primitive) generalised Reed-Muller codes, but the version presented here has been 'translated' into a directly applicable form. For the orginal statement of the result, see Theorem 8 in [6] (and see Chapter 5 of [3] for a nice treatment of the coding-theoretic background).

Theorem 2.5 *Let* U *be a* G_n*-submodule of* $\mathbb{F}_p[[V_n]]^1$ *such that* U *is not contained in* $\ker \beta_{1,0}^n$. *Then* $U = \operatorname{im} \beta_{k,1}^n$ *for some* k *satisfying* $2 \leq k \leq n$, *and moreover these are all distinct.* $\square$

We now have enough to describe the submodule structure of $\mathbb{F}_p[[V_n]]^1$, so we collect our results together as:

Theorem 2.6 *Let* $V_n = V(n, \mathbb{F}_p)$ *and* $G_n = GL(V_n)$. *Then the proper* G_n*-submodules of* $\mathbb{F}_p[[V_n]]^1$ *are:*

- $\ker \beta_{1,0}^n$

- $\operatorname{im} \beta_{k,1}^n$ *for* $k = 2, 3, \ldots, n$

- $\operatorname{im} \beta_{k,1}^n \cap \ker \beta_{1,0}^n$ *for* $k = 2, 3, \ldots, n$,

and these are all distinct. Moreover, we have that

$$0 < \operatorname{im} \beta_{n,1}^n < \operatorname{im} \beta_{n-1,1}^n < \cdots < \operatorname{im} \beta_{2,1}^n < \mathbb{F}_p[[V]]^1$$

is a composition series for $\mathbb{F}_p[[V_n]]^1$.

Proof. Delsarte's theorem gives the G_n-submodules of $\mathbb{F}_p[[V_n]]^1$ not contained in $\ker \beta_{1,0}^n$. If U is a submodule contained in $\ker \beta_{1,0}^n$ then (by Lemma 2.4) $U + \operatorname{im} \beta_{n,1}^n$ must therefore equal $\operatorname{im} \beta_{k,1}^n$ for some k and (as $\ker \beta_{1,0}^n \cap \operatorname{im} \beta_{k,1}^n$ contains U and is of codimension 1 in $\operatorname{im} \beta_{k,1}^n$) we have $U = \ker \beta_{1,0}^n \cap \operatorname{im} \beta_{k,1}^n$. $\square$

We can now give the corresponding result for $\mathbb{F}_p W$.

Theorem 2.7 *Let $V = V(\aleph_0, \mathbb{F}_p)$ and $G = GL(V)$. Let $W = [[V]]^1$. Then the G-submodules of $\mathbb{F}_p W$ are:*

- $\ker \beta_{1,0}$

- $\operatorname{im} \beta_{k,1}$ *for $k = 2, 3, \ldots$*

- $\operatorname{im} \beta_{k,1} \cap \ker \beta_{1,0}$ *for $k = 2, 3, \ldots$,*

and these are all distinct. Moreover, we have that

$$0 < \ldots < \operatorname{im} \beta_{n,1} < \cdots < \operatorname{im} \beta_{2,1} < \mathbb{F}_p W$$

is a composition series for $\mathbb{F}_p W$.

Proof. Let $x \in \mathbb{F}_p W$. Then for large enough n we have that $x \in \mathbb{F}_p[[V_n]]^1$. So by Theorem 2.6, the G_n-submodule generated by x equals $\operatorname{im} \beta_{k,1}^n$ or $\operatorname{im} \beta_{k,1}^n \cap \ker \beta_{1,0}^n$, for some k. In the former case we may assume that $x = \beta_{k,1}(w)$ for some $w \in [[V_n]]^k$, and then clearly the G-submodule of $\mathbb{F}_p W$ generated x is $\operatorname{im} \beta_{k,1}$. In the latter case we may assume that $x = \beta_{k,1}(w - w')$ for some distinct $w, w' \in [[V_n]]^k$ (as this is a generator for $\operatorname{im} \beta_{k,1}^n \cap \ker \beta_{1,0}^n$) and then the G-submodule generated by x is seen to be $\operatorname{im} \beta_{k,1} \cap \ker \beta_{1,0}$. $\square$

So we have described all the G-submodules of $\mathbb{F}_p W$. Now we use Theorem 1.10 to describe all the closed G-invariant subgroups of $K_0 = \mathbb{F}_p{}^W$, by taking annihilators. Corresponding to the submodule $\operatorname{im} \beta_{k+1,1}$ of $\mathbb{F}_p W$ we have the folllowing submodule of $\mathbb{F}_p{}^W$:

$$Pol_k = \left\{ f \in \mathbb{F}_p{}^W : \Sigma_{x \in W} a_x f(x) = 0 \text{ for all } \Sigma_{x \in W} a_x x \in \operatorname{im} \beta_{k+1,1} \right\}$$

$$= \left\{ f \in \mathbb{F}_p{}^W : \sum_{x \in [[w]]^1} f(x) = 0 \text{ for all } w \in [[V]]^{k+1} \right\},$$

and the annihilator of $\ker \beta_{1,0}$ is Con, which is the submodule of constant functions. Thus we have:

Corollary 2.8 *The closed, G-invariant submodules of $\mathbb{F}_p{}^W$ are*

- Con

- Pol_k *for $k = 1, 2, \ldots$*

- $Pol_k + Con$ *for $k = 1, 2, \ldots$,*

and these are all distinct. $\square$

Remark 2.1 The above is proved in ([1], Theorem 1.11) for the case $p = 2$. We have used the same notation for the submodules as in [2].

We now provide a different description of the composition factors of the finite-dimensional G_n-module $\mathbb{F}_p[[V_n]]^1$.

Definition 2.9 Suppose E is any field of characteristic p and consider the ring $E[x_1, \ldots, x_n]$ of polynomials in variables $x_1, \ldots, x_n$ as a module for G_n (with G_n acting by substitutions in the usual way). The G_n-action preserves the total degree of an element of the ring, and leaves invariant the ideal I generated by $x_1^p, \ldots, x_n^p$. We let $X_E(i)$ be the image in $E[x_1, \ldots, x_n]/I$ of the homogeneous polynomials of degree i, and regard this as a G_n-module. This is called the module of *truncated polynomials* of degree i.

Remarks 2.2 (i) When $p = 2$, the module $X_{\mathbb{F}_p}(i)$ is just the i-th exterior power of V_n.

(ii) If $i \leq n(p-1)$ there is a natural basis of $X_E(i)$ consisting of $x_1^{a_1} \ldots x_n^{a_n} + I$ where the $a_j \leq p - 1$ and $\Sigma_j a_j = i$. It is easy to see that $X_E(i)$ is naturally isomorphic with $X_{\mathbb{F}_p}(i) \otimes_{\mathbb{F}_p} E$.

(iii) If i is divisible by $p - 1$ then the scalar transformations in G_n act trivially on $X_{\mathbb{F}_p}(i)$.

Some of the results about these modules which we wish to quote are stated in the literature for the case where E is algebraically closed. The following trivial lemma allows us to deduce the corresponding results for the prime field.

Lemma 2.10 *Let $E \subseteq F$ be a field extension and G a group. Suppose M is a finite dimensional EG-module. Extend this to an FG-module $M' = F \otimes_E M \geq M$.*

(i) If M' is an irreducible FG-module, then M is an irreducible EG-module.

(ii) If M has a series of submodules with m non-zero factors $M_1, \ldots, M_m$ and M' has a composition series with m composition factors, then $M_1, \ldots, M_m$ are irreducible EG-modules and the composition factors of M' are $F \otimes_E M_1, \ldots, F \otimes_E M_m$.

(iii) If M is irreducible and N is an EG-module such that M' and $F \otimes_E N$ are isomorphic as FG-modules, then N and M are isomorphic EG-modules.

Proof. (i) If H is an E-subspace of M of E-dimension k, then the F-subspace H' of M' spanned by H has F-dimension k. Moreover, if H is G-invariant, then so is H'.

(ii) As in (i), taking the F-subspaces of M' generated by the terms of the series of submodules of M gives a series of submodules of M' with m non-zero factors isomorphic to $F \otimes_E M_1, \ldots, F \otimes_E M_m$. The statement now follows from the Jordan-Hölder theorem and (i).

(iii) Without loss of generality, we may assume that N is an EG-submodule of M' of E-dimension n. Let $x_1, \ldots, x_n$ be an E-basis for M and $y_1, \ldots, y_n$ an E-basis for N. For each $g \in G$ the matrix representing g with respect to each of these two bases has entries in E. Moreover there is an invertible matrix A with entries in F which for each $g \in G$ conjugates the one matrix to the other. The entries of A therefore give a non-zero solution to a system of linear equations with coefficients in E, and so we can actually find such a solution in E. This matrix of elements of E can then be used to set up a non-zero EG-homomorphism from M to N. The rest follows from irreducibility of M, and comparison of dimensions. $\square$

Now we let S_n denote the elements of G_n of determinant 1. Clearly we can regard any G_n-module as an S_n-module, just by restriction. Let F denote an algebraically closed field of characteristic p.

Theorem 2.11 *Suppose $n \geq 3$. Then the composition factors of $\mathbb{F}_p[[V_n]]^1$, viewed as an S_n-module, are $\left\{ X_{\mathbb{F}_p}(i(p-1)) \; : \; i = 0, 1, \ldots, n-1 \right\}$.*

Proof. By a result of I. Suprunenko and A. Zalesskii (Theorem 1.8(a) of [14], together with the description of the highest weights of the modules $X_F(i(p-1))$ in, for example, the fourth paragraph of [13]) the composition factors of $F[[V_n]]^1$, considered as FS_n-modules, are $X_F(i(p-1))$ for $i = 0, \ldots, n-1$. In particular, there are n composition factors. Also Theorem 2.6 gives a series of submodules of the $\mathbb{F}_pS_n$-module $\mathbb{F}_p[[V_n]]^1$ with n non-zero factors. The result now follows from Lemma 2.10. $\square$

Remark 2.3 The case $p = 2$ (where the composition factors are exterior powers of V_n) is proved directly in [1].

2.2   Cohomology groups

As above, we let $V = V(\aleph_0, p)$ be a countably infinite dimensional vector space over the field with p elements, and $G = \mathrm{GL}(\aleph_0, p)$ its automorphism group. Let $W = [[V]]^1$, considered as a permutation structure with automorphisms those permutations induced by G. Let $\pi_0 : C_0 \to W$ be the principal finite cover of W with fibre groups cyclic of order p (and each fibre of size p). Let K_0 be the kernel of this. In Theorem 2.8 we described the closed G-invariant subgroups K of K_0: a result which was deduced from the parallel situation of finite-dimensional V.

We now show:

Theorem 2.12 *For each possible kernel K we have $H_c^1(G, K_0/K) = \{0\}$.*

Applying (1.12) we get:

Corollary 2.13 *All covering expansions of π_0 split. Any such covering expansion is determined (up to conjugacy in $\mathrm{Aut}(C_0)$) by its kernel, and the possibilities for the kernels are given in Theorem 2.8.* $\square$

Remarks 2.4 For the case $p = 2$ Ahlbrandt and Ziegler ([2]) deduce Theorem 2.12 from results of G. Bell ([4]) about the vanishing of the first cohomology groups of the finite general linear groups $GL(n, 2)$ acting on exterior powers of $V(n, 2)$ (if $n \geq 4$), together with results on envelopes in totally categorical structures. Instead of using envelopes, we use Lemma 1.13. In place of Bell's results, we shall use the following. The cases not covered by Bell's work are due to A. Kleschev. The notation is as in Theorem 2.11.

Theorem 2.14 *For $n \geq 4$ and $i = 0, \ldots, n-1$ we have*

$$H^1(S_n, X_{\mathbb{F}_p}(i(p-1))) = 0.$$

Proof. For $p = 2$ this is in [4]. For $p = 3$ it is ([12], Theorem 4.8). For $p > 3$ it is ([12], Theorem 4.6). $\square$

Corollary 2.15 *If $n \geq 4$ and M is a submodule or quotient module of the $\mathbb{F}_p S_n$-module $\mathbb{F}_p[[V_n]]^1$, then $H^1(S_n, M) = \{0\}$.*

Proof. This follows from Theorem 2.11, Lemma 1.14 (for discrete groups), and Theorem 2.14. $\square$

Proof of 2.12. We use Lemma 1.13 with $\Gamma = GL(V)$ and $M = K_0/K$. Remember that $K_0 = \mathbb{F}_p{}^W$ and K is a closed, Γ-invariant subgroup of K_0. Let $(V_i : 5 \leq i < \omega)$ be an increasing chain of finite dimensional subspaces of V (with V_i of dimension i) with union the whole of V. Let T_i be a complement to V_i in V, and choose these so that $T_i \geq T_{i+1}$ for all i. Let

$$G_i = \{g \in \Gamma : gV_i = V_i, \ g|V_i \text{ has determinant } 1 \text{ and } gx = x \ \forall x \in T_i\}.$$

Then the G_i form an increasing chain whose union is dense in Γ, and G_i is naturally isomorphic to the special linear group $SL(V_i)$ (called S_i in the above). Let K_i be those functions in K_0 which are zero on $[[V_i]]^1$. Thus, K_0/K_i is isomorphic to $\mathbb{F}_p{}^{[[V_i]]^1}$. Let $M_i = (K + K_i)/K$. Then

$$M/M_i = (K_0/K)/(K + K_i/K) \cong K_0/(K + K_i) \cong (K_0/K_i)/(K + K_i/K_i)$$

and all these isomorphisms hold as isomorphisms of G_i-modules. So M/M_i is isomorphic to a quotient module of $\mathbb{F}_p{}^{[[V_i]]^1}$. But the latter is isomorphic to $\mathbb{F}_p[[V_i]]^1$ (this module is self-dual) and so by Corollary 2.15 we get $H^1(G_i, M/M_i) = \{0\}$. Lemma 1.13 is now applicable, and this finishes the proof of 2.12. $\square$

3 Grassmannians of a disintegrated set

Throughout this section p will be a prime number. Let $D = \mathbb{N}$, $G = \text{Sym}(D)$, let $k \in \mathbb{N}$ and let $W = [D]^k$, the Grassmannian of k-sets from the disintegrated set D. We shall describe the finite covers of W with fibre group of order p (and fibres of size p). So we wish to find all the closed G-submodules K of $K_0 = \mathbb{F}_p{}^W$, and compute the cohomology groups $H_c^1(G, K_0/K)$ in each case. We shall use facts about the representation theory of the finite symmetric groups for which we refer to [11]. We use the following notation throughout.

We denote by M^k the G-module $\mathbb{F}_p[D]^k$. If $n \in \mathbb{N}$ let $[n]^k$ be the set of k-subsets from $\{1, \ldots, n\}$, and let $M^k[n] = \mathbb{F}_p[n]^k$, which we regard as a $\mathrm{Sym}(n)$-module. The *Specht submodule* S^k of M^k is the G-submodule generated by the element:

$$e = \sum_{r=0}^{k} \sum_{A \in [k]^r} (-1)^r \{x \leq k : x \notin A\} \cup \{x + k : x \in A\}.$$

Also, if $n \geq 2k$ then the $\mathrm{Sym}(n)$-submodule of $M^k[n]$ generated by this is called the Specht submodule of $M^k[n]$ and is denoted by $S^k[n]$. There is a natural $\mathrm{Sym}(n)$-invariant inner product on $M^l[n]$ (with $[n]^l$ as an orthonormal basis), and any submodule of $M^l[n]$ either contains $S^l[n]$ or is orthogonal to it (see [11], Theorem 4.8). The Specht modules can also be characterised as the intersections of the kernels of the module homomorphisms $\beta_{k,l}$ (and $\beta_{k,l}^n$) definied below (cf. [11], Corollary 17.18). In particular, this shows that $S^k[n] = M^k[n] \cap S^k$. It can be shown that S^k is irreducible ([9], Corollary 3.3) (although this is not necessarily true of $S^k[n]$ for arbitrary n).

3.1   Kernels

As in the previous section, if $0 \leq l < k$ we define a map $\beta_{k,l} : M^k \to M^l$ by setting, for $w \in W$,

$$\beta_{k,l}(w) = \sum_{x \in [w]^l} x$$

and extending linearly. This is clearly a G-homomorphism. Similarly, we define maps $\beta_{k,l}^n : M^k[n] \to M^l[n]$ for $0 \leq l \leq k \leq n$. The following in proved in [9]:

Theorem 3.1 *Any proper, non-zero G-submodule of M^k is an intersection of kernels of homomorphisms $\beta_{k,l}$ for $0 \leq l < k$. The composition factors of M^k are Specht modules $S^0, \ldots, S^k$ (where S^0 is the one-dimensional trivial module $\mathbb{F}_p$).* $\square$

Remark 3.1 Note that there are only finitely many submodules of M^k. In fact, there is an algorithm which enables one to write down the full submodule lattice of M^k (and the only computation involved in this is checking divisibility by p of a finite number of binomial coefficients). See [9] for further details.

We can now use Theorem 1.10 to describe the closed G-submodules of $\mathbb{F}_p{}^W$. Consider the maps $\alpha_{l,k} : \mathbb{F}_p{}^{[D]^l} \to \mathbb{F}_p{}^W$ given by

$$(\alpha_{l,k}(f))(w) = \sum_{x \in [w]^l} f(x).$$

It is easy to see that these are continuous G-module homomorphisms and that $f \in \mathbb{F}_p{}^W$ annihilates $\ker(\beta_{k,l})$ if and only if f is in the image of $\alpha_{l,k}$. Thus we have:

Corollary 3.2 *Any closed G-submodule of $\mathbb{F}_p{}^W$ is a sum of images of homomorphisms $\alpha_{l,k}$, for some $0 \leq l \leq k$. There is a (topological) composition series of $\mathbb{F}_p{}^W$ by closed G-submodules, where the composition factors are duals of Specht modules $(S^l)^*$ for $0 \leq l \leq k$, each appearing with multiplicity one.* $\square$

Remark 3.2 The dual $(S^l)^*$ consists of all linear functions $f : S^l \to \mathbb{F}_p$. We can regard this as the quotient module of $\mathbb{F}_p{}^{[D]^l}$ by the annihilator in $\mathbb{F}_p{}^{[D]^l}$ of S^l.

3.2   Cohomology groups

Lemma 3.3 *If $a \in \mathbb{N}$ is such that $p^a \geq l$ and $n = p^a + 2l - 1$ then the Specht module $S^l[n]$ is self-dual and irreducible.*

Proof. This follows from ([11], Theorem 23.13). $\square$

Lemma 3.4 *If $S^l[n]$ is irreducible, then the map obtained by considering a derivation into $S^l[n]$ as a derivation into $M^l[n]$ gives an embedding of $H^1(\mathrm{Sym}(n), S^l[n])$ into $H^1(\mathrm{Sym}(n), M^l[n])$*

Proof. Applying the long exact sequence of cohomology (Lemma 1.15) to the exact sequence

$$0 \to S^l[n] \to M^l[n] \to M^l[n]/S^l[n] \to 0$$

we get the exact sequence

$$H^0(\mathrm{Sym}(n), M^l[n]) \xrightarrow{\phi} H^0(\mathrm{Sym}(n), M^l[n]/S^l[n]) \to H^1(\mathrm{Sym}(n), S^l[n])$$
$$\xrightarrow{\psi} H^1(\mathrm{Sym}(n), M^l[n])$$

where ψ is as described in the statement of the lemma. So it is only necessary to show that ϕ is surjective. Let T be the subspace of vectors of $M^l[n]$ orthogonal to $S^l[n]$ under the natural form on $M^l[n]$. As $S^l[n]$ is irreducible, we have $M^l[n] = S^l[n] \oplus T$. Now suppose $x + S^l[n] \in M^l[n]/S^l[n]$ is fixed by all elements of $\mathrm{Sym}(n)$. Then there exists a unique $y \in T$ such that $y - x \in S^l[n]$. It follows that y is fixed by $\mathrm{Sym}(n)$ and $\phi(y) = x + S^l[n]$, as required. $\square$

Corollary 3.5 *If $p \neq 2$, $n = p^a + 2l - 1$ and $p^a \geq l$ then $H^1(\mathrm{Sym}(n), S^l[n])$ $= \{0\}$.*

Proof. By Shapiro's lemma ([5], III.6.2) we have

$$H^1(\mathrm{Sym}(n), M^l[n]) \cong H^1(\mathrm{Sym}(l) \times \mathrm{Sym}(n - l), \mathbb{F}_p)$$

which is trivial, as any derivation into a trivial module is a homomorphism, and $\mathrm{Sym}(l) \times \mathrm{Sym}(n - l)$ has no homomorphic image of order p. The result now follows from the above two lemmas. □

Theorem 3.6 *If $p \neq 2$ then for all closed G-submodules K of K_0 we have $H_c^1(G, K_0/K) = \{0\}$. All finite covers of W with fibre group of order p split and (assuming the fibres have order p) each such cover is determined by its kernel. The possibilities for the kernels are described in Corollary 3.2.*

Proof. By Corollary 3.2 all of this follows exactly as in Corollary 2.13 once we show that $H_c^1(G, (S^l)^*) = \{0\}$ for $l = 0, \ldots, k$. We use Lemma 1.13 (with $\Gamma = G$ and $M = (S^l)^*$), as in the proof of Theorem 2.12. Let $n_1 < n_2 < \cdots$ be such that $S^l[n_i]$ is irreducible and self dual for all $i \in \mathbb{N}$. Let G_i consist of permutations in G fixing each $n > n_i$, and let M_i consist of linear functions $S^l \to \mathbb{F}_p$ which are zero on $S^l[n_i]$. This is an open G_i-invariant subgroup of M and M/M_i is isomorphic (as a G_i-module) to $(S^l[n_i])^*$, which by choice of n_i is isomorphic (as G_i-module) to $S^l[n_i]$. The result now follows from Lemma 1.13 and Corollary 3.5. □

3.3   The case $p = 2$

Throughout we assume that $p = 2$ and $n = 2^a + 2l - 1$ where $2^a > l$. Thus $S^l[n]$ is self-dual and irreducible. We denote by Z_2 the cyclic group of order 2.

As before, $\pi_0 : C_0 \to W$ is a principal finite cover with fibres of size 2 and fibre groups Z_2. The difficulty with this case is that the cohomology groups $H_c^1(G, (S^l)^*)$ are not all zero.

Lemma 3.7 *If $k \geq 2$ there are covering expansions $\pi_1 : C_1 \to W$ and $\pi_2 : C_2 \to W$ of π_0 with trivial kernels, whose automorphism groups T_1 and T_2 are not conjugate in $\mathrm{Aut}(C_0)$. Thus, $H_c^1(G, K_0) \neq \{0\}$.*

Proof. The connection between the two parts of the claim is given by Corollary 1.11. It follows from a version of Shapiro's lemma ([8]) that $H_c^1(G, K_0) = Z_2$, but we shall describe explicitly π_1 and π_2.

We take π_1 to be a covering expansion with trivial fibre group: pick a transversal of the fibres and let π_1 be the expansion of π_0 by this (as a unary relation). Now let $w = \{1, \ldots, k\} \in W$ and let $H \leq G$ be those elements of the stabiliser of w which induce an even permutation on w. This is of index 2 in the stabiliser of w. Let C_2 be the set of left cosets of H in G (regarded as a permutation structure with automorphisms those permutations induced by left multiplication by elements of G) and $\pi_2 : C_2 \to W$ be given by $\pi_2(gH) = gw$. This is a finite cover with fibres of size 2 and fibre group Z_2. It is clear that the only element of G fixing each element of W is the identity. So π_2 may be regarded as a covering expansion of π_0 with trivial kernel and $\mathrm{Aut}(C_1)$ and $\mathrm{Aut}(C_2)$ are not conjugate in $\mathrm{Aut}(C_0)$ (as they have different fibre groups). $\square$

By Shapiro's lemma ([5], III.6.2) we have the following.

Lemma 3.8

$$H^1(\mathrm{Sym}(n), M^l[n]) = \begin{cases} Z_2 & \text{if } l = 0, 1 \\ Z_2 \times Z_2 & \text{if } l \geq 2. \end{cases} \quad \square$$

Shapiro's lemma can be used to calculate the derivations $\mathrm{Sym}(n) \to M^l[n]$. If $l \geq 2$ then (modulo inner derivations) these are as follows. Note that to specify such a derivation, it is enough to give its value on each of the transpositions $(i, i+1) \in \mathrm{Sym}(n)$ for $i = 1, \ldots, n-1$ (as these generate $\mathrm{Sym}(n)$).

- 0

- $\delta_0 : \mathrm{Sym}(n) \to M^l[n]$ given by

$$\delta_0(g) = \begin{cases} 0 & \text{if } g \text{ is even} \\ \mathbf{j} = \sum_{w \in [n]^l} w & \text{if } g \text{ is odd.} \end{cases}$$

- $\delta_1 : \mathrm{Sym}(n) \to M^l[n]$ given by

$$\delta_1((i, i+1)) = \sum \{w \in [n]^l \ : \ i, i+1 \in w\}.$$

- $\delta_2 : \mathrm{Sym}(n) \to M^l[n]$ given by

$$\delta_2((i, i+1)) = \sum \{w \in [n]^l \ : \ i, i+1 \notin w\}.$$

If $l = 0$ or 1 then δ_0 gives the non-zero element of $H^1(\mathrm{Sym}(n), M^l[n])$.

Lemma 3.9 *Let $l \neq 0$. There exists an inner derivation $d_x : \mathrm{Sym}(n) \to M^l[n]$ with $\mathrm{im}(\delta_j + d_x) \subseteq S^l[n]$ if and only if $l = 2$ and $j = 2$.*

Proof. **Case $j = 1$.** Suppose $l > 2$. Let $x \in M^l[n]$. Recall that

$$S^l[n] = \bigcap_{r=0}^{l-1} \ker \beta_{l,r}^n.$$

So it is enough to find $0 \le r < l$ such that $\beta_{l,r}^n \circ (\delta_1 + d_x) \ne 0$. Now,

$$\beta_{l,l-2}^n(\delta_1((i,i+1))) + x + (i,i+1)x) = \beta_{l,l-2}^n(\sum \{w \in [n]^l : i, i+1 \in w\})$$
$$+ \beta_{l,l-2}^n(x + (i,i+1)x).$$

Moreover, $x + (i,i+1)x$ is a sum of terms of the form $\{i\} \cup w + \{i+1\} \cup w$ for some $w \in [n\backslash\{i,i+1\}]^{l-1}$. Thus no element of the form $w' \in [n\backslash\{i,i+1\}]^{l-2}$ appears in the support of $\beta_{l,l-2}^n(x+(i,i+1)x)$. However any such w' appears in $\beta_{l,l-2}^n(\sum\{w \in [n]^l : i, i+1 \in w\})$ with coefficient 1. Thus $\delta_1 + d_x \not\le$ ker $\beta_{l,l-2}^n$.

Case $j = 0$. The image of δ_0 is a one-dimensional submodule of $M^l[n]$, so is not contained in $S^l[n]$ (which is irreducible). If $\text{im}(\delta_0 + d_x) \subseteq S^l[n]$ for some $x \in M^l[n]$ then as in the previous case $\beta_{l,r}^n(\mathbf{j} + (i,i+1)x + x) = 0$ for $0 \le r < l$. It is now easy to argue that this implies $\beta_{l,r}^n(\mathbf{j}) = 0$, and so $\mathbf{j} \in S^l[n]$, a contradiction.

Case $j = 2$. If $0 \le r < l$ then for $x \in M^l[n]$:

$$\beta_{l,r}^n(\delta_2((i,i+1)) + x + (i,i+1)x) =$$

$$\binom{n-2-r}{l-r} \sum_{w \in [n\backslash\{i,i+1\}]^r} w + \sum_{y \in Y}(\{i\} \cup y + \{i+1\} \cup y)$$

for some $Y \subseteq [n \backslash \{i,i+1\}]^{r-1}$.

So, this is zero for all r only if each of the binomial coefficients

$$\binom{n-2}{l}, \binom{n-2-1}{l-1}, \ldots, \binom{n-2-(l-1)}{1}$$

is divisible by 2. By ([11], Corollary 22.5) this happens if and only if

$$n - 2 - l \equiv -1 \bmod 2^s$$

where 2^s is the smallest power of 2 greater than l. Recalling that $n = 2^a + 2l - 1$ and $2^a > l$, this says that $l = 2$. Note that if $l = 2$ then the above shows that $\text{im}(\delta_2) \subseteq S^2[n]$.

It remains to show that (in the case $l = 2$) there is no inner derivation d_x such that $\text{im}(\delta_1 + d_x) \subseteq S^2[n]$. But in this case $\delta_1 = \delta_0 + \delta_2$ and $\text{im}(\delta_2) \subseteq S^2[n]$, which implies $\text{im}(\delta_0 + d_x) \subseteq S^2[n]$. This contradicts the result already established for $j = 0$. $\square$

Corollary 3.10 *For* $l \ge 1$ *we have*

$$H^1(\text{Sym}(n), S^l[n]) = \begin{cases} 0 & \text{if } l \ne 2 \\ Z_2 & \text{if } l = 2. \end{cases}$$

Proof. This follows directly from the above lemma and Lemma 3.4. □

Corollary 3.11 *(i) We have*

$$H^1_c(G, (S^l)^*) = \begin{cases} 0 & \text{if } l \neq 2 \\ Z_2 & \text{if } l = 2. \end{cases}$$

(ii) If K is a closed G-submodule of $K_0 = \mathbb{F}_2^W$ then $H^1_c(G, K_0/K)$ is trivial if $(S^2)^*$ is not a composition factor of K_0/K, otherwise, it is of order 2.

Proof. (i) The case $l = 0$ follows from the fact that S^0 is the trivial module $\mathbb{F}_2$, and G has no subgroup of index 2. In the other cases where $l \neq 2$ the result follows from Corollary 3.10 and Lemma 1.13 exactly as in Corollary 3.6.

Now suppose that $l = 2$. First note that there is a non-inner derivation $d :$ $G \to (S^2)^*$ (otherwise $H^1_c(G, \mathbb{F}_2^{[D]^2})$ is trivial, by Corollary 3.2 and Lemma 1.14, contradicting Lemma 3.7). Now suppose that d' is another non-inner derivation. Then by Lemma 1.13 (or rather, its proof) for infinitely many choices of n, the derivations induced by d and d' on $(S^2[n])^*$ are non-inner. Thus by Corollary 3.10, and the fact that these $S^2[n]$ are self-dual, we deduce that $d - d'$ induces an inner derivation on $(S^2[n])^*$ for infinitely many n. It follows from the proof of Lemma 1.13 that $d - d'$ is inner.

(ii) If $(S^2)^*$ is not a composition factor of K_0/K this follows from Corollary 3.2 and Lemma 1.14. More generally, K_0 has the property that the trivial G-module appears as a (topological) composition factor with multiplicity 1, and this appears as a submodule of K_0 (the constant functions). Thus if M is a closed submodule of a continuous homomorphic image of K_0 and N is a closed submodule of this then any fixed point of G on M/N comes from a fixed point of G on M. So then the long exact sequence (Lemma 1.15) shows that the sequence

$$0 \to H^1_c(G, N) \to H^1_c(G, M) \to H^1_c(G, M/N)$$

is exact. It is now easy to deduce (ii) from (i) (and Corollary 3.2). □

Remark 3.3 By the dual version of Corollary 3.16 of [9] we have that $(S^2)^*$ is a composition factor of K_0/K if and only if $\mathrm{im}(\alpha_{2,k})$ is not contained in K (cf. Corollary 3.2).

We use the notation of Lemma 3.7.

Theorem 3.12 *Let $\pi : C \to W$ be a covering expansion of π_0 with kernel K. Then $\mathrm{Aut}(C)$ is conjugate in $\mathrm{Aut}(C_0)$ to KT_1 or to KT_2 (and only the first of these if $k = 1$). These are non-conjugate in $\mathrm{Aut}(C_0)$ if and only if $\mathrm{im}(\alpha_{2,k}) \not\subseteq K$. In any case, π is split.*

Proof. Let $d : G \to K_0$ be the derivation corresponding to π_2 of Lemma 3.7. Then the derivation corresponding to the split cover with automorphism group KT_2 and kernel K is $\bar{d}$, the result of composing d with the natural map $K_0 \to K_0/K$. By Corollary 3.11 and Corollary 1.11 the theorem follows once we show that if $\mathrm{im}(\alpha_{2,k}) \not\leq K$ then $\bar{d}$ is not inner. But if $\bar{d}$ is inner, there exists $a \in K_0$ such that $d - d_a$ has image in K, so is a derivation into K. Now, if $\mathrm{im}(\alpha_{2,k}) \not\leq K$ then $(S^2)^*$ is a composition factor of K_0/K, so not a composition factor of K and thus all continuous derivations into K are inner (by Corollary 3.11). So d is inner, contradicting Lemma 3.7. $\square$

4   REFERENCES

[1] G. Ahlbrandt, M. Ziegler, 'Invariant subspaces of V^V', J. Algebra 151 (1992), 26–38.

[2] G. Ahlbrandt, M. Ziegler, 'What's so special about $(\mathbb{Z}/4\mathbb{Z})^\omega$?', Archive for Mathematical Logic 31 (1991), 115–132.

[3] E. F. Assmus Jr. and J. Key, Designs and Their Codes, Cambridge University Press, Cambridge, 1992.

[4] G. B. Bell, 'On the cohomology of the finite special linear groups I, II' , J. Algebra 54 (1978), 216–238, 239–259.

[5] Kenneth S. Brown, Cohomology of Groups, Springer GTM 87, Springer Verlag, Berlin 1982.

[6] Philippe Delsarte, 'On cyclic codes that are invariant under the general linear group', IEEE Trans. Information Theory, Vol. IT-16 (1970), 760–769.

[7] D. M. Evans, 'Finite covers with finite kernels', Ann. Pure Appl. Logic, *to appear.*

[8] D. M. Evans, A. A. Ivanov and D. Macpherson, 'Finite covers', *to appear in* Model Theory of Groups and Automorphism Groups, Proceedings of a Summer School in Blaubeuren, 1995 (ed. D. M. Evans), London Mathematical Society Lecture Notes 244, Cambridge University Press, 1997.

[9] D. G. D. Gray, 'The structure of some permutation modules for the symmetric group of infinite degree', J. Algebra 193 (1997), 122–143.

[10] W. A. Hodges, A. Pillay, 'Cohomology of structures and some problems of Ahlbrandt and Ziegler', J. London Math. Soc. (2) 50 (1994), 1-16.

[11] G. D. James, The Representation Theory of the Finite Symmetric Groups, Springer LNM 682, Springer, Berlin, 1978.

[12] A. S. Kleschev, '1-cohomology of a special linear group with coefficients in a module of truncated polynomials', Math. Zametkii 49 (1992), 63–71.

[13] A. E. Zalesskii and I. D. Suprunenko, 'Reduced symmetric powers of natural realizations of the groups $SL_m(P)$ and $Sp_m(P)$ and their restrictions to subgroups', Siberian Math. Journal 31 (1990), 33–46.

[14] A. E. Zalesskii and I. D. Suprunenko, 'Permutation representations and a fragment of the decomposition matrix of symplectic and special linear groups over a finite field', Siberian Math. Journal 31 (1990), 46–60.

Authors address

School of Mathematics,
UEA, Norwich NR4 7TJ,
England
e-mail: d.evans@uea.ac.uk
     d.gray@uea.ac.uk

On "star" schemata of Kossak and Paris

Vladimir Kanovei[2] [3]

ABSTRACT Kossak and Paris introduced the "star" versions of the Induction and Collection schemata for Peano arithmetic, in which one admits, as extra parameters, subsets of a given nonstandard Peano model coded in a fixed elementary end extension of the model. We prove that the "star" schemata are not finitely axiomatizable over recursively saturated models. A partial solution of a conjecture of Kossak and Paris is obtained.

Introduction

Kossak and Paris [2] have suggested the study of properties of second–order **PA** structures of the form $\langle M; N/M \rangle$, where M and N are nonstandard models of the Peano arithmetic, **PA**, N being an end extension of M (so that M is an initial segment of N), and N/M is the collection of all sets $X \subseteq M$ of the form $X = X' \cap M$, where $X' \subseteq N$ is an N-finite set (*i. e.* X' is coded in N as a finite set by some $a \in N$).

Let $\Sigma_n[N/M]$ denote the extension of the class of Σ_n formulas of the **PA** language by elements of M occuring in the usual way and sets $X \in N/M$ used as extra second–order parameters (with no quantification over them allowed).

This enrichment of the language leads us to the question: are the Induction and Collection schemata, restricted to the class of $\Sigma_{n+1}[N/M]$ formulas, really stronger than those restricted to $\Sigma_n[N/M]$ formulas ? Kossak and Paris obtained (see [2]) positive answers for the case when $n = 1$ or 2, and formulated it as a conjecture that the result should be true for all n.

This note is written to present a partial answer. We prove that, at least

[1] Received September 2, 1996; revised version April 4, 1997.

[2] The author thanks the organizers of the Warsaw meeting on Peano arithmetic (April 1996) and LC'96 (San Sebastian, July 1996) for the opportunity of giving preliminary talks on the main results of the paper.

[3] Partially supported by AMS and DFG grants and visiting appointments from IPM (Tehran), Caltech, and universities of Amsterdam, Wisconsin (Madison), and Wuppertal.

in the case when M is countable and recursively saturated, there exists a countable elementary end extension N of M such that M models the schemata for $\Sigma_{n-1}[N/M]$ formulas but does not model those for $\Sigma_{n+1}[N/M]$ formulas.

The level n is still missing. The other open problem is to eliminate the requirement that M is recursively saturated.

The proof involves a coding technique for subsets of **PA** models. In particular, we prove that, given a model $M \models$ **PA** and a set $X \subseteq M$, for any n there exists a set $A \subseteq M$ such that M still models both Induction and Collection for $\Sigma_n(A)$ formulas (where A can occur as an extra second–order parameter), but X is $\Delta_{n+1}(A)$ in M.

1   Preliminaries

We give Kaye [1] as a general reference in matters of notation, but take some space to introduce more special notation which reflects the scope of the paper.

Let M be a countable **PA** model, fixed for the remainder.

An M-*finite set* will mean: a set $X \subseteq M$ coded in M as a finite set. The notion of an M-*finite sequence* (of elements of M) is understood similarly.

A set $X \subseteq M$ is M-*piecewise definable*, M-p. df. in brief, iff $X \cap u$ is M-finite for every M-finite set u.

Σ_n and Π_n will denote the ordinary classes of formulas in the **PA** language.

By $\mathit{\Sigma_n}$ (slanted !) we shall denote the collection of all Σ_n-formulas of the **PA** language, with elements of M allowed as parameters.

Let $\mathcal{X} \subseteq \mathcal{P}(M)$. By $\mathit{\Sigma_n}[\mathcal{X}]$ we shall denote the collection of all formulas obtained from $\mathit{\Sigma_n}$ by the permission to use terms composed from characteristic functions of sets $X \in \mathcal{X}$ to substitute **PA** variables. We write $\mathit{\Sigma_n}(X)$ or $\mathit{\Sigma_n}(X, Y)$ instead of resp. $\mathit{\Sigma_n}[\{X\}]$ or $\mathit{\Sigma_n}[\{X, Y\}]$.

By Σ_n we shall also denote the collections of all subsets of M, $M \times M$ etc. definable in M by $\mathit{\Sigma_n}$ formulas (where, by definition, elements of M may occur as parameters). We define $\Sigma_\infty = \bigcup_{1 \leq n \in \omega} \Sigma_n$.

Other similar notation, like $\Pi_n(X)$, has the corresponding meaning.

Finally, $\Delta... = \Sigma... \cap \Pi...$, in all cases.

It will always be the case that the subsets X of M involved as extra set parameters are M-piecewise definable.

2   The main results

Let Γ be a definability class. We shall consider the following schemata of axioms, where Φ is assumed to be a formula in Γ:

Γ-Collection:
$$\forall a_0\,[\,\forall a < a_0\,\exists b\,\Phi(a,b)\ \Longrightarrow\ \exists b_0\,\forall a < a_0\,\exists b < b_0\,\Phi(a,b)\,],$$

Γ-Induction: $\Phi(0)\,\&\,\forall a\,[\,\Phi(a)\ \Longrightarrow\ \Phi(a+1)\,]\ \Longrightarrow\ \forall a\,\Phi(a).$

Let $N \models \mathbf{PA}$ be an end extension of M. Following Kossak and Paris [2], we consider the schemata for the classes $\Gamma = \Sigma_n[N/M]$. (Notation $B\Sigma_n^*$ and $I\Sigma_n^*$ was used in [2] to denote $\Sigma_n[N/M]$-Collection and $\Sigma_n[N/M]$-Induction.)

Working with a hierarchy, one naturally wants to figure out whether a given property on a level $n+1$ is strictly stronger than it is on level n. Regarding the Induction and Collection schemata, Kossak and Paris obtained the following results (see [2]). First, every countable model $M \models \mathbf{PA}$ has an elementary end extension N such that M models $\Sigma_1[N/M]$-Induction but does not model $\Sigma_2[N/M]$-Collection. Second, every countable model $M \models \mathbf{PA}$ has an elementary end extension N such that M models $\Sigma_2[N/M]$-Induction but does not model $\Sigma_3[N/M]$-Induction. They conjectured that the results generalize to higher levels.

We do not know how to prove this conjecture even in the case of recursively saturated models M. The following theorem gives a partial result.

Theorem 2.1 (Main theorem)
Let $n \geq 2$. Assume that M is a countable recursively saturated model of **PA**. *There exists a countable elementary end extension N of M such that M models $\Sigma_{n-1}[N/M]$-Induction but does not model $\Sigma_{n+1}[N/M]$-Collection.*

Induction usually implies Collection; for instance Σ_n-Induction implies Σ_n-Collection for any particular n, see *e.g.* Proposition 4.1 in Sieg [3], so the theorem formally yields the result for either of the schemata separately. However, to make the exposition self-contained, we shall prove independently that M also models $\Sigma_{n-1}[N/M]$-Collection and does not model $\Sigma_{n+1}[N/M]$-Induction.

The level n is still missing. On the other hand, the theorem implies that for any n we have "essential" gap at least for one of the successive pairs, $n-1, n$ and $n, n+1$.

The proof is based on two ideas concerning how to code subsets of Peano models. The first idea appears in the following theorem, perhaps of separate interest.

Theorem 2.2 *Let $n \geq 1$. Suppose that M is a countable model of* **PA** *and $T \subseteq M$ is an inductive set for M. Let finally $X \subseteq M$ be an M-p. df. set. Then there exists an M-p. df. set $A \subseteq M$ such that M models both Induction and Collection for $\Sigma_n(T,A)$, but X is $\Delta_{n+1}(T,A)$ in M.*

(A set $T \subseteq M$ is *inductive* iff M models $\Sigma_m(T)$-Induction for all m.

The set T enters the result and the proof as a uniform parameter.) Thus any M-p. df. set $X \subseteq M$ ($e.\,g.$ X may effectively code a cofinal map from some $M_{\leq a}$ to M, violating the Collection schema) can be coded in $\langle M\,;T,A\rangle$ at level $n+1$ in such a way that the schemata still hold in M at level n and below.

To prove Theorem 2.2, we introduce the notion of a $\Sigma_n(T)$-generic matrix in Section 3. A *matrix* here is essentially a sequence $\mu = \langle \mu_a : a \in M \rangle$ of functions $\mu_a \in 2^M$. We show that $\Sigma_n(T)$-generic matrices do not violate $\Sigma_n(T)$-Induction and $\Sigma_n(T)$-Collection in M.

Then we use, in Section 4, a "double" $\Delta_{n+1}(T)$ matrix μ, which is essentially a double sequence $\langle \mu_{ai} : a \in M,\ i \in \{0,1\} \rangle$ of functions $\mu_{ai} \in 2^M$, satisfying the property that, for any M-p. df. set $X \subseteq M$, putting $\mu_a = \mu_{a1}$ iff $a \in X$ and $\mu_a = \mu_{a0}$ otherwise, one obtains a $\Sigma_n(T)$-generic matrix $\mu = \langle \mu_a : a \in M \rangle$ independently on the choice of X. On the other hand, μ codes X in such a way that X is $\Delta_{n+1}(T,\mu)$ in M.

To prove Theorem 2.2, we apply this construction for a given M-p. df. set $X \subseteq M$. This results in a $\Sigma_n(T)$-generic matrix μ (so M models the schemata for the class $\Sigma_n(T,\mu)$) such that X is $\Delta_{n+1}(T,\mu)$ in M. It remains to convert μ to a set $A \subseteq M$.

Let us describe how this theorem works in the proof of Theorem 2.1. We consider a countable recursively saturated model M of **PA**. There exists an inductive satisfaction class $T \subseteq M$. Note that M models $\Sigma_m(T)$-Induction for all $m \in \omega$, and T satisfies the Tarski rules for a class of true formulas provided elements of M are adequately treated as Gödel numbers of **PA** formulas.

We then fix a cofinal M-p. df. map $\beta : M_{<a_0} \longrightarrow M$, a_0 being an arbitrary nonstandard element of M. Applying Theorem 2.2, we obtain an M-p. df. set $A \subseteq M$ such that β is $\Delta_{n+1}(T,A)$ in M and M satisfies the Induction and Collection schemata for $\Sigma_n(T,A)$.

As the second part of the proof of Theorem 2.1, we define in Section 5 a countable elementary end extension N of M (an ultrapower of M) such that

(1) Both T and A belong to N/M, therefore β is $\Delta_{n+1}[N/M]$ in M.

(2) Every element of N/M belongs to $\Delta_2(T,A)$ in M.

Now (1) implies that $\Sigma_{n+1}[N/M]$-Collection and $\Sigma_{n+1}[N/M]$-Induction fail in M. On the other hand, it follows from (2) that Collection and Induction for the class $\Sigma_{n-1}[N/M]$ hold in M by the choice of A.

We do not know how to reduce $\Delta_2(T,\mu)$ to $\Delta_1(T,\mu)$ in (2), that would improve Σ_{n-1} to Σ_n in Theorem 2.1. The other open problem is to eliminate the assumption that the given model M is recursively saturated. (This property is used in the ultrapower construction of N.)

3 Generic matrices

This section starts the proof of Theorem 2.2. Thus let us suppose that M is a nonstandard countable model of **PA**, and T is an inductive set for M (so that M models $\Sigma_n(T)$-Induction for all n), but not necessarily a satisfaction class. For instance, this includes the case when T is the empty set; then the classes $\Sigma_n(T)$ etc. below become equal to Σ_n etc.

Generic matrices

We wish to consider generic sequences of maps from M into $2 = \{0,1\}$. Technically, this can be realized in the form of generic matrices.

A *matrix* is an arbitrary function $\mu : M \times M \longrightarrow \{0,1\}$. Alternatively, a matrix μ can be seen as the indexed family $\langle \mu_a : a \in M \rangle$, where every $\mu_a \in 2^M$ is defined by $\mu_a(l) = \mu(a, l)$ for all l.

Let COND denote the set of all M-finite functions p such that

i) the domain $\mathrm{dom}\, p$ is an M-finite subset of $M \times M$;

ii) all values of p are among 0 and 1.

Elements of COND, called *conditions* below, are identified with their codes in M, so that COND is understood as a definable class in M.

The set COND is ordered by inclusion: $p \leq p'$ iff p' extends p as a function. In this case, we say that p' is *stronger than* p. A set $D \subseteq$ COND is *dense* iff every $p \in$ COND is extended by some $p' \in D$.

Let $C \subseteq$ COND. A condition p *decides* C iff either $p \in C$ or there is no stronger condition $p' \in C$. We observe that the set $\{p : p$ decides $C\}$ is dense; and if a set C is dense then *deciding* C is equivalent to *belonging to* C.

A matrix μ *extends* a condition $p \in$ COND iff $p \subseteq \mu$, *i.e.* $p(a,l) = \mu(a,l)$ for all $\langle a, l \rangle \in \mathrm{dom}\, p$. A matrix μ *decides* a set $C \subseteq$ COND iff μ extends a condition which decides C. As above if C is dense then μ decides C iff μ extends a condition in C.

Now we introduce the notion of a $\Sigma_n(T)$-generic matrix. The definition intends to meet the following two requirements of opposite character:

1. Any $\Sigma_n(T)$-generic matrix μ has to decide $\Delta_{n+1}(T)$ sets,   *and*:

2. One would be able to define a $\Sigma_n(T)$-generic matrix μ of class $\Delta_{n+1}(T)$.

The latter requirement implies that *some*, and even some *dense*, $\Delta_{n+1}(T)$ sets cannot be decided by μ. However we can decide a reasonably large subfamily of dense $\Delta_{n+1}(T)$ sets. For example this subfamily will contain all sets of the form $\{p : p$ decides $C\}$ where C is an arbitrary $\Sigma_n(T)$ subset of COND.

Suppose that $E \subseteq M \times \text{COND}$. We put $E^b = \{m : \langle b, m \rangle \in E\}$ for all elements $b \in M$, and

$$\mathcal{D}_{cE} = \{p \in \text{COND} : p \text{ decides every } E^b, \ b < c\}.$$

Evidently $\mathcal{D}_{cE}$ is dense in COND provided E is $\Sigma_n(T)$ in M for some n.

Definition 3.1 Let Γ be a definability type. A matrix μ is Γ-*generic* iff for every Γ-set E and every $c \in M$, μ extends a condition $p \in \mathcal{D}_{cE}$. $\square$

Proposition 3.2 *Let* $n \geq 1$. *There exists a* $\Sigma_n(T)$-*generic matrix* μ *of class* $\Delta_{n+1}(T)$ *in* M.

Proof Using a $\Sigma_n(T)$ set universal for all $\Sigma_n(T)$ sets in M get an appropriate $\Delta_{n+1}(T)$ enumeration of all relevant sets $\mathcal{D}_{cE}$, and define μ as the limit of a certain increasing $\Delta_{n+1}(T)$ sequence of conditions. $\square$

Forcing

Let us consider the extension of the language of **PA** by the set $T \subseteq M$ as an extra second order parameter, as above, and a one more constant, $\breve{\mu}$, for a generic matrix. In other words, now "terms" of the form $\breve{\mu}(a, k)$ are admitted to substitute **PA** variables. In particular, $\Sigma_n(\breve{\mu}, T)$ will denote the collection of all $\Sigma_n(T)$ formulas where in addition $\breve{\mu}$ may occur in the mentioned way. The notation $\Pi_n(\breve{\mu}, T)$ is treated similarly.

For a condition p, let p^+ be the matrix which extends p by zeros, that is,

$$p^+(a, l) = \begin{cases} p(a, l) & \text{whenewer} \quad \langle a, l \rangle \in \text{dom}\, p \\ 0 & \text{otherwise} \end{cases}$$

Definition 3.3 The forcing relation $p \, \text{forc} \, \varphi$ is introduced; here $p \in$ COND while φ is a closed formula of one of the classes $\Sigma_n(\breve{\mu}, T)$, $\Pi_n(\breve{\mu}, T)$; $n \geq 1$

1. Let $\varphi(\breve{\mu}, T)$ be a closed $\Sigma_1(\breve{\mu}, T)$ formula. We set $p \, \text{forc} \, \varphi(\breve{\mu}, T)$ iff the computation of the truth value of $\varphi(p^+, T)$ in M gives the result true after an M-finite number of steps, in such a way that every value $p^+(a, l)$ which factually occurs in the computation satisfy $\langle a, l \rangle \in \text{dom}\, p$.

2. $p \, \text{forc} \, \exists a \, \varphi(a)$ iff there exists $a \in M$ such that $p \, \text{forc} \, \varphi(a)$.

3. Let Φ be a closed $\Pi_m(\breve{\mu}, T)$ formula, $m \geq 1$. Then $p \, \text{forc} \, \Phi$ iff none among the conditions p' extending p forces Φ^-. (Here Φ^- denotes the result of straightforward transformation of $\neg \Phi$ to $\Sigma_n(\breve{\mu}, T)$ form.) $\square$

Proposition 3.4 *Let* $\Phi(a_1, ..., a_m)$ *be a* $\Sigma_n(\breve{\mu}, T)$ *formula,* $n \geq 1$. *Then the set* $\{\langle a_1, ..., a_m, p \rangle \in M^m \times \text{COND} : p \text{ forc } \Phi(a_1, ..., a_m)\}$ *is* $\Sigma_n(T)$ *in* M.

Proof The statement in the case $n = 1$ follows from item 1 of Definition 3.3; then the result extends to the general case by induction. $\square$

In particular the set $\{p \in \text{COND} : p \text{ forc } \Phi\}$ is $\Sigma_n(T)$ in M for any closed $\Sigma_n(T)$-formula Φ.

Corollary 3.5 *Assume that* $n \geq 1$. *Let* μ *be an* $\Sigma_n(T)$-*generic matrix. Then for any* $m \leq n$ *and any closed* $\Sigma_m(T)$ *formula* Φ *there exists a condition* p, *extended by* μ, *which decides* Φ (*i.e. forces* Φ *or forces* Φ^-.) $\square$

The following lemma connects the truth of **PA** formulas, having T and a generic matrix μ as extra parameters, with the forcing.

Lemma 3.6 *Assume that* $n \geq 1$. *Let* μ *be an* $\Sigma_n(T)$-*generic matrix. Let* $\varphi(\breve{\mu})$ *be a* $\Sigma_m(\breve{\mu}, T)$ *formula,* $1 \leq m \leq n+1$. *Then* $\varphi(\mu)$ *is true in* M *iff some condition* p *extended by* μ *forces* $\varphi(\breve{\mu})$.

Proof The proof goes on by induction on m. The case $m = 1$ is easy.

To carry out the step, suppose that $m \leq n$. Consider a $\Sigma_{m+1}(\breve{\mu}, T)$ formula $\varphi(\breve{\mu})$ of the form $\exists a \, \psi(a, \breve{\mu})$, where ψ is a $\Pi_m(\breve{\mu}, T)$ formula.

Assume that $\varphi(\mu)$ is true. Then $\psi(a, \mu)$ holds in M for some $a \in M$, so that the $\Sigma_m(T)$ formula $\psi^-(a, \mu)$ is false and, by the induction hypothesis, none among conditions p extended by μ forces $\psi^-(a, \breve{\mu})$. By Corollary 3.5, there exists a condition $p \subset \mu$ which forces $\psi(a, \breve{\mu})$. Therefore $p \text{ forc } \varphi(\breve{\mu})$.

Conversely suppose that a condition $p \subset \mu$ forces $\varphi(\breve{\mu})$, that is, forces $\psi(a, \breve{\mu})$ for some a. We prove that $\psi(a, \mu)$ is true in M. Assume on the contrary that $\psi(a, \mu)$ is false, that is, $\psi^-(a, \mu)$ is true in M. Applying the induction hypothesis, we obtain a condition $p' \subset \mu$ which forces $\psi^-(a, \breve{\mu})$. One may assume that $p \subseteq p'$ since p also is extended by μ. This is a contradiction because $p \text{ forc } \psi(a, \breve{\mu})$. $\square$

Lemma 3.7 *Let* $n \geq 1$. *Suppose that* μ *is a* $\Sigma_n(T)$-*generic matrix. Then* M *satisfies both* Induction *and* Collection *for formulas in* $\Sigma_n(T, \mu)$.

Proof Induction. Consider a $\Sigma_n(\breve{\mu}, T)$ formula $\Phi(\breve{\mu}, a)$. It suffices to prove that if the set $A = \{a \in M : \neg \, \Phi(\mu, a) \text{ in } M\}$ is nonempty then it contains a least element in M. Consider an arbitrary $a' \in A$. By Proposition 3.4 and the genericity, μ extends a condition p which decides every sentence $\Phi(\breve{\mu}, a)$, $a \leq a'$, in M. We pick the M-least $a \leq a'$ such that p forces $\neg \, \Phi(\breve{\mu}, a)$, and use Lemma 3.6 having in mind that Π_n is convertible to Σ_{n+1}.

Collection. Consider a $\Sigma_n(\breve{\mu}, T)$ formula $\Phi(\breve{\mu}, a, b)$. Let $a_0 \in M$. It suffices to find $b_0 \in M$ such that the following holds in M :

$$\forall a < a_0 \, [\, \exists b \, \Phi(\mu, a, b) \implies \exists b < b_0 \, \Phi(\mu, a, b) \,].$$

By the genericity there exists a condition p, extended by μ, which decides the formula $\exists b \, \Phi(\breve{\mu}, a, b)$ for all $a < a_0$. By Proposition 3.4 the forcing relation is definable in $\langle M; T \rangle$; hence for any $a < a_0$ there exists the M-least element $b = b(a) \in M$ such that either $p \operatorname{forc} \Phi(\breve{\mu}, a, b)$ or $b = 0$ and p does not force $\exists b \, \Phi(\breve{\mu}, a, b)$. Moreover there exists $b_0 = \max_{a < a_0} b(a) \in M$, as required.   □

4 Coding sets by generic matrices

In this section, we complete the proof of Theorem 2.2; a **PA** model M and an inductive for M set $T \subseteq M$ remain fixed. Let us also fix a number $n \geq 1$.

Let a *double matrix* be any function $\mu : M \times 2 \times M \longrightarrow 2 = \{0, 1\}$. A double matrix μ can be seen as the indexed family $\langle \mu_{ai} : a \in M \, \& \, i \in \{0, 1\} \rangle$ where every "row" $\mu_{ai} \in 2^M$ is defined by $\mu_{ai}(l) = \mu(a, i, l)$ for all $l \in M$.

In this case for any set $X \subseteq M$ we define a matrix $\mu = \mu * X = \langle \mu_a : a \in M \rangle$ by $\mu_a = \mu_{a1}$ whenever $a \in X$ and $\mu_a = \mu_{a0}$ otherwise. Matrices of the form $\mu * X$ generated by M-p. df. sets $X \subseteq M$ will be called *M-flips* of μ.

Lemma 4.1 *There exists a double matrix μ which is $\Delta_{n+1}(T)$ in M, $\mu_{a0} \neq \mu_{a1}$ for any $a \in M$, and all M-flips of μ are $\Sigma_n(T)$-generic.*

Proof Let DCOND denote the set of all M-finite functions $\mathbf{p}$ such that

1) The domain $\operatorname{dom} \mathbf{p}$ is an M-finite subset of $M \times 2 \times M$ satisfying the following requirement: $\langle a, 0, l \rangle \in \operatorname{dom} \mathbf{p} \iff \langle a, 1, l \rangle \in \operatorname{dom} \mathbf{p}$.

2) All values of $\mathbf{p}$ are among the numbers 0 and 1.

Elements of DCOND, called *double conditions*, are identified with their codes in M so that DCOND is understood as a subset of M. We put

$$|\mathbf{p}| = \operatorname{dom} \operatorname{dom} \operatorname{dom} \mathbf{p} = \{a : \exists i \, \exists b \, (\langle a, i, b \rangle \in \operatorname{dom} \mathbf{p})\}$$

for any double condition $\mathbf{p}$; this is an M-finite subset of M, of course. The set DCOND is ordered by inclusion.

Now we introduce flips of double conditions. Let $\mathbf{p} \in$ DCOND, $u = |\mathbf{p}|$, and $U \subseteq u$ is an M-finite set. We define $p = \mathbf{p} * U \in$ COND, an *M-flip* of $\mathbf{p}$, as follows: $p(a, l) = \mathbf{p}(a, 1, l)$ whenever $a \in U$, and $p(a, l) = \mathbf{p}(a, 0, l)$ otherwise. Thus, an M-flip of a double condition is a condition in COND.

Assertion Let $\mathbf{p} \in \text{DCOND}$, $E \subseteq M \times \text{COND}$ be a $\Sigma_m(T)$ set in M for some m, and $c \in M$. There exists a double condition $\mathbf{p}'$ extending $\mathbf{p}$ such that every M-flip of $\mathbf{p}'$ decides each of the sets E^b, $b < c$.

Proof Let us fix an enumeration $\{q_k : 1 \leq k \leq k_0\}$ (where $k_0 \in M$) of all M-flips of $\mathbf{p}$ in M. We need one more definition. Let $q \subseteq p$ be a pair of conditions in COND, q' a condition satisfying $\text{dom}\, q' = \text{dom}\, q$. We define the substitution $p' = p[q/q']$ as follows:

$$p'(a,l) = \begin{cases} q'(a,l) & \text{iff } \langle a,l \rangle \in \text{dom}\, q = \text{dom}\, q' \\ p(a,l) & \text{iff } \langle a,l \rangle \in \text{dom}\, p \setminus \text{dom}\, q \end{cases}$$

Let now q be one of the conditions q_k. One can construct an increasing M-finite sequence of conditions, $q = p_0 \subseteq p_1 \subseteq p_2 \subseteq \cdots \subseteq p_{k_0} = p'$ such that every condition $p'_k = p_k[q/q_k]$ $(1 \leq k \leq k_0)$ decides each of the sets E^b, $b < c$. In particular every $p'[q/q_k]$ decides each E^b. We define $\mathbf{p}' \in \text{DCOND}$ as follows:

$$\mathbf{p}'(a,i,l) = \begin{cases} \mathbf{p}(a,i,l) & \text{if } \langle a,i,l \rangle \in \text{dom}\, \mathbf{p} \\ p'(a,l) & \text{if } \langle a,l \rangle \in \text{dom}\, p' \text{ and } \langle a,i,l \rangle \notin \text{dom}\, \mathbf{p} \end{cases}$$

Then every M-flip of $\mathbf{p}'$ is equal to some $p'[q/q_k]$.   □ (the assertion)

Now, using the assertion, one ends the proof of Lemma 4.1 in the way outlined above, for the proof of Proposition 3.2, in addition taking care of requirement $\mu_{a0} \neq \mu_{a1}$. (The latter would easily follow from a very moderate amount of genericity of μ itself, which indeed we shall not exploit.)

Note that the assertion can capture only the flips generated by M-finite sets (the reasoning essentially proceeds in M). This is why X was required to be an M-p. df. set in the lemma.   □ (Lemma 4.1)

Let us complete the proof of Theorem 2.2. Suppose that X is an M-p. df. set. Let μ be the double matrix given by Lemma 4.1. Then $\mu = \mu * X$ is a $\Sigma_n(T)$-generic matrix, so that M models both Induction and Collection for $\Sigma_n(T, \mu)$ by Lemma 3.7. On the other hand, since $\mu_{a0} \neq \mu_{a1}$ for all a, we have

$$a \in X \iff \forall l\, [\mu(a,l) = \mu(a,1,l)] \iff \exists l\, [\mu(a,l) \neq \mu(a,0,l)],$$

so X is $\Delta_{n+1}(T,\mu)$ in M because μ is chosen to be $\Delta_{n+1}(T)$ in M. It remains to convert the matrix $\mu \in 2^{M \times M}$ to a set $A \subseteq M$.   □ (Theorem 2.2)

5   The extension

This section proves Theorem 2.1. Thus we suppose that $n \geq 2$ and M is a countable recursively saturated model of **PA**. Then there exists an inductive satisfaction class $T \subseteq M$ for M. In particular, M models $\Sigma_m(T)$-Induction for all m.

Since M is countable, there exists a cofinal increasing sequence $\langle b_k : k \in \omega \rangle$ in M. We put $d_k = 2^k 3^{b_k}$. Let us fix a number $a_0 \in M \setminus \omega$ and a $1-1$ map $\beta : M_{<a_0}$ onto the set $\Delta = \{d_k : k \in \omega\}$.

Note that β is M-p. df. as Δ is a cofinal subset of M of order type ω. Hence by Theorem 2.2 there exists an M-p. df. set $A \subseteq M$ such that M models both Induction and Collection for $\Sigma_n(T, A)$, and β is $\Delta_{n+1}(T, A)$ in M.

The continuation of the proof involves the following lemma.

Lemma 5.1   *There is a countable elementary end extension N of M such that*

(a) *Both A and T belong to N/M.*

(b) *Every element of N/M is $\Delta_2(T, A)$ in M.*

Let us demonstrate that the lemma implies Theorem 2.1. Requirement (b) guarantees that every $\Sigma_{n-1}[N/M]$ subset of M belongs to $\Sigma_n(T, A)$ in M. So M models $\Sigma_{n-1}[N/M]$-Collection and $\Sigma_{n-1}[N/M]$-Induction by Lemma 3.7.

Requirement (a) implies $\beta \in \Delta_{n+1}[N/M]$ in M by the choice of A. It immediately follows that $\Sigma_{n+1}[N/M]$-Collection fails in M by the choice of β. To see that $\Sigma_{n+1}[N/M]$-Induction fails as well, consider a $\Sigma_{n+1}[N/M]$ formula $\Phi(k)$ which says that there exist numbers $a < a_0$ and b satisfying $\beta(a) = 2^k 3^b$. It is clear that $\Phi(k)$ is true in M iff $k \in \omega$. $\square$ (Theorem 2.1)

Proof of the lemma. The required extension N will be defined as an ultrapower of M of the form $N = \text{Ult}_\mathcal{U}\, \mathfrak{F}$, where

$$\mathfrak{F} = \{f \in M^M : f \text{ is definable in } M \text{ by a } \mathbf{PA} \text{ formula}$$
$$\text{with parameters in } M\}$$

and $\mathcal{U}$ is an ultrafilter in the algebra $\mathcal{A}$ of all subsets of M definable in M by a **PA** formula with parameters in M.

Since T and A may not be **PA** definable in M, we have to use an ultrafilter as the principal coding tool in order to fulfill requirement (a). The ultrafilter $\mathcal{U}$ will be defined in two steps.

Coding T and A

We put $P_a = \{x \in M : a\text{-th prime divides } x \text{ in } M\}$ and $R_a = M \setminus P_a$. Let $Z = \{2b : b \in T\} \cup \{2b+1 : b \in A\}$; we now define

$$Q_a = \begin{cases} P_a & \text{iff} \quad a \in Z \\ R_a & \text{otherwise} \end{cases}$$

We set finally $\mathcal{U}_0 = \{Q_a : a \in M\}$.

Notice that all M-finite intersections of sets Q_a are unbounded in M. Furthermore, if $\mathcal{U} \subseteq \mathcal{A}$ is an ultrafilter such that $\mathcal{U}_0 \subseteq \mathcal{U}$, and $N = \mathrm{Ult}_{\mathcal{U}} \mathcal{F}$, then both T and A belong to N/M, which yields requirement (a) of the lemma.

Trying to expand $\mathcal{U}_0$ to an ultrafilter

For any set $B \in \mathcal{A}$, it must be determined which among the sets B, $M \backslash B$, belongs to $\mathcal{U}$. We have two restrictions on this expansion, of somewhat opposite direction. First we must satisfy requirement (b) of the lemma; second we have to guarantee that $N = \mathrm{Ult}_{\mathcal{U}} \mathcal{F}$ is an *end* extension.

Let $\{\phi_p(x)\}_p$ be a formal recursive enumeration in **PA** of all formulas of the language of **PA** having x as the only free variable. We are willing to set $A_p = \{x \in M : M \models \phi_p(x)\}$ for all $p \in M$. This is inconsistent, generally speaking, since many "formulas" ϕ_p may not have a definite external meaning. The satisfaction class T converts the definition to legitimate form. We put

$$A_p = \{x \in M : \ulcorner \phi_p(x) \urcorner \in T\} \quad \text{and} \quad C_p = M \setminus A_p \, ;$$

where $\ulcorner \cdot \urcorner$ denotes the Gödel number of $\cdot$. [4] We define an auxiliary ultrafilter $\mathcal{U}' = \mathcal{U}_0 \cup \{B_p : p \in M\}$ where each B_p is either A_p or C_p. The definition of B_p goes on in $\langle M ; T, A \rangle$ by the (internal) induction on p as follows:

$$B_p \; = \; \begin{cases} A_p & \text{iff} \quad A_p \text{ is compatible with } \mathcal{U}_0 \cup \{B_q : q < p\} \, ; \\ C_p & \text{otherwise} \end{cases}$$

We say that a set $B \subseteq M$ is *compatible* with a family $\mathcal{X}$ of subsets of M iff $B \cap \bigcap \mathcal{X}'$ is unbounded in M for any M-finite $\mathcal{X}' \subseteq \mathcal{X}$. We say that $\mathcal{X}$ is *compatible* iff M is compatible with $\mathcal{X}$. (In the way how the proof goes on, the notion of an M-finite family of subsets of M is well defined.)

Justification

The inductive definition of B_p is carried out in $\langle M ; T, A \rangle$ (meaning M with T and A as extra second-order parameters). Therefore to see that the definition is legitimate we have to justify it in the frameworks of our assumptions.

We recall that M satisfies both Induction and Collection for $\Sigma_n(T, A)$, where $n \geq 2$ is a fixed number (from Theorem 2.1).

---

[4] Notice that every subset of M definable in M by a **PA** formula with parameters in M is equal to some A_p, $p \in M$, *but not vice versa*, because the family of all sets A_p contains sets of nonstandard M-finite levels of the arithmetical hierarchy in M, available via T .

Let a *good sequence* mean an M-finite binary sequense $s = \langle i_0, ..., i_r \rangle \in M$ corresponding to the construction of A_p and C_p in the sense that, for all $p \le r$,

$$i_p = \begin{cases} 1 & \text{iff} \quad (*) \quad A_p \text{ is compatible with } \mathcal{U}_0 \cup \{B_q : q < p\} \\ 0 & \text{otherwise} \end{cases}$$

In other words $i_p = 1$ iff $B_p = A_p$

We now explore the "complexity" of the requirement $(*)$ (saying that the set $B_{ps} = A_p \cap \bigcap_{q<p,\, s(q)=1} A_q \cap \bigcap_{q<p,\, s(q)=0} C_q$ has unbounded, in M, intersection with any among sets $Q_{<b} = \bigcap_{a<b} Q_a$).

Since A, T are M-p. df. sets, we can associate with any $b \in M$ a particular **PA** formula $\gamma_b(x)$ with parameters in M such that $Q_{<b} = \{x : \gamma_b(x)\}$ in M and the map $b \longmapsto \ulcorner\gamma_b\urcorner$ is $\Delta_1(T, A)$, i.e. recursive w. r. t. T and A, in M. Let Φ_{pbs} denote the following perhaps infinite but M-finite sequence of symbols in M, which looks like a **PA** formula from the M-th point of view:

$$\forall x \, \exists y \ge x \, [\, \phi_p(y) \,\&\, \gamma_b(y) \,\&\, \forall q < p \, (s(q) = 1 \iff \phi_q(y))\,]\,.$$

(By the way we cannot add the quantifier $\forall b$ because this would involve T and A as parameters.) Since T is a satisfaction class for M, we have, for all $p, b \in M$ and an M-finite sequence s,

$$\ulcorner\Phi_{pbs}\urcorner \in T \iff \text{the intersection } B_{ps} \cap Q_{<b} \text{ is unbounded}.$$

Then $(*) \iff \forall b \, (\ulcorner\Phi_{pbs}\urcorner \in T)$, so that the property of "being a good sequence" can be expressed by a formula of the type

(bounded quantifier) ($\Sigma_1(T, A)$-formula $\&$ $\Pi_1(T, A)$-formula),

(because the function $p, b, s \longmapsto \ulcorner\Phi_{pbs}\urcorner$ is $\Delta_1(T, A)$ in M), which is within both $\Sigma_2(T, A)$ and $\Pi_2(T, A)$. It follows that the formula "there exists a good sequence of length k" is $\Sigma_2(T, A)$ as well.

Therefore we can apply $\Sigma_2(T, A)$-**Induction** (a good sequence obviously cannot be maximal) getting a good sequence s_k in M of length k for any $k \in M$. (The uniqueness of a good sequence for any fixed length is easily verified.) This conclusion justifies the construction of sets B_p, $p \in M$.

One more important consequence from our consideration is that the set $S = \{p \in M : B_p = A_p\}$ is $\Delta_2(T, A)$ in M; this will be used below.

The ultrapower

Thus the sets B_p are well defined, and so is $\mathcal{U}' = \mathcal{U}_0 \cup \{B_p : p \in M\}$, the auxiliary ultrafilter, therefore $\mathcal{U} = \mathcal{U}' \cap \mathcal{A}$ is an ultrafilter in $\mathcal{A}$, and $\mathcal{U}$ is compatible in the sense above. We shall prove that $N = \mathtt{Ult}_{\mathcal{U}}\,\mathcal{F}$ is the required extension of M. There are just two points which we have to check: first, N is an end extension of M, second, requirement (b) of Lemma 5.1.

End extension

By the choice of $\mathcal{F}$, to guarantee that N is an end extension of M, it suffices to prove that if $W \subseteq M \times M$ is definable in M by a **PA** formula (parameters in M, but not T or A, allowed), $c_0 \in M$, $W_k = \{a : \langle k, a \rangle \in W\}$ for all k, and $X = \bigcup_{c < c_0} W_c \in \mathcal{U}$ then there exists $c < c_0$ such that $W_c \in \mathcal{U}$.

To prove this fact assume on the contrary that $W_c \notin \mathcal{U}$ for all $c < c_0$. Let us verify that $W_{<k} = \bigcup_{c<k} W_c \notin \mathcal{U}$ for all $k \le c_0$ by induction on k; this immediately leads to contradiction. Since we have $\Sigma_2(T, A)$-Induction in M, it suffices to check that the property $W_{<k} = \bigcup_{c<k} W_c \notin \mathcal{U}$ can be expressed in M by a $\Sigma_2(T, A)$ formula. Such a formula can be defined as follows:

$$\exists p \left[(p \notin S \,\&\, W_{<k} = A_p) \bigvee (p \in S \,\&\, W_{<k} = C_p) \right],$$

where, we recall, $S = \{p \in M : B_p = A_p\}$ is a $\Delta_2(T, A)$ set in M.

It remains to replace the equality $W_{<k} = A_p$ by something like the formula $\ulcorner \forall x (x \in W_{<k} \iff \phi_p(x)) \urcorner \in T$ and accordingly replace $W_{<k} = C_p$.

Requirement (b) of Lemma 5.1

The following is sufficient: if $W \subseteq M \times M$ and $W_k = \{a : \langle k, a \rangle \in W\}$ for all $k \in M$ are as above then $Y = \{k : W_k \in \mathcal{U}\} \in \Delta_2(T, A)$. We observe that

$$k \in Y \iff \exists p \left[p \in S \,\&\, W_k = A_p \right] \iff \forall p \left[p \in S \implies W_k \ne C_p \right].$$

Both the equality $W_k = A_p$ and the inequality $W_k \ne C_p$ can be reduced to $\Delta_1(T, A)$ as above, therefore $Y \in \Delta_2(T, A)$, as required. $\square$
(Lemma 5.1)

Acknowledgements. The author acknowledges with pleasure useful discussions with A. Enayat, R. Kaye, R. Kossak, J. Schmerl regarding the content of this paper. The author thanks the referee for important improvements.

6 References

[1] R. W. Kaye. *Models of Peano Arithmetic*, Volume 15 of *Oxford Logic Guides*, Oxford Science Publications, 1991.

[2] R. Kossak and J. B. Paris. Subsets of models of arithmetic. *Arch. Math. Logic*, 1992, 32, P. 65 – 73.

[3] W. Sieg. Fragments of arithmetic. *Annals of Pure and Applied Logic*, 1985, 28, P. 33 – 72.

Added in proof: It was in September 1997, after the final version of this paper had been submitted, that Richard Kaye let me know a nice improvement of the reasonning in Section 5, which seems to close the gap between $n - 1$ and $n + 1$ in Theorem 2.1.

Author address

Moscow Transport Engineering Institute
e-mail: kanovei@mech.math.msu.su
         kanovei@math.uni-wuppertal.de

Arithmetizing proofs in analysis

Ulrich Kohlenbach

1 Introduction

In this paper we continue our investigations started in [15] and [16] on the question:

What is the impact on the growth of extractable uniform bounds the use of various analytical principles Γ in a given proof of an $\forall\exists$–sentence might have?

To be more specific, we are interested in analyzing proofs of sentences having the form

$$(1) \quad \forall u^1, k^0 \forall v \leq_\rho tuk \exists w^0 A_0(u, k, v, w),$$

where A_0 is a quantifier–free formula[2] (containing only u, k, v, w as free variables) in the language of a suitable subsystem $\mathcal{T}^\omega$ of arithmetic in all finite types, t is a closed term and $\leq_\rho$ is defined pointwise (ρ being an arbitrary finite type).

From a proof of (1) carried out in $\mathcal{T}^\omega$ one can extract an effective uniform bound Φuk on $\exists w$, i.e.

$$(2) \quad \forall u^1, k^0 \forall v \leq_\rho tuk \exists w \leq_0 \Phi uk\, A_0(u, k, v, w),$$

where the complexity (and in particular the growth) of Φ is limited by the complexity of the system $\mathcal{T}^\omega$ (see [13],[15]).

By the predicate 'uniform' we refer to the fact that the bound Φ does not depend on $v \leq_\rho tuk$.

In [13] we have discussed in detail, how sentences (1) arise naturally in analysis and why such uniform bounds are of numerical interest (e.g. in the context of approximation theory).

---

[1] Received September 96; revised version February 97.
[2] Throughout this paper $A_0, B_0, C_0, \ldots$ always denote quantifier–free formulas.

Proofs in analysis can be formalized in a suitable base theory $\mathcal{T}^\omega$ plus certain (in general non–constructive) analytical principles Γ (usually not derivable in $\mathcal{T}^\omega$). In order to determine faithfully the contribution of the use of Γ to the growth of extractable bounds Φ we introduced in [15] a hierarchy of weak subsystems $G_n A^\omega$ of arithmetic in all finite types whose definable type–1–objects correspond to the well–known Grzegorczyk hierarchy of functions.

As the essential proof–theoretic tool, monotone functional interpretation (which was introduced in [13]) was used to extract bounds Φ (given by closed term of $G_n A^\omega$) from proofs

$$(3) \quad G_n A^\omega + \Delta + \text{AC–qf} \vdash (1),$$

where

$$\text{AC}^{\rho,\tau}\text{–qf} \; : \; \forall x^\rho \exists y^\tau A_0(x,y) \to \exists Y^{\tau\rho} \forall x^\rho A_0(x,Yx)$$

is the schema of choice for quantifier–free formulas and Δ is a set of 'axioms' having the form

$$(4) \quad \forall x^\delta \exists y \leq_\rho s x \forall z^\tau G_0(x,y,z),$$

where G_0 is a quantifier–free formula containing only x,y,z free and s is a closed term.

In particular for $n = 2$ (resp. $n = 3$) the extractability of a bound $\Phi u k$ which is a polynomial (resp. a finitely iterated exponential function) in $u^M x := \max_{i \leq x} u(i)$ and k is guaranteed (see [15] for details).

In [14] we have shown that for suitable Δ already $G_2 A^\omega + \Delta + \text{AC–qf}$ covers a substantial part of standard analysis. In fact essentially only analytical axioms (4) having types $\delta, \rho \leq 1, \tau = 0$ are sufficient.

The proof of the verification of the extracted bound Φ also relies on these non–constructive principles Δ, in fact even on their strengthened versions

$$(5) \quad \tilde{\Delta} := \left\{ \exists Y \leq_{1(1)} s \forall x, z G_0(x, Yx, z) | \forall x^1 \exists y \leq_1 s x \forall z^0 G_0(x,y,z) \in \Delta \right\}$$

relatively to the intuitionistic variant $G_n A_i^\omega$ of $G_n A^\omega$.

However combining the methods from [15] with techniques from [12] one can replace the use of (5) by the use of the 'ε–weakenings' of (5) thereby achieving

$$(6) \quad G_n A_i^\omega + \Delta_\varepsilon \vdash \forall u^1, k^0 \forall v \leq_\rho t u k \exists w \leq_0 \Phi u k A_0(u,k,v,w),$$

where

$$(7) \Delta_\varepsilon := \left\{ \forall x^1, z^0 \exists y \leq_1 s x \bigwedge_{i=0}^{z} G_0(x,y,i) | \forall x^1 \exists y \leq_1 s x \forall z^0 G_0(x,y,z) \in \Delta \right\}$$

The ε–weakening Δ_ε of Δ usually is constructively provable in suitable subsystems of intuitionistic arithmetic in all finite types. This passage from

$\tilde{\Delta}$ to Δ_ε – which may be viewed as an ε-arithmetization of the original proof – however is not necessary for the extraction of Φ but only for a **constructive verification** of Φ.

Whereas a number of important analytical principles can be expressed directly as axioms (4) – in particular relatively to systems like $\widehat{\text{PA}}^\omega\!\!\restriction$ or $G_n A^\omega$ for $n \geq 3$ the binary König's lemma WKL can be expressed in this form (see [12] for details) – there are many theorems not having this form but which can be proved from WKL relatively to base systems like $\widehat{\text{PA}}^\omega\!\!\restriction +$ AC–qf which essentially is a finite type extension of the second-order theory RCA_0 known from reverse mathematics. Examples of such theorems are the following principles:

- Every pointwise continuous function $f : [0,1]^d \to \mathbb{R}$ is uniformly continuous.

- The attainment of the maximum value of $f \in C([0,1]^d, \mathbb{R})$ on $[0,1]^d$.[3]

- The sequential form of the Heine–Borel covering property for $[0,1]^d$.

- Dini's theorem.

- The existence of a uniformly continuous inverse function for every strictly increasing continuous function $f : [0,1] \to \mathbb{R}$.

The problem in treating these principles relative to weak base theories as $G_2 A^\omega$ is that their usual proofs (using WKL) are not formalizable within e.g. $G_2 A^\omega$. In particular WKL can not even be expressed in its usual formulation in this system, since this involves the coding functional $f_{\langle\rangle}x := \langle f0, \ldots, f(x-1)\rangle$ which is available in $G_n A^\omega$ only for $n \geq 3$. In order to treat the principles above faithfully we introduced in [15] the axiom (having the form (4))

$$F^- :\equiv \forall \Phi^{2(0)}, y^{1(0)} \exists y_0 \leq_{1(0)} y \forall k^0, z^1, n^0$$
$$\Big(\bigwedge_{i <_0 n} (zi \leq_0 yki) \to \Phi k(\overline{z,n}) \leq_0 \Phi k(y_0 k) \Big),$$

where, for $z^{\rho 0}, (\overline{z,n})(k^0) :=_\rho zk$, if $k <_0 n$ and $:= 0^\rho$, otherwise.

This axiom implies (already relatively to $G_2 A^\omega + \text{AC}^{1,0}$–qf) the following **principle of uniform Σ_1^0–boundedness**

$$\Sigma_1^0\text{-UB}^- :\equiv \left\{ \begin{array}{l} \forall y^{1(0)} \big(\forall k^0 \forall x \leq_1 yk \exists z^0\, A(x,y,k,z) \to \exists \chi^1 \forall k^0, x^1, n^0 \\ \qquad \big(\bigwedge_{i <_0 n} (xi \leq_0 yki) \to \exists z \leq_0 \chi k\, A((\overline{x,n}), y, k, z) \big) \big), \end{array} \right.$$

---

[3]This statement can be expressed as an axiom (4) (if f is endowed with a modulus of uniform continuity). However this requires a very complicated representation of the elements $f \in C([0,1]^d, \mathbb{R})$ which can be avoided using the principle of uniform boundedness discussed below.

where $A \equiv \exists l^0 A_0(l)$ is a purely existential formula (see [15] for a detailed discussion of this principle).

In $G_2 A^\omega + \Sigma_1^0\text{-}UB^-$ and hence in $G_2 A^\omega + F^- + AC^{1,0}\text{-}qf$ one can give very short and perspicuous proofs of the analytical theorems listed above and since F^- has the form of an axiom Δ we can extract a polynomial bound from such a proof (see [17] for details). The verification of this so far still depends on the non–standard axiom F^- which does not hold classically, i.e. it does not hold in the full set–theoretic type structure $\mathcal{S}^\omega$ (but only in the type structure of all so–called strongly majorizable functionals $\mathcal{M}^\omega$). Nevertheless, using the ε–arithmetization technique mentioned above, one can replace the use of F^- by its ε–weakening and this ε–weakening is provable e.g. in $G_3 A_i^\omega$ (see [15]). In this case ε-arithmetization still is not needed for the extraction of an uniform bound but now it is needed even for a **classical verification**.

On the other hand there are central theorems in analysis whose proofs use arithmetical comprehension, more precisely instances of

$$AC_{ar} : \forall x^0 \exists y^0 A(x,y) \to \exists f^1 \forall x^0 A(x, fx),$$

where $A \in \Pi_\infty^0$ (A may contain parameters of arbitrary type), and which are not covered by the results mentioned above.

Examples are the following theorems

1) The principle of convergence for bounded monotone sequences of real numbers (or equivalently: every bounded monotone sequence of reals has a Cauchy modulus (PCM)).

2) For every sequence of real numbers which is bounded from above there exists a least upper bound.

3) The Bolzano–Weierstraß property for bounded sequences in $\mathbb{R}^d$ (for every fixed d).

4) The Arzelà–Ascoli lemma.

5) The existence of the limit superior for bounded sequences of real numbers.

Using a convenient representation of real numbers, (PCM) can be formalized as follows:

$$(\text{PCM}) : \begin{cases} \forall a_{(\cdot)}^{1(0)}, c^1 \left(\forall n^0 (c \leq_\mathbb{R} a_{n+1} \leq_\mathbb{R} a_n) \right. \\ \qquad\qquad \left. \to \exists h^1 \forall k^0 \forall m, \tilde{m} \geq_0 hk(|a_m -_\mathbb{R} a_{\tilde{m}}| \leq_\mathbb{R} \tfrac{1}{k+1})\right). \end{cases}$$

(PCM) immediately follows from its arithmetical weakening

$$(\text{PCM}^-) : \begin{cases} \forall a_{(\cdot)}^{1(0)}, c^1 \left(\forall n^0 (c \leq_\mathbb{R} a_{n+1} \leq_\mathbb{R} a_n) \right. \\ \qquad\qquad \left. \to \forall k^0 \exists n^0 \forall m, \tilde{m} \geq_0 n(|a_m -_\mathbb{R} a_{\tilde{m}}| \leq_\mathbb{R} \tfrac{1}{k+1})\right) \end{cases}$$

by an application of AC_{ar} to

$$A :\equiv \forall m, \tilde{m} \geq n(|a_m -_{\mathbb{R}} a_{\tilde{m}}| \leq_{\mathbb{R}} \frac{1}{k+1}) \in \Pi_1^0$$

($\leq_{\mathbb{R}} \in \Pi_1^0$ follows from the fact that real numbers are given as Cauchy sequences of rationals with fixed rate of convergence in our theories).

It is well–known that a constructive functional interpretation of the negative translation of AC_{ar} requires so–called bar–recursion and cannot be carried out e.g. in Gödel's term calculus T (see [23] and [18]). AC_{ar} is (using classical logic) equivalent to $CA_{ar}+AC^{0,0}$–qf, where

$$CA_{ar} : \exists g^1 \forall x^0 (g(x) =_0 0 \leftrightarrow A(x)) \text{ with } A \in \Pi_\infty^0,$$

and therefore causes an immense rate of growth (when added to e.g. G_2A^ω). From the work in the context of 'reverse mathematics' (see e.g. [6],[22]) it is known that 1)–5) imply CA_{ar} relatively to (a second–order version of) $\widehat{PA}^\omega \upharpoonright +AC^{0,0}$–qf (see [5] for the definition of $\widehat{PA}^\omega \upharpoonright$). In [14] it is shown that this holds even relatively to G_2A^ω.

In contrast to these general facts on huge growth we prove in this paper a theorem which in particular implies that if (PCM) is applied in a proof only to sequences (a_n) which are given explicitly in the parameters of the proposition (which is proved) then this proof can be (effectively) transformed (without causing new growth) into a proof of the same conclusion which uses only (PCM$^-$) for these sequences. By this transformation the use of AC_{ar} is eliminated and the determination of the growth caused (potentially by (PCM)) reduces to the determination of the growth caused by (PCM$^-$). This reduction is achieved using the method of elimination of Skolem function for monotone formulas (developed in [16]).

In difference to (PCM) the (negative translation of the) principle (PCM$^-$) has a simple constructive monotone functional interpretation which is fulfilled by a functional Ψ which is primitive recursive in the sense of [9]. Because of the nice behaviour of the monotone functional interpretation with respect to the modus ponens one obtains (by applying Φ to Ψ) a monotone functional interpretation of (1) and so, using tools from [13],[15], a uniform bound ξ for $\exists w$, i.e.

$$\forall u^1, k^0 \forall v \leq_\rho tuk \exists w \leq_0 \xi uk A_0(u, k, v, w),$$

where ξ is **primitive recursive in the sense of Kleene [9]** (and not only in the generalized sense of Gödel's calculus T). X-Mozilla-Status: 0000

(This conclusion also holds for sequences of instances $\forall n^0 PCM(\chi uvn)$ of $PCM(a)$ instead of $PCM(\chi uv)$.)

In this case ε-arithmetization – namely the reduction of the use of instances of (PCM) to corresponding instances of its arithmetical weakening (PCM^-) – is necessary already for the **construction of the bound Φ.**

In our treatment of the Bolzano–Weierstraß theorem (as well as the Arzelà–Ascoli lemma) in section 5 below the use of the method of elimination of Skolem functions is combined with the use of the non–standard axiom F^- mentioned above: Single (sequences of) instances of the Bolzano–Weierstraß theorem can be proved (relatively to $G_2A^\omega + AC^{1,0}$–qf) from single instances of the second–order axiom Π^0_1–CA plus F^-. Π^0_1–CA is studied in [16] where it is shown that single instances of this principle (in contrast to its full second–order universal closure, which is equivalent to full arithmetical comprehension over numbers) also contribute at most by a **primitive recursive functional in the sense of Kleene**. By the method of F^-–elimination discussed above, the resulting bound from a proof which uses single instances of the Bolzano–Weierstraß theorem then can be classically (and even constructively) verified. Here ε-arithmetization of a given proof is used twice for the construction of a bound (by elimination of Skolem functions) and for a classical verification (by elimination of the non–standard axiom F^-).

Finally we investigate the principle of the existence of the limit superior of a bounded sequence of real numbers. It turns out that the use of single instances of this principle in the proof of a theorem (1) can be reduced to an arithmetical Π^0_5–principle whose monotone functional interpretation can be fulfilled by a functional from the fragment T_1 of Gödels calculus T with the recursor constants R_ρ for $\rho \leq 1$ (this fragment of T is sufficient to define the Ackermann function but no functions of essentially greater rate of growth).

In section 2 we present the theorems from [16] on which our investigations in the present paper are based in order to make this paper independent from the reading of [16]. However we assume the reader to be familiar with [15] and all undefined notions in this paper are used in the sense of [15].

2   Proof–theoretic tools

In this section we recall some of our proof–theoretic results from [16] which will be used in section 5 below.

Definition 2.1 ([16]) *Let $A \in \mathcal{L}(G_nA^\omega)$ be a formula having the form*

$$A \equiv \forall u^1 \forall v \leq_\tau tu \exists y_1^0 \forall x_1^0 \ldots \exists y_k^0 \forall x_k^0 \exists w^\gamma A_0(u, v, y_1, x_1, \ldots, y_k, x_k, w),$$

where A_0 is quantifier–free and contains only $u, v, \underline{y}, \underline{x}, w$ free. Furthermore let t be $\in G_nR^\omega$ and τ, γ are arbitrary finite types.

1) A is called (arithmetically) **monotone** *if*

$$
Mon(A) :\equiv \left\{ \begin{array}{l} \forall u^1 \forall v \leq_\tau tu \forall x_1, \tilde{x}_1, \ldots, x_k, \tilde{x}_k, y_1, \tilde{y}_1, \ldots y_k, \tilde{y}_k \\[1mm] \left(\bigwedge_{i=1}^{k} (\tilde{x}_i \leq_0 x_i \wedge \tilde{y}_i \geq_0 y_i) \wedge \exists w^\gamma A_0(u, v, y_1, x_1, \ldots, y_k, x_k, w) \right. \\[3mm] \qquad\qquad \left. \to \exists w^\gamma A_0(u, v, \tilde{y}_1, \tilde{x}_1, \ldots, \tilde{y}_k, \tilde{x}_k, w) \right). \end{array} \right.
$$

2) The **Herbrand normal form** A^H *of A is defined to be*

$$
A^H :\equiv \forall u^1 \forall v \leq_\tau tu \forall h_1^{\rho_1}, \ldots, h_k^{\rho_k} \exists y_1^0, \ldots, y_k^0, w^\gamma
$$

$$
\underbrace{A_0(u, v, y_1, h_1 y_1, \ldots, y_k, h_k y_1 \ldots y_k, w)}_{A_0^H :\equiv}, \; \text{where } \rho_i = 0 \underbrace{(0) \ldots (0)}_{i}.
$$

Theorem 2.2 ([16]) *Let* $n \geq 1$ *and* $\Psi_1, \ldots, \Psi_k \in G_n R^\omega$. *Then*

$$
G_n A^\omega + Mon(A) \vdash \forall u^1 \forall v \leq_\tau tu \forall h_1, \ldots, h_k \left(\bigwedge_{i=1}^{k} (h_i \text{ monotone}) \right.
$$

$$
\left. \to \exists y_1 \leq_0 \Psi_1 u \underline{h} \ldots \exists y_k \leq_0 \Psi_k u \underline{h} \exists w^\gamma A_0^H \right) \to A,
$$

where

$$
(h_i \text{ monotone}) :\equiv \forall x_1, \ldots, x_i, y_1, \ldots, y_i \left(\bigwedge_{j=1}^{i} (x_j \geq_0 y_j) \to h_i \underline{x} \geq_0 h_i \underline{y} \right).
$$

Definition 2.3 (Bounded choice) $b\text{-}AC := \bigcup_{\delta, \rho \in T} \left\{ (b\text{-}AC^{\delta, \rho}) \right\}$ *denotes the schema of bounded choice*

$$
(b\text{-}AC^{\delta, \rho}) \; : \; \forall Z^{\rho\delta} \left(\forall x^\delta \exists y \leq_\rho Zx \; A(x, y, Z) \to \exists Y \leq_{\rho\delta} Z \forall x A(x, Yx, Z) \right).
$$

Theorem 2.4 ([16]) *Let A be as in thm.2.2 and* Δ *be a set of sentences* $\forall x^\delta \exists y \leq_\rho sx \forall z^\eta G_0(x, y, z)$ *where s is a closed term of* $G_n A^\omega$ *and* G_0 *a quantifier-free formula, and let* A' *denote the negative translation[4] of A. Then the following rule holds:*

$$
\left\{ \begin{array}{l} G_n A^\omega + AC\text{-}qf + \Delta \vdash A^H \wedge Mon(A) \Rightarrow \\[2mm] G_n A^\omega + \tilde{\Delta} \vdash A \text{ and by monotone functional interpretation} \\[2mm] \text{one can extract a tuple } \underline{\Psi} \in G_n R^\omega \text{ such that} \\[2mm] G_n A_i^\omega + \tilde{\Delta} \vdash \underline{\Psi} \text{ satisfies the monotone functional interpretation of } A', \end{array} \right.
$$

---

[4]Here we can use Gödel's [7] translation or any of the various negative translations. For a systematical treatment of negative translations see [18].

where $\tilde{\Delta} := \{\exists Y \leq_{\rho\delta} s \forall x^\delta, z^\eta G_0(x, Yx, z) : \forall x^\delta \exists y \leq_\rho sx \forall z^\eta G_0(x, y, z) \in \Delta\}$. (In particular the second conclusion can be proved in $G_n A_i^\omega + \Delta + b\text{-}AC$).

Remark 2.1 In theorems 2.2,2.4 one may also have tuples '$\exists \underline{w}$' instead of '$\exists w^\gamma$' in Δ.

For our applications in paragraph 5 we need the following corollary of theorem 2.4:

Corollary 2.5 ([16]) Let $\forall x^0 \exists y^0 \forall z^0 A_0(u^1, v^\tau, x, y, z) \in \mathcal{L}(G_n A^\omega)$ be a formula which contains only u, v as free variables and satisfies provably in $G_n A^\omega + \Delta + AC\text{-}qf$ the following monotonicity property:

$$(*) \quad \forall u, v, x, \tilde{x}, y, \tilde{y}(\tilde{x} \leq_0 x \wedge \tilde{y} \geq_0 y \wedge \forall z^0 A_0(u, v, x, y, z) \to$$
$$\forall z^0 A_0(u, v, \tilde{x}, \tilde{y}, z)),$$

(i.e. $Mon(\exists x \forall y \exists z \neg A_0)$). Furthermore let $B_0(u, v, w^\gamma) \in \mathcal{L}(G_n A^\omega)$ be a (quantifier-free) formula which contains only u, v, w as free variables and $\gamma \leq 2$. Then from a proof

$$G_n A^\omega + \Delta + AC\text{-}qf \vdash$$

$$\forall u^1 \forall v \leq_\tau tu\big(\exists f^1 \forall x, z\, A_0(u, v, x, fx, z) \to \exists w^\gamma B_0(u, v, w)\big) \wedge (*)$$

one can extract a term $\chi \in G_n R^\omega$ such that

$$G_n A_i^\omega + \Delta + b\text{-}AC \vdash \forall u^1 \forall v \leq_\tau tu \forall \Psi^*\big((\Psi^* \text{ satisfies the mon.funct.}$$

$$interpr.of\, \forall x^0, g^1 \exists y^0 A_0(u, v, x, y, gy)) \to \exists w \leq_\gamma \chi u \Psi^*\, B_0(u, v, w)\big)^5.$$

In the conclusion $\Delta + b\text{-}AC$ can be replaced by $\tilde{\Delta}$ as defined in thm.2.4. If $\tau \leq 1$ and the types of existential quantifiers in the axioms Δ are ≤ 1, then $G_n A^\omega + \Delta + AC\text{-}qf$ may be replaced by $E\text{-}G_n A^\omega + \Delta + AC^{\alpha,\beta}\text{-}qf$, where $(\alpha = 0 \wedge \beta \leq 1)$ or $(\alpha = 1 \wedge \beta = 0)$, since elimination of extensionality applies in this case.

The mathematical significance of corollary 2.5 for the extraction of bounds from given proofs by arithmetization rests on the following fact: Direct monotone functional interpretation of

$$G_n A^\omega + \Delta + AC\text{-}qf \vdash$$

$$\forall u^1 \forall v \leq_\tau tu\big(\exists f^1 \forall x, z\, A_0(u, v, x, fx, z) \to \exists w^\gamma B_0(u, v, w)\big)$$

---

[5]'Ψ^* satisfies the mon. funct.interpr. of $\forall x, g \exists y A_0(u, v, x, y, gy)$' is meant here for fixed u, v (and not uniformly as a functional in u, v), i.e. $\exists \Psi\big(\Psi^*\text{ s–maj } \Psi \wedge \forall x, g\, A_0(u, v, x, \Psi xg, g(\Psi xg))\big)$.

provides only a bound on $\exists w$ which depends on a functional which satisfies the monotone functional interpretation of (1) $\exists f \forall x, z\, A_0$ or if we let remain the double negation in front of $\exists$ (which comes from the negative translation) (2) $\neg\neg\exists f \forall x, z\, A_0$. However in our applications the monotone functional interpretation of (1) would require non–computable functionals (since f in general is not recursive). The monotone functional interpretation of (2) can be carried out only using bar-recursive functionals (see [23]). In contrast to this the bound χ only depends on a functional which satisfies the monotone functional interpretation of the negative translation of $\forall x \exists y \forall z\, A_0(x, y, z)$: In our applications in section 5 such a functional can be constructed in $\widehat{PR}^{\omega}$ except for the existence of the limit superior of a bounded sequence of real numbers where the fragment T_1 of Gödel's calculus T with R_ρ for $\rho \leq 1$ is needed (note that the Ackermann function is definable in T_1).

In particular by arithmetizing the original proof the use of the **analytical** premise $\exists f^1 \forall x, z A_0$ has been replaced by the use of the **arithmetical** premise $\forall x^0 \exists y^0 \forall z^0 A_0$.

3 Real numbers in $G_2 A_i^\omega$

Suppose that a proposition $\forall x \exists y A(x, y)$ is proved in one of the theories T^ω from [16], where the variables x, y may range over $\mathbb{N}, \mathbb{Z}, \mathbb{Q}, \mathbb{R}$ or e.g. $C[0,1]$ etc. What sort of numerical information on '$\exists y$' relatively to the 'input' x can be extracted from a given proof depends in particular on how x is represented, i.e. on the numerical data by which x is given:
Suppose e.g. x that is a variable on $\mathbb{R}$ and real numbers are represented by **arbitrary** Cauchy sequences of rational numbers x_n, i.e.

$$(1) \quad \forall k^0 \exists n^0 \forall m, \tilde{m} \geq n \big(|x_m - x_{\tilde{m}}| \leq \frac{1}{k+1} \big).$$

Let us consider the (obviously true) proposition

$$(2) \quad \forall x \in \mathbb{R} \exists l \in \mathbb{N}(x \leq l).$$

Given x by a representative (x_n) in the sense of (1) it is not possible to compute an l which satisfies (2) on the basis of this representation, since this would involve the computation of a number n which fulfils a (in general undecidable) universal property like $\forall m, \tilde{m} \geq n(|x_m - x_{\tilde{m}}| \leq 1)$ to define l as $\lceil |x_n| \rceil + 1$.

If however real numbers are represented by Cauchy sequences with a **fixed Cauchy modulus**, e.g. $1/(k+1)$, i.e.

$$(3) \quad \forall m, \tilde{m} \geq k \big(|x_m - x_{\tilde{m}}| \leq \frac{1}{k+1} \big),$$

then the computation of l is trivial: $l := \Phi\left((x_n)\right) := \lceil |x_0| \rceil + 1$. Φ is not a function : $\mathbb{R} \to \mathbb{N}$ since it is not extensional: Different Cauchy sequences $(x_n), (\tilde{x}_n)$ which represent the same real number, i.e. $\lim_{n\to\infty}(x_n - \tilde{x}_n) = 0$, yield in general different numbers $\Phi\left((x_n)\right) \neq \Phi\left((\tilde{x}_n)\right)$. Following E. Bishop [3] , [4] we call Φ an **operation** : $\mathbb{R} \to \mathbb{N}$. This phenomenon is a general one (and not caused by the special definition of Φ): The only computable operations $\mathbb{R} \to \mathbb{N}$, which are extensional, are operations which are constant, since the computability of Φ implies its continuity as a functional[6] : $\mathbb{N}^{\mathbb{N}} \to \mathbb{N}$ and therefore (if it is extensional w.r.t. $=_{\mathbb{R}}$) the continuity as a function $\mathbb{R} \to \mathbb{N}$.

The importance of the representation of complex objects as e.g. real numbers is also indicated by the fact that the logical form of properties of these objects depends essentially on the representation: If $(x_n), (\tilde{x}_n)$ are arbitrary Cauchy sequences (in the sense of (1)) then the property that both sequences represent the same real number is expressed by the Π_3^0–formula

$$(4) \quad \forall k \exists n \forall m, \tilde{m} \geq n\left(|x_m - \tilde{x}_m| \leq \frac{1}{k+1}\right).$$

For Cauchy sequences with fixed Cauchy modulus as in (2) this property can be expressed by the (logically much simpler) Π_1^0–formula

$$(5) \quad \forall k\left(|x_k - \tilde{x}_k| \leq \frac{3}{k+1}\right).$$

For Cauchy sequences with modulus $1/(k+1)$ (4) and (5) are equivalent (provably in $G_2 A_i^\omega$). But for arbitrary Cauchy sequences (4) does not imply (5) in general.

If $(x_n) \subset \mathbb{Q}$ is an arbitrary Cauchy sequence then $AC^{0,0}$ applied to

$$\forall k \exists n \forall m, \tilde{m} \geq n\left(|x_m - x_{\tilde{m}}| \leq \frac{1}{k+1}\right)$$

yields the existence of a function f^1 such that
$\forall k \forall m, \tilde{m} \geq fk\left(|x_m - x_{\tilde{m}}| \leq \frac{1}{k+1}\right)$.

For $m, \tilde{m} \geq k$ this implies $|x_{fm} - x_{f\tilde{m}}| \leq \frac{1}{k+1}$ (choose $k' \in \{m, \tilde{m}\}$ with $fk' \leq fm, f\tilde{m}$ and apply the Cauchy property to $m' := fm, \tilde{m}' := f\tilde{m}$), i.e. the sequence $(x_{fn})_{n\in\mathbb{N}}$ is a Cauchy sequence with modulus $1/(k+1)$ which has the same limit as $(x_n)_{n\in\mathbb{N}}$.

---

[6]An operation $\Phi : \mathbb{R} \to \mathbb{N}$ is given by a functional : $\mathbb{N}^{\mathbb{N}} \to \mathbb{N}$ (which is extensional w.r.t. $=_1$!) since sequences of rational numbers are coded as sequences of natural numbers.

Thus in the presence of $AC^{0,0}$ (or more precisely the restriction $AC^{0,0}-\forall$ of $AC^{0,0}$ to Π_1^0-formulas) both representations (1) and (2) equivalent. However $AC^{0,0}-\forall$ is not provable in any of our theories and the addition of this schema to the axioms would yield an explosion of the rate of growth of the provably recursive functions. In fact every $\alpha(< \varepsilon_0)$-recursive function is provably recursive in $G_2A^\omega + AC^{0,0}-\forall$. This follows from the fact that iterated use of $AC^{0,0}-\forall$ combined with classical logic yields full arithmetical comprehension

$$CA_{ar} \ : \ \exists f^1 \forall x^0 (fx =_0 0 \leftrightarrow A(x)),$$

where A is an arithmetical formula, i.e. a formula containing only quantifiers of type 0. CA_{ar} applied to QF–IA proves the induction principle for every arithmetical formula. Hence full Peano–arithmetic PA is a subsystem of $G_2A^\omega + AC^{0,0}-\forall$.

As a consequence of this situation we have to specify the representation of real numbers we choose:

Definition 3.1 *A real number is given by a Cauchy sequence of rational numbers with modulus* $1/(k+1)$.

The reason for this representation is two–fold:

1) As we have seen above any numerically interesting application of the extraction of a bound presupposes that the input is given as a numerically reasonable object. This is also the reason why in constructive analysis (in the sense of Bishop) as well as in complexity theory for analysis (in the sense of H. Friedman and K.-I. Ko, see [11]) real numbers are always endowed with a rate of convergence, continuous functions with a modulus of continuity and so on. Also in the work by H. Friedman, S. Simpson (see e.g. [22]) and others on the program of so–called 'reverse mathematics', real numbers are always given with a fixed rate of convergence.

2) For our representation of real numbers we can achieve that quantification over real numbers is nothing else then quantification over $\mathbb{N}^{\mathbb{N}}$, i.e. $\forall x^1, \exists y^1$. Because of this many interesting theorems in analysis have the logical form $\forall \exists F_0$ (see [13] for a discussion on that) so that our method of extracting feasible bounds applies.

1) and 2) are in fact closely related: If real numbers would be represented as arbitrary Cauchy sequences then a proposition $\forall x \in \mathbb{R} \exists y \in \mathbb{N} \ A(x,y)$ would have the logical form

$$\forall x^1 (\forall k \exists n \forall m F_0 \rightarrow \exists y^0 A),$$

where $(*) \ \forall k \exists n \forall m F_0$ expresses the Cauchy property of the sequence of rational numbers coded by x^1. By our reasoning in [15] we know that in general we can only obtain an effective bound on y which depends on x

together with a Skolem function for (∗). But this just means that the computation of the bound requires that x is given with a Cauchy modulus. As concerned with provability in our theories like $G_n A^\omega + AC$-qf the representation with fixed modulus is no real restriction: In section 5 we will show in particular that the a proof of

$$\forall(x_n)\left(\exists f^1 \forall k \forall m, \tilde{m} \geq fk(|x_m - \tilde{x}_m| \leq \frac{1}{k+1}) \to \exists y^0 A\right)$$

can be transformed into a proof of

$$\forall(x_n)\left(\forall k \exists n \forall m, \tilde{m} \geq n(|x_m - \tilde{x}_m| \leq \frac{1}{k+1}) \to \exists y^0 A\right).$$

within the same theory (i.e. without any use of $AC^{0,0}$) for a large class of formulas A.

The representation of $\mathbb{R}$ presupposes a **representation of** $\mathbb{Q}$: Rational numbers are represented as codes $j(n,m)$ of pairs (n,m) of natural numbers n, m. $j(n,m)$ represents

the rational number $\frac{\frac{n}{2}}{m+1}$, if n is even, and

the negative rational $-\frac{\frac{n+1}{2}}{m+1}$ if n is odd.

By the surjectivity of our pairing function j from [15] every natural number can be conceived as code of a uniquely determined rational number. On the codes of $\mathbb{Q}$, i.e. on $\mathbb{N}$, we define an equivalence relation by

$$n_1 =_\mathbb{Q} n_2 := \frac{\frac{j_1 n_1}{2}}{j_2 n_1 + 1} = \frac{\frac{j_1 n_2}{2}}{j_2 n_2 + 1} \text{ if } j_1 n_1, j_1 n_2 \text{ both are even}$$

and analogously in the remaining cases, where $\frac{a}{b} = \frac{c}{d}$ is defined to hold iff $ad =_0 cb$ (for $bd > 0$).
On $\mathbb{N}$ one easily defines functions $|\cdot|_\mathbb{Q}, +_\mathbb{Q}, -_\mathbb{Q}, \cdot_\mathbb{Q} :_\mathbb{Q}, \max_\mathbb{Q}, \min_\mathbb{Q} \in G_2 R^\omega$ and (quantifier–free) relations) $<_\mathbb{Q}, \leq_\mathbb{Q}$ which represent the corresponding functions and relations on $\mathbb{Q}$. In the following we sometimes omit the index $\mathbb{Q}$ if this does not cause any confusion.

Notational convention: For better readability we often write e.g. $\frac{1}{k+1}$ instead of its code $j(2,k)$ in $\mathbb{N}$. So e.g. we write $x^0 \leq_\mathbb{Q} \frac{1}{k+1}$ for $x \leq_\mathbb{Q} j(2,k)$.

By the coding of rational numbers as natural numbers, **sequences of rationals** are just functions f^1 (and every function f^1 can be conceived as a sequence of rational numbers in a unique way). In particular representatives of real numbers are functions f^1 modulo this coding. We now show that **every** function can be conceived as an representative of a uniquely determined Cauchy sequence of rationals with modulus $1/(k+1)$ and therefore can be conceived as an representative of a uniquely determined real number.[7]

To achieve this we need the following functional $\widehat{f}$.

Definition 3.2 *The functional* $\lambda f^1.\widehat{f} \in G_2 R^\omega$ *is defined such that*

$$\widehat{f}n = \begin{cases} fn, & \text{if } \forall k,m,\tilde{m} \leq_0 n\big(m,\tilde{m} \geq_0 k \to |fm -_{\mathbb{Q}} f\tilde{m}| \leq_{\mathbb{Q}} \frac{1}{k+1}\big) \\ f(n_0-1) & \text{for } n_0 := \min l \leq_0 n[\exists k,m,\tilde{m} \leq_0 l\big(m,\tilde{m} \geq_0 k \wedge \\ & \qquad |fm -_{\mathbb{Q}} f\tilde{m}| >_{\mathbb{Q}} \frac{1}{k+1}\big)], \text{ otherwise.} \end{cases}$$

One easily verifies (within $G_2 A_i^\omega$) that

1) if f^1 represents a Cauchy sequence of rational numbers with modulus $1/(k+1)$, then $\forall n^0(fn =_0 \widehat{f}n)$,

2) for every f^1 the function $\widehat{f}$ represents a Cauchy sequence of rational numbers with modulus $1/(k+1)$.

Hence every function f gives a uniquely determined real number, namely that number which is represented by $\widehat{f}$. Quantification $\forall x \in \mathbb{R} A(x)$ ($\exists x \in \mathbb{R} A(x)$) so reduces to the quantification $\forall f^1 A(\widehat{f})$ ($\exists f^1 A(\widehat{f})$) for properties A which are extensional w.r.t. $=_{\mathbb{R}}$ below (i.e. which are really properties of real numbers). **Operations** $\Phi : \mathbb{R} \to \mathbb{R}$ are given by functionals $\Phi^{1(1)}$ (which are extensional w.r.t.$=_1$). A real function : $\mathbb{R} \to \mathbb{R}$ is given by a functional $\Phi^{1(1)}$ which (in addition) is extensional w.r.t. $=_{\mathbb{R}}$. Following the usual notation we write (x_n) instead of fn and $(\widehat{x}_n)$ instead of $\widehat{f}n$.

In the following we define various relations and operations on functions which correspond to the usual relations and operations on $\mathbb{R}$ for the real numbers represented by the respective functions:

Definition 3.3 *1)* $(x_n) =_{\mathbb{R}} (\tilde{x}_n) :\equiv \forall k^0\big(|\widehat{x}_k -_{\mathbb{Q}} \widehat{\tilde{x}}_k| \leq_{\mathbb{Q}} \frac{3}{k+1}\big);$

2) $(x_n) <_{\mathbb{R}} (\tilde{x}_n) :\equiv \exists k^0\big(\widehat{\tilde{x}}_k - \widehat{x}_k >_{\mathbb{Q}} \frac{3}{k+1}\big);$

3) $(x_n) \leq_{\mathbb{R}} (\tilde{x}_n) :\equiv \neg(\widehat{\tilde{x}}_n) <_{\mathbb{R}} (\widehat{x}_n);$

4) $(x_n) +_{\mathbb{R}} (\tilde{x}_n) := (\widehat{x}_{2n+1} +_{\mathbb{Q}} \widehat{\tilde{x}}_{2n+1});$

5) $(x_n) -_{\mathbb{R}} (\tilde{x}_n) := (\widehat{x}_{2n+1} -_{\mathbb{Q}} \widehat{\tilde{x}}_{2n+1});$

6) $|(x_n)|_{\mathbb{R}} := (|\widehat{x}_n|_{\mathbb{Q}});$

7) $(x_n) \cdot_{\mathbb{R}} (\tilde{x}_n) := (\widehat{x}_{2(n+1)k} \cdot_{\mathbb{Q}} \widehat{\tilde{x}}_{2(n+1)k}),$
  where $k := \lceil \max_{\mathbb{Q}}(|x_0|_{\mathbb{Q}} + 1, |\tilde{x}_0|_{\mathbb{Q}} + 1)\rceil;$

---

[7]A related representation of real numbers is sketched in [1] .

8) For (x_n) and l^0 we define

$$(x_n)^{-1} := \begin{cases} (\max_{\mathbb{Q}}(\widehat{x}_{(n+1)(l+1)^2}, \frac{1}{l+1})^{-1}), & \text{if } \widehat{x}_{2(l+1)} >_{\mathbb{Q}} 0 \\ (\min_{\mathbb{Q}}(\widehat{x}_{(n+1)(l+1)^2}, \frac{-1}{l+1})^{-1}), & \text{otherwise;} \end{cases}$$

9) $\max_{\mathbb{R}}\big((x_n),(\tilde{x}_n)\big) := \big(\max_{\mathbb{Q}}(\widehat{x}_n, \widehat{\tilde{x}}_n)\big),$
$\min_{\mathbb{R}}\big((x_n),(\tilde{x}_n)\big) := \big(\min_{\mathbb{Q}}(\widehat{x}_n, \widehat{\tilde{x}}_n)\big).$

One easily verifies the following

Lemma 3.4 *1)* $(x_n) =_{\mathbb{R}} (\tilde{x}_n)$ *resp.* $(x_n) <_{\mathbb{R}} (\tilde{x}_n)$, $(x_n) \le_{\mathbb{R}} (\tilde{x}_n)$ *hold iff the corresponding relations hold for those real numbers which are represented by* $(x_n),(\tilde{x}_n)$.

2) Provably in $G_2 A_i^\omega$, $(x_n) +_{\mathbb{R}} (\tilde{x}_n)$, $(x_n) -_{\mathbb{R}} (\tilde{x}_n)$, $(x_n) \cdot_{\mathbb{R}} (\tilde{x}_n)$, $\max_{\mathbb{R}}$ $((x_n),(\tilde{x}_n))$, $\min_{\mathbb{R}} ((x_n),(\tilde{x}_n))$ *and* $|(x_n)|_{\mathbb{R}}$ *also represent Cauchy sequences with modulus* $1/(k+1)$ *which represent the real number obtained by addition (subtraction,...) of those real numbers which are represented by* $(x_n),(\tilde{x}_n)$. *This also holds for* $(x_n)^{-1}$ *if* $|(x_n)|_{\mathbb{R}} \ge_{\mathbb{R}}$ $\frac{1}{l+1}$ *for the number* l *used in the definition of* $(x_n)^{-1}$. *In particular the operations* $+_{\mathbb{R}}, -_{\mathbb{R}}$ *etc. are extensional w.r.t. to* $=_{\mathbb{R}}$ *and therefore represent functions[8].*

3) The functionals $+_{\mathbb{R}}, -_{\mathbb{R}}, \cdot_{\mathbb{R}}, \max_{\mathbb{R}}, \min_{\mathbb{R}}$ *of type* $1(1)(1)$, $|\cdot|_{\mathbb{R}}$ *of type* $1(1)$ *and* $()^{-1}$ *of type* $1(1)(0)$ *are definable in* $G_2 R^\omega$.

Remark 3.1 *Since our theories* $G_n A_i^\omega$ *contain all* $\mathbb{N}, \mathbb{N}^{\mathbb{N}}$*-true purely universal sentences* $\forall \underline{x}^{0/1} A_0(\underline{x})$ *as axioms (because they do not contribute to the growth of extractable bounds at all, see [15] for details), it is easy to check that the basic properties of* $=_{\mathbb{R}}, \le_{\mathbb{R}}, +_{\mathbb{R}}, \ldots$ *can be proved in* $G_2 A_i^\omega$. *They are either directly purely universal or can be strengthened to universal statements, e.g.*

$x =_{\mathbb{R}} y \wedge y =_{\mathbb{R}} z \to x =_{\mathbb{R}} z$ *follows from the universal axiom*

$\forall x^1, y^1, k^0\big(|\widehat{x}(6(k+1)) -_{\mathbb{Q}} \widehat{y}(6(k+1))| \le_{\mathbb{Q}} \frac{3}{6(k+1)+1} \wedge$
$|\widehat{y}(6(k+1)) -_{\mathbb{Q}} \widehat{z}(6(k+1))| \le_{\mathbb{Q}} \frac{3}{6(k+1)+1} \to |\widehat{x}(k) -_{\mathbb{Q}} \widehat{z}(k)| \le_{\mathbb{Q}} \frac{3}{k+1}\big).$

Rational numbers q coded by r_q have as canonical representative in $\mathbb{R}$ (besides other representatives) the constant function $\lambda n^0.r_q$. One easily shows that $\forall k\big(|(x_n) -_{\mathbb{R}} \lambda n.\hat{x}_k| \le_{\mathbb{R}} \frac{1}{k+1}\big)$ for every function (x_n).

Notational convention: For notational simplicity we often omit the embedding $\mathbb{Q} \hookrightarrow \mathbb{R}$, e.g. $x^1 \le_{\mathbb{R}} y^0$ stands for $x \le_{\mathbb{R}} \lambda n.y^0$. From the type of the objects it will be always clear what is meant.

---

[8]The functional $()^{-1}$ is extensional for all l and $(x_n),(y_n)$ such that $|(x_n)|_{\mathbb{R}}, |(y_n)|_{\mathbb{R}} \ge \frac{1}{l+1}$.

If $(f_n)_{n \in \mathbb{N}}$ of type $1(0)$ represents a $\frac{1}{k+1}$-Cauchy sequence of **real** numbers, then

$f(n) := \widehat{f}_{3(n+1)}(3(n+1))$ represents the limit of this sequence, i.e. $\forall k (|f_k -_{\mathbb{R}} f| \leq_{\mathbb{R}} \frac{1}{k+1})$. One easily verifies this fact in $G_2 A_i^\omega$.

Representation of $\mathbb{R}^d$ in $G_2 A_i^\omega$:

For every fixed d we represent $\mathbb{R}^d$ as follows: Elements of $\mathbb{R}^d$ are represented by functions f^1 in the following way: Using the construction $\widehat{f}$ from above, every f^1 can be conceived as a representative of such a d-tuple of Cauchy sequences of real numbers, namely the sequence which is represented by

$$\big(\widehat{\nu_1^d(f)}, \ldots, \widehat{\nu_d^d(f)}\big), \text{ where } \nu_i^d(f) := \lambda x^0 . \nu_i^d(fx),$$

(ν_i^d are the coding functions $\in G_2 R^\omega$ from [15]).

Since the $\widehat{\nu_i^d(f)}$ represent Cauchy sequences of rationals with Cauchy modulus $\frac{1}{k+1}$, elements of $\mathbb{R}^d$ are so represented as Cauchy sequences of elements in $\mathbb{Q}^d$ which have the Cauchy modulus $\frac{1}{k+1}$ w.r.t. the maximum norm $\|f^1\|_{\max} := \max_{\mathbb{R}} \big(|\nu_1^d(f)|_{\mathbb{R}}, \ldots, |\nu_d^d(f)|_{\mathbb{R}}\big)$.

Quantification $\forall (x_1, \ldots, x_d) \in \mathbb{R}^d$ so reduces to $\forall f^1 A(\widehat{\nu_1^d(f)}, \ldots, \widehat{\nu_d^d(f)})$ for $\mathbb{R}^d$–extensional properties A (likewise for $\exists$).

The operations $+_{\mathbb{R}^d}, -_{\mathbb{R}^d}, \ldots$ are defined via the corresponding operations on the components, e.g. $x^1 +_{\mathbb{R}^d} y^1 :\equiv \nu^d(\nu_1^d x +_{\mathbb{R}} \nu_1^d y, \ldots, \nu_d^d x +_{\mathbb{R}} \nu_d^d y)$.

Sequences of elements in $\mathbb{R}^d$ are represented by (f_n) of type $1(0)$.

Representation of $[0,1] \subset \mathbb{R}$ in $G_2 A_i^\omega$

We now show that every element of $[0, 1]$ can be represented already by a bounded function $f \in \{f : f \leq_1 M\}$, where M is a fixed function from $G_2 R^\omega$ and that every function from this set can be conceived as an (representative of an) element in $[0,1]$: Firstly we define a function $q \in G_2 R^\omega$ by

$$q(n) := \begin{cases} \min l \leq_0 n[l =_{\mathbb{Q}} n], & \text{if } 0 \leq_{\mathbb{Q}} n \leq_{\mathbb{Q}} 1 \\ 0^0, & \text{otherwise.} \end{cases}$$

It is clear that every rational number $\in [0, 1] \cap \mathbb{Q}$ has a unique code by a number $\in q(\mathbb{N})$ and $\forall n^0 (q(q(n)) =_0 q(n))$. Also every such number codes an element of $\in [0, 1] \cap \mathbb{Q}$ (0^0 codes $0 \in \mathbb{Q}$ since $j(0, 0) = 0$). We may conceive every number n as a representative of a rational number $\in [0, 1] \cap \mathbb{Q}$, namely of the rational coded by $q(n)$.

In contrast to $\mathbb{R}$ we can restrict the set of representing functions for $[0,1]$ to the compact (in the sense of the Baire space) set $f \in \{f : f \leq_1 M\}$, where $M(n) := j(6(n+1), 3(n+1) - 1)$ (here j is the Cantor pairing function): Each fraction r having the form $\frac{i}{3(n+1)}$ (with $i \leq 3(n+1)$) is represented by a number $k \leq M(n)$, i.e. $k \leq M(n) \wedge q(k)$ codes r. Thus $\{k : k \leq M(n)\}$ contains (modulo this coding) an $\frac{1}{3(n+1)}$-net for $[0,1]$.

We define a functional $\lambda f.\tilde{f} \in G_2 R^\omega$ such that

$$\tilde{f}(k) = q(i_0), \text{ where } i_0 = \mu i \leq_0 M(k)[\forall j \leq_0 M(k)(|\hat{f}(3(k+1)) -_{\mathbb{Q}} q(j)|$$
$$\geq_{\mathbb{Q}} |\hat{f}(3(k+1)) -_{\mathbb{Q}} q(i)|)].$$

$\tilde{f}$ has (provably in $G_2 A_i^\omega$) the following properties:

1) $\forall f^1(\tilde{f} \leq_1 M)$.

2) $\forall f^1(\hat{\tilde{f}} =_1 \tilde{f})$.

3) $\forall f^1(0 \leq_{\mathbb{R}} \tilde{f} \leq_{\mathbb{R}} 1)$.

4) $\forall f^1(0 \leq_{\mathbb{R}} f \leq_{\mathbb{R}} 1 \to f =_{\mathbb{R}} \tilde{f})$.

5) $\forall f^1(\tilde{\tilde{f}} =_{\mathbb{R}} \tilde{f})$.

By this construction quantification $\forall x \in [0,1]\ A(x)$ and $\exists x \in [0,1]\ A(x)$ reduces to quantification having the form $\forall f \leq_1 M\ A(\tilde{f})$ and $\exists f \leq_1 M\ A(\tilde{f})$ for properties A which are $=_{\mathbb{R}}$-extensional (for f_1, f_2 such that $0 \leq_{\mathbb{R}} f_1, f_2 \leq_{\mathbb{R}} 1$), where $M \in G_2 R^\omega$. Similarly one can define a representation of $[a, b]$ for variable a^1, b^1 such that $a <_{\mathbb{R}} b$ by bounded functions $\{f^1 : f \leq_1 M(a,b)\}$. However by remark 3.2 below one can easily reduce the quantification over $[a, b]$ to quantification over $[0,1]$ so that we do not need this generalization. But on some occasions it is convenient to have an explicit representation for $[-k, k]$ for all natural numbers k. This representation is analogous to the representation of $[0, 1]$ except that we now define $M_k(n) := j(6k(n+1), 3(n+1) - 1)$ as the bounding function. The construction corresponding to $\lambda f.\tilde{f}$ is also denoted by $\tilde{f}$ since it will be always clear from the context what interval we have in mind.

Representation of $[0, 1]^d$ in $G_2 A_i^\omega$

Using the construction $f \mapsto \tilde{f}$ from the representation of $[0,1]$ we also can represent $[0, 1]^d$ for every fixed number d by a bounded set $\{f^1 : f \leq_1 M_d\}$ of functions, where $M_d : \nu^d(M, \ldots, M) \in G_2 R^\omega$ for every fixed d: $\widetilde{f(\leq M_d)}$ represents the vector in $[0, 1]^d$ which is represented by $(\widetilde{(\nu_1^d f)}, \ldots, \widetilde{(\nu_d^d f)})$. If (in the other direction) $f_1, \ldots, f_d$ represent real numbers $x_1, \ldots, x_d \in [0, 1]$, then $f := \nu^d(\tilde{f}_1, \ldots, \tilde{f}_d) \leq_1 \nu^d(M, \ldots, M)$ represents $(x_1, \ldots, x_d) \in [0,1]^d$ in this sense.

Remark 3.2 *For $a, b \in \mathbb{R}$ with $a \leq_{\mathbb{R}} b$, quantification $\forall x \in [a, b]\ A(x)$ ($\exists x \in [a, b]\ A(x)$) reduces to quantification over $[0,1]$ (and therefore – modulo our representation– over $\{f : f \leq_1 M\}$) by $\forall \lambda \in [0,1]\ A((1-\lambda)a + \lambda b)$ and analogously for $\exists x$. This transformation immediately generalizes to $[a_1, b_1] \times \cdots \times [a_d, b_d]$ using $\lambda_1, \ldots, \lambda_d$.*

4   Sequences and series in $G_2A_i^\omega$: Convergence with moduli involved

By our representation of real numbers by functions f^1 developed in the previous section, sequences of real numbers are given as functions $f^{1(0)}$ in $G_2A_i^\omega$. We will use the usual notation (a_n) instead of f. In this section we are concerned with the following properties of sequences of real numbers:

1) (a_n) is a Cauchy sequence, i.e.
$$\forall k^0 \exists n^0 \forall m, \tilde{m} \geq_0 n \left(|a_m -_{\mathbb{R}} a_{\tilde{m}}| \leq_{\mathbb{R}} \tfrac{1}{k+1} \right).$$

2) (a_n) is convergent, i.e. $\exists a^1 \forall k^0 \exists n^0 \forall m \geq_0 n \left(|a_m -_{\mathbb{R}} a| \leq_{\mathbb{R}} \tfrac{1}{k+1} \right).$

3) (a_n) is convergent with a modulus of convergence, i.e.
$$\exists a^1, h^1 \forall k^0 \forall m \geq_0 hk(|a_m -_{\mathbb{R}} a| \leq_{\mathbb{R}} \frac{1}{k+1}).$$

4) (a_n) is a Cauchy sequence with a Cauchy modulus, i.e.
$$\exists h^1 \forall k^0 \forall m, \tilde{m} \geq_0 hk(|a_m -_{\mathbb{R}} a_{\tilde{m}}| \leq_{\mathbb{R}} \frac{1}{k+1}).$$

One easily shows within $G_2A_i^\omega$ that 4) $\leftrightarrow$ 3) $\rightarrow$ 2) $\rightarrow$ 1). Using $AC^{0,0}\text{-}\forall^0$ one can prove that 1) $\rightarrow$ 4) (and therefore 1) $\leftrightarrow$ 2) $\leftrightarrow$ 3) $\leftrightarrow$ 4)).
However, as we already have discussed in the previous section, the addition of $AC^{0,0}\text{-}\forall^0$ to G_2A^ω would make all $\alpha(< \varepsilon_0)$–recursive functions provably recursive.

Thus since we are working in (extensions of) G_2A^ω we have to distinguish carefully between e.g. 1) and 4). In the next section we will study the relationship between 1) and 4) in detail and show in particular that the use of sequences of single instances of 4) in proofs of $\forall u^1 \forall v \leq_\rho tu \exists w^2 A_0$–sentences relatively to e.g. $G_2A^\omega + \Delta + AC\text{-}qf$ (where Δ is defined as in thm.2.4) can be reduced the use of the same instances of 1).

For **monotone** sequences (a_n) the equivalence of 2) and 3) (and hence that of 2) and 4)) is already provable using only the **quantifier–free** choice

$AC^{0,0}$–qf:

Let (a_n) be say increasing, i.e.

$$(i)\ \forall n^0(a_n \leq_{\mathbb{R}} a_{n+1}),$$

and a^1 be such that

$$(ii)\ \forall k^0 \exists n^0 \forall m \geq_0 n \left(|a_m - a| \leq_{\mathbb{R}} \frac{1}{k+1}\right).$$

$AC^{0,0}$–qf applied to $\forall k^0 \exists n^0 \underbrace{\left(|a_n - a| <_{\mathbb{R}} \frac{1}{k+1}\right)}_{\in \Sigma_1^0}$ yields

$\exists h^1 \forall k^0 \left(|a_{hk} - a| <_{\mathbb{R}} \frac{1}{k+1}\right)$, which gives $\exists h^1 \forall k^0 \forall m \geq_0 hk\left(|a_m - a| <_{\mathbb{R}} \frac{1}{k+1}\right)$, since –by (i),(ii)– $a_{hk} \leq a_m \leq a$ for all $m \geq_0 hk$. (Here we use the fact that $\forall n(a_n \leq_{\mathbb{R}} a_{n+1}) \to \forall m, \tilde{m}(m \geq \tilde{m} \to a_{\tilde{m}} \leq_{\mathbb{R}} a_m)$. This follows in G_2A^ω from the universal sentence

$(+)\ \forall a_{(\cdot)}^{1(0)}, n, l \left(\forall k < n\left(\hat{a}_k(l) \leq_{\mathbb{Q}} \hat{a}_{k+1}(l) + \frac{3}{l+1}\right) \to \forall m, \tilde{m} \leq n \left(m \geq \tilde{m} \to a_{\tilde{m}} \leq_{\mathbb{R}} a_m + \frac{5n}{l+1}\right)\right).\ (+)$ is true (and hence an axiom of G_2A^ω) since $\hat{a}_k(l) \leq_{\mathbb{Q}} \hat{a}_{k+1}(l) + \frac{3}{l+1} \to a_k \leq_{\mathbb{R}} a_{k+1} + \frac{5}{l+1}$.)

If one of the properties 1), . . . ,4) –say $i \in \{1, \ldots, 4\}$– is fulfilled for two sequences $(a_n), (b_n)$, then $i)$ is also fulfilled (provably in $G_2A_i^\omega$) for $(a_n +_{\mathbb{R}} b_n), (a_n -_{\mathbb{R}} b_n), (a_n \cdot_{\mathbb{R}} b_n)$ and (if $b_n \neq 0$ and $b_n \to b \neq 0$) for $\left(\frac{a_n}{b_n}\right)$, where in the later case the modulus in 3),4) depends on an estimate $l \in \mathbb{N}$ such that $|b| \geq \frac{1}{l+1}$ (The construction of the moduli for $(a_n +_{\mathbb{R}} b_n), (a_n -_{\mathbb{R}} b_n), (a_n \cdot_{\mathbb{R}} b_n), \left(\frac{a_n}{b_n}\right)$ from the moduli for $(a_n), (b_n)$ (for i=3,4) is similar to our definition of $+_{\mathbb{R}}, -_{\mathbb{R}}, \cdot_{\mathbb{R}}, (\cdot)^{-1}$ given in the previous section.

The most important property of bounded monotone sequences (a_n) of real numbers is their convergence. We call this fact 'principle of convergence for monotone sequences' (PCM). Because of the difference between 1) and 4) above we have in fact to consider two versions of this principle:

$$(PCM1)\ :\ \left\{ \begin{array}{l} \forall a_{(\cdot)}^{1(0)}, c^1 \left(\forall n^0 (c \leq_{\mathbb{R}} a_{n+1} \leq_{\mathbb{R}} a_n)\right. \\ \qquad \left. \to \forall k^0 \exists n^0 \forall m, \tilde{m} \geq_0 n(|a_m -_{\mathbb{R}} a_{\tilde{m}}| \leq_{\mathbb{R}} \frac{1}{k+1})\right), \end{array} \right.$$

$$(PCM2)\ :\ \left\{ \begin{array}{l} \forall a_{(\cdot)}^{1(0)}, c^1 \left(\forall n^0 (c \leq_{\mathbb{R}} a_{n+1} \leq_{\mathbb{R}} a_n)\right. \\ \qquad \left. \to \exists h^1 \forall k^0 \forall m, \tilde{m} \geq_0 hk(|a_m -_{\mathbb{R}} a_{\tilde{m}}| \leq_{\mathbb{R}} \frac{1}{k+1})\right), \end{array} \right.$$

Both principles cannot be derived in any of the theories $G_nA^\omega + \Delta + AC$–qf. In fact $(PCM1)$ is equivalent (relatively to G_3A^ω) to the second–order axiom of Σ_1^0–induction whereas $(PCM2)$ is equivalent (relatively to $G_3A^\omega + AC^{0,0}$–qf) even to arithmetical comprehension over numbers (see [14]; for the system RCA_0, known from reverse mathematics, the equivalence between $(PCM2)$ and arithmetical comprehension is due to [6]). We

3

now determine the contribution of the use of $(PCM1)$ to the growth of extractable uniform bounds. This will be used in the next section to determine the growth which may be caused be single sequences of instances of $(PCM2)$.

Using the construction $\tilde{a}(n) := \max_{\mathbb{R}}(0, \min_{i\leq n}(a(i)))$, we can express $(PCM1)$ in the following logically more simple form[9]

$$(1)\ \forall a^{1(0)}\forall k^0\exists n^0\forall m >_0 n\big(\tilde{a}(n) -_{\mathbb{R}} \tilde{a}(m) \leq_{\mathbb{R}} \frac{1}{k+1}\big).$$

(If $a^{1(0)}$ fulfils $\forall n(0 \leq_{\mathbb{R}} a(n+1) \leq_{\mathbb{R}} a(n))$, then $\forall n(\tilde{a}(n) =_{\mathbb{R}} a(n))$. Furthermore $\forall n(0 \leq_{\mathbb{R}} \tilde{a}(n+1) \leq_{\mathbb{R}} \tilde{a}(n))$ for all $a^{1(0)}$. Thus by the transformation $a \mapsto \tilde{a}$, quantification over all decreasing sequences $\subset \mathbb{R}_+$ reduces to quantification over all $a^{1(0)}$).
By $AC^{0,0}$-qf (1) is equivalent to

$$(2)\ \forall a^{1(0)}, k^0, g^1\exists n^0\big(gn >_0 n \to \tilde{a}(n) -_{\mathbb{R}} \tilde{a}(gn) \leq_{\mathbb{R}} \frac{1}{k+1}\big).$$

We now construct a functional Ψ which provides a bound for $\exists n$, i.e.

$$(3)\ \forall a^{1(0)}, k^0, g^1\exists n \leq_0 \Psi akg\big(gn >_0 n \to \tilde{a}(n) -_{\mathbb{R}} \tilde{a}(gn) \leq_{\mathbb{R}} \frac{1}{k+1}\big).$$

Let $C(a) \in \mathbb{N}$ ($C(a) \geq 1$) be an upper bound for the real number represented by $\tilde{a}(0)$ (with C s-maj C), e.g. $C(a) := (a(0))(0) + 1$. We show that
$\Psi akg := \max_{i<C(a)k'}\big(\Phi_{it}i0g\big)(= \max_{i<C(a)k'}\big(g^i(0)\big))$ satisfies (3) (provably in PRA$^\omega$):
Claim: $\exists i < C(a)k'\big(g(g^i0) > g^i0 \to \tilde{a}(g^i0) -_{\mathbb{R}} \tilde{a}(g(g^i0)) \leq_{\mathbb{R}} \frac{1}{k+1}\big)$.
Case 1: $\exists i < C(a)k'(g(g^i0) \leq g^i0)$: Obvious!
Case 2: $\forall i < C(a)k'(g(g^i0) > g^i0)$:
Assume $\forall i < C(a)k'(\tilde{a}(g^i0) -_{\mathbb{R}} \tilde{a}(g(g^i0))) >_{\mathbb{R}} \frac{1}{k+1})$.
Then $\tilde{a}(0) -_{\mathbb{R}} \tilde{a}(g^{C(a)k'}0) > C(a)$, contradicting $\tilde{a}(n) \in [0, C(a)]$ for all n.

---

[9]Here we use that $\forall n^0\big(a(n+1) \leq_{\mathbb{R}} an\big) \to \forall n^0\big(\Phi_{\min_{\mathbb{R}}}(a,n) =_{\mathbb{R}} an\big)$, where $\Phi_{\min_{\mathbb{R}}}$ is a functional from G_2R^ω which computes the minimum of the real numbers $a(0),\ldots,a(n)$ (such a functional can be defined similarly to $\min_{\mathbb{R}}$ in section 3 noting that $\Phi_{\min_{\mathbb{Q}}}(f^1, n^0) = \min_{\mathbb{Q}}(f0,\ldots,fn)$ is definable in G_2R^ω). This follows in G_2A^ω from the purely universal sentence
$(+)\ \forall a^{1(0)}, n, k\Big(\forall l < n\big(\big(\widehat{a(l+1)}\big)(k) \leq_{\mathbb{Q}} (\widehat{al})(k) + \frac{3}{k+1}\big) \to |\Phi_{\min_{\mathbb{R}}}(a,n) -_{\mathbb{R}} an| \leq_{\mathbb{R}} \frac{5n}{k+1}\Big)$. $(+)$ is true (and hence an axiom of G_2A^ω) since
$\big(\widehat{a(l+1)}\big)(k) \leq_{\mathbb{Q}} (\widehat{al})(k) + \frac{3}{k+1} \to a(l+1) \leq_{\mathbb{R}} al + \frac{5}{k+1}$.

In contrast to (2) the bounded proposition (3) has the form of an axiom Δ in the theorems from [15] and section 2. Hence the monotone functional interpretation of (3) requires just a majorant for Ψ. In particular we may use $\Psi \in \widehat{PR}^\omega$ itself since Ψ s–maj Ψ.

Thus from a proof of e.g. a sentence $\forall x^0 \forall y \leq_\rho sx \exists z^0 A_0(x, y, z)$ in $G_n A^\omega +$ $\Delta + (PCM1)+AC$–qf we can (in general) extract only a bound t for z (i.e. $\forall x \forall y \leq sx \exists z \leq tx\, A_0(x, y, z)$) which is defined in $\widehat{PR}^\omega$ since the definition of Ψ uses the functional Φ_{it} which is not definable in $G_\infty R^\omega$ (see [15]). If however the proof uses (3) above only for functions g which can be bounded by terms in $G_k R^\omega$, then we can extract a $t \in G_{\max(k+1,n)} R^\omega$ since the iteration of a function $\in G_k R^\omega$ is definable in $G_{k+1} R^\omega$ (for $k \geq 2$).

The monotone functional interpretation of the negative translation of (1) requires (taking the quantifier hidden in $\leq_{\mathbb{R}}$ into account) a majorant for a functional Φ which bounds '$\exists n$' in

$$(3)' \; \forall a^{1(0)}, k^0, g^1, h^1 \exists n \big(gn > n \to \widehat{\bar{a}(n)}(hn) -_{\mathbb{Q}} \widehat{\bar{a}(gn)}(hn) \leq_{\mathbb{Q}}$$
$$\frac{1}{k+1} + \frac{3}{h(n)+1} \big).$$

However every Φ which provides a bound for (2) a fortiori yields a bound for $(3)'$ (which does not depend on h). Hence Ψ satisfies (provably in PRA_i^ω) the monotone functional interpretation of the negative translation of (1), i.e. $(PCM1)$.

5   The rate of growth caused by sequences of instances of analytical principles whose proofs rely on arithmetical comprehension

In this section we apply the results presented in section 2 in order to determine the impact on the rate of growth of uniform bounds for provably $\forall u^1 \forall v \leq_\tau tu \exists w^\gamma A_0$–sentences which may result from the use of sequences (which however may depend on the parameters of the proposition to be proved) of instances of:

1) $(PCM2)$ and the convergence of bounded monotone sequences of real numbers.

2) The existence of a greatest lower bound for every sequence of real numbers which is bounded from below.

3) Π_1^0–CA and Π_1^0–AC.

4) The Bolzano–Weierstraß property for bounded sequences in $\mathbb{R}^d$ (for every fixed d).

5) The Arzelà–Ascoli lemma.

6) The existence of lim sup and lim inf for bounded sequences in $\mathbb{R}$.

5.1 $(PCM2)$ and the convergence of bounded monotone sequences of real numbers

Let $a^{1(0)}$ be such that $\forall n^0(0 \leq_\mathbb{R} a(n+1) \leq_\mathbb{R} an)$[10]
$(PCM2)$ implies

$$\exists h^1 \forall k^0, m^0(m \geq_0 hk \to a(hk) -_\mathbb{R} a(m) \leq_\mathbb{R} \frac{1}{k+1}).$$

$\big(a(hk)\big)_k$ is a Cauchy sequence with modulus $\frac{1}{k+1}$ whose limit equals the limit of $\big(a(m)\big)_{n\in\mathbb{N}}$. The existence of a limit a_0 of $(a(m))_m$ now follows from the remarks below lemma 3.4 : $a_0 k := (a(h(\widehat{3(k+1)})))(3(k+1))$. Thus we only have to consider $(PCM2)$. In order to simplify the logical form of $(PCM2)$ we use the construction $\tilde{a}(n) := \max_\mathbb{R}(0, \min_{i\leq n}(a(i)))$ from the previous section (recall that this construction ensures that $\tilde{a}$ is monotone decreasing and bounded from below by 0. If a already fulfils these properties nothing is changed by the passage from a to $\tilde{a}$).

$$(PCM2)(a^{1(0)}) :\equiv \exists h^1 \forall k^0, m^0\big(m \geq_0 hk \to \tilde{a}(hk) -_\mathbb{R} \tilde{a}(m) \leq_\mathbb{R} \frac{1}{k+1}\big).$$

We now show that the contribution of single instances $(PCM2)(a)$ of $(PCM2)$ to the growth of uniform bounds is (at most) given by the functional $\Psi akg := \max_{i<C(a)k'}\big(\Phi_{it}i0g\big)$ (where $\mathbb{N}^* \ni C(a) \geq \tilde{a}(0)$) as above:

Proposition 5.1 *Let $n \geq 2$ and $B_0(u^1, v^\tau, w^\gamma) \in \mathcal{L}(G_n A^\omega)$ be a quantifier-free formula which contains only u^1, v^τ, w^γ free, where $\gamma \leq 2$. Furthermore*

---

[10]The restriction to the lower bound 0 is (convenient but) not essential: If $\forall n^0(c \leq_\mathbb{R} a(n+1) \leq_\mathbb{R} an)$ we may define $a'(n) := a(n) -_\mathbb{R} c$. $(PCM2)$ applied to a' implies $(PCM2)$ for a. Everything holds analogously for increasing sequences which are bounded from above.

let $\xi, \iota \in G_n R^\omega$ and Δ be as in thm.2.4. Then the following rule holds

$$
\begin{cases}
G_n A^\omega + \Delta + AC\text{-}qf \vdash \forall u^1 \forall v \leq_\tau tu\big((PCM2)(\xi uv) \to \exists w^\gamma B_0(u,v,w)\big) \\
\Rightarrow \exists (eff.)\chi, \tilde\chi \in G_n R^\omega \text{ such that} \\
G_n A_i^\omega + \Delta + b\text{-}AC \vdash \forall u^1 \forall v \leq_\tau tu \forall \tilde\Psi^* \big((\tilde\Psi^* \text{ satisfies the mon.funct.} \\
\quad \text{interpr. of } \forall k^0, g^1 \exists n^0 (gn > n \to (\widetilde{\xi uv})(n) -_{\mathbb{R}} (\widetilde{\xi uv})(gn) \leq_{\mathbb{R}} \frac{1}{k+1})) \\
\hspace{6cm} \to \exists w \leq_\gamma \tilde\chi u \tilde\Psi^* B_0(u,v,w)\big) \\
\text{and} \\
G_n A_i^\omega + \Delta + b\text{-}AC \vdash \forall u^1 \forall v \leq_\tau tu \forall \Psi^* \big((\Psi^* \text{ satisfies the mon. funct.} \\
\quad \text{interpr. of } \forall a^{1(0)}, k^0, g^1 \exists n^0 (gn > n \to \tilde a(n) -_{\mathbb{R}} \tilde a(gn) \leq_{\mathbb{R}} \frac{1}{k+1})) \\
\hspace{6cm} \to \exists w \leq_\gamma \chi u \Psi^* B_0(u,v,w)\big) \\
\text{and therefore} \\
PRA_i^\omega + \Delta + b\text{-}AC \vdash \forall u^1 \forall v \leq_\tau tu \exists w \leq_\gamma \chi u \Psi \, B_0(u,v,w),
\end{cases}
$$

where $\Psi := \lambda a, k, g. \max\limits_{i < C(a)k'} \big(\Phi_{it} i 0 g\big) = \max\limits_{i < C(a)k'} \big(g^{(i)}(0)\big)$
and $C(a) := (a(0))(0) + 1$.
In the conclusion, $\Delta + b\text{-}AC$ can be replaced by $\tilde\Delta$, where $\tilde\Delta$ is defined as
in theorem 2.4. If $\Delta = \emptyset$, then $b\text{-}AC$ can be omitted from the proof of the
conclusion. If $\tau \leq 1$ and the types of the $\exists$-quantifiers in Δ are ≤ 1, then
$G_n A^\omega + \Delta + AC\text{-}qf$ may be replaced by $E\text{-}G_n A^\omega + \Delta + AC^{\alpha,\beta}\text{-}qf$, where α, β
are as in cor.2.5.

Proof: The existence of $\tilde\chi$ follows from cor.2.5 since

$$
G_2 A^\omega \vdash \forall a^{1(0)} \forall k, \tilde k, n, \tilde n \big(\tilde k \leq_0 k \wedge \tilde n \geq_0 n \wedge
$$
$$
\forall m \geq_0 n(\tilde a(n) -_{\mathbb{R}} \tilde a(m) \leq_{\mathbb{R}} \tfrac{1}{k+1}) \to \forall m \geq_0 \tilde n (\tilde a(\tilde n) -_{\mathbb{R}} \tilde a(m) \leq_{\mathbb{R}} \tfrac{1}{k+1})).
$$

Ψ fulfils the monotone functional interpretation of
$\forall a^{1(0)}, k^0, g^1 \exists n^0 (gn > n \to \tilde a(n) -_{\mathbb{R}} \tilde a(gn) \leq_{\mathbb{R}} \frac{1}{k+1})$ (see the end of section
4) and hence (using lemma 2.2.11 from [15]) $\Psi(\xi^*(u^M, t^* u^M))$ satisfies the
monotone functional interpretation of

$$
\forall k^0, g^1 \exists n^0 (gn > n \to (\widetilde{\xi uv})(n) -_{\mathbb{R}} (\widetilde{\xi uv})(gn) \leq_{\mathbb{R}} \tfrac{1}{k+1}),
$$

where ξ^* s–maj $\xi \wedge t^*$ s–maj t.

χ is defined by $\chi := \lambda u, \Psi^*. \tilde\chi u \big(\Psi^*(\xi^*(u^M, t^* u^M))\big)$.

Remark 5.1   *1) The computation of the bound $\tilde\chi$ in the proposition
above needs only a functional $\tilde\Psi^*$ which satisfies the monotone func-*

tional interpretation of

$$(+) \; \forall k^0, g^1 \exists n^0 (gn > n \to (\widetilde{\xi uv})(n) -_{\mathbb{R}} (\widetilde{\xi uv})(gn) \leq_{\mathbb{R}} \frac{1}{k+1}).$$

For special ξ such a functional may be constructable without the use of Φ_{it}. Furthermore for fixed u the number of iterations of g only depends on the k–instances of $(+)$ which are used in the proof.

2) *If the given proof of the assumption of this proposition applies Ψ only to functions g of low growth, then also the bound $\chi u \Psi$ is of low growth: e.g. if only $g := S$ is used and type/$w = 0$, then $\chi u \Psi$ is a polynomial in u^M (in the sense of [15]).*

Corollary to the proof of prop.5.1:
The rule

$$
\begin{cases}
G_n A^\omega + \Delta + \text{AC-qf} \vdash \forall u^1 \forall v \leq_\tau tu \\
(\exists f^0 \forall k \forall m, \tilde{m} > fk(|(\xi uv)(\tilde{m}) -_{\mathbb{R}} (\xi uv)(m)| \leq \frac{1}{k+1}) \to \exists w^\gamma B_0(u,v,w)) \\
\Rightarrow \\
G_n A^\omega + \tilde{\Delta} \vdash \forall u^1 \forall v \leq_\tau tu \\
(\forall k \exists n \forall m, \tilde{m} > n(|(\xi uv)(\tilde{m}) -_{\mathbb{R}} (\xi uv)(m)| \leq \frac{1}{k+1}) \to \exists w^\gamma B_0(u,v,w))
\end{cases}
$$

holds for arbitrary sequences $(\xi uv)^{1(0)}$ of real numbers (this also extends to more general monotone formulas $\forall u^1 \forall v \leq_\tau tuB(u,v)$ in the sense of thm.2.4). The restriction to bounded monotone sequences $\tilde{\xi} uv$ is used only to ensure the existence of a functional Ψ which satisfies the monotone functional interpretation of $(+)$ above.
We now consider a generalization $(PCM2^*)(a_{(\cdot)}^{1(0)(0)})$ of $(PCM2)(a^{1(0)})$ which asserts the existence of a sequence of Cauchy moduli for a sequence $\tilde{a}_l$ of bounded monotone sequences:

$$(PCM2^*)(a_{(\cdot)}^{1(0)(0)}) :\equiv \exists h^{1(0)} \forall l^0, k^0 \forall m \geq_0 hkl (\widetilde{(a_l)}(hkl) -_{\mathbb{R}} \widetilde{(a_l)}(m)$$
$$\leq_{\mathbb{R}} \frac{1}{k+1}).$$

Proposition 5.2 *Let $n, B_0(u,v,w), t, \Delta$ be as in prop.5.1. $t, \xi \in G_n R^\omega$.*

Then the following rule holds

$$
\begin{cases}
G_n A^\omega + \Delta + AC\text{-}qf \vdash \forall u^1 \forall v \leq_\tau tu\big((PCM2^*)(\xi uv) \rightarrow \exists w^\gamma B_0(u,v,w)\big) \\
\Rightarrow \exists(eff.)\chi \in G_n R^\omega \ such\ that \\
G_n A_i^\omega + \Delta + b\text{-}AC \vdash \\
\quad \forall u^1 \forall v \leq_\tau tu \forall \Psi^*\big((\Psi^* \ satisfies\ the\ mon.\ funct.\ interpr.\ of \\
\quad\quad \forall a^{1(0)(0)}, k^0, g^1 \exists n^0(gn > n \rightarrow \forall l \leq k(\widetilde{(a_l)}(n) -_\mathbb{R} \widetilde{(a_l)}(gn) \leq_\mathbb{R} \frac{1}{k+1}))) \\
\quad\quad\quad \rightarrow \exists w \leq_\gamma \chi u \Psi^* B_0(u,v,w)\big) \\
and\ in\ particular \\
PRA_i^\omega + \Delta + b\text{-}AC \vdash \forall u^1 \forall v \leq_\tau tu \exists w \leq_\gamma \chi u \Psi' B_0(u,v,w),
\end{cases}
$$

where $\Psi' := \lambda a, k, g. \displaystyle\max_{i < C(a,k)(k+1)^2} (\Phi_{it} i 0 g)$ *and*

$\mathbb{N}^* \ni C(a,k) \geq \max_\mathbb{R}(\widetilde{(a_0)}(0), \dots, \widetilde{(a_k)}(0))$ *(with* C *s-maj* $C)^{11}$.
In the conclusion, $\Delta + b\text{-}AC$ *can be replaced by* $\tilde\Delta$, *where* $\tilde\Delta$ *is defined as in theorem 2.4. If* $\Delta = \emptyset$, *then* $b\text{-}AC$ *can be omitted from the proof of the conclusion. If* $\tau \leq 1$ *and the types of the* $\exists$*-quantifiers in* Δ *are* ≤ 1, *then* $G_n A^\omega + \Delta + AC\text{-}qf$ *may be replaced by* $E\text{-}G_n A^\omega + \Delta + AC^{\alpha,\beta}\text{-}qf$, *where* α, β *are as in cor.2.5.*
As in prop.5.1 we also have a term $\tilde\chi$ *which needs only a* $\tilde\Psi^*$ *for the instance* $a := \xi uv$.

Proof: The first part of the proposition follows from corollary 2.5 since $(PCM2^*)(a)$ is implied by

$$
\exists h^1 \forall k^0 \forall m \geq_0 hk \forall l \leq_0 k\big(\widetilde{(a_l)}(hk) -_\mathbb{R} \widetilde{(a_l)}(m) \leq_\mathbb{R} \frac{1}{k+1}\big)
$$

and

$$
\begin{aligned}
G_2 A^\omega \vdash \forall a_{(\cdot)}^{1(0)(0)} \forall k, \tilde k, n, \tilde n\big(\tilde k \leq_0 k \wedge \tilde n \geq_0 n \wedge \\
\forall m \geq_0 n \forall l \leq_0 k\big(\widetilde{(a_l)}(n) -_\mathbb{R} \widetilde{(a_l)}(m) \leq_\mathbb{R} \frac{1}{k+1}\big) \\
\rightarrow \forall m \geq_0 \tilde n \forall l \leq_0 \tilde k\big(\widetilde{(a_l)}(\tilde n) -_\mathbb{R} \widetilde{(a_l)}(m) \leq_\mathbb{R} \frac{1}{k+1}\big)\big).
\end{aligned}
$$

It remains to show that Ψ' satisfies the monotone functional interpretation of
$\forall a^{1(0)(0)}, k^0, g^1 \exists n^0\big(gn > n \rightarrow \forall l \leq k(\widetilde{(a_l)}(n) - \widetilde{(a_l)}(gn) \leq \frac{1}{k+1})\big)$:
Assume

$$
\forall i < C(a,k)(k+1)^2\big(g(g^i 0) > g^i 0 \wedge \exists l \leq k(\widetilde{(a_l)}(g^i 0) - \widetilde{(a_l)}(g(g^i 0)) > \frac{1}{k+1}\big).
$$

---

[11] E.g. take $C(a,k) := \displaystyle\max_{i \leq k}(a_i(0)(0) + 1)$.

Then

$$\forall i < C(a,k)(k+1)^2\big(g(g^i 0) > g^i 0\big) \text{ and}$$

$$\exists l \le k \exists j \big(\forall i < C(a,k)(k+1) \dot- 1\big((j)_i < (j)_{i+1} < C(a,k)(k+1)^2\big)\wedge$$
$$\forall i < C(a,k)(k+1)\big(\widetilde{(a_l)}(g^{(j)_i}0) - \widetilde{(a_l)}(g(g^{(j)_i}0)) > \tfrac{1}{k+1}\big)\big)$$

and therefore

$$\exists l \le k \exists j \Big(\forall i < C(a,k)(k+1) \dot- 1$$
$$\big(g^{(j)_{i+1}}0 > g^{(j)_i}0 \wedge \widetilde{(a_l)}(g^{(j)_i}0) - \widetilde{(a_l)}(g^{(j)_{i+1}}0) > \tfrac{1}{k+1}\big)$$
$$\wedge g(g^{(j)}C(a,k)(k+1)\dot- 1(0)) > g^{(j)}C(a,k)(k+1)\dot- 1(0)$$
$$\wedge \widetilde{(a_l)}(g^{(j)}C(a,k)(k+1)\dot- 1(0)) - \widetilde{(a_l)}(g(g^{(j)}C(a,k)(k+1)\dot- 1(0))) > \tfrac{1}{k+1}\Big).$$

Hence

$$\exists l \le k \exists j \forall i < C(a,k)(k+1)$$
$$\big(g^{(j)_{i+1}}0 > g^{(j)_i}0 \wedge \widetilde{(a_l)}(g^{(j)_i}0) - \widetilde{(a_l)}(g^{(j)_{i+1}}0) > \frac{1}{k+1}\big),$$

which contradicts $\widetilde{(a_l)} \subset [0, C(a,k)]$.

5.2   The principle (GLB) 'every sequence of real numbers in $\mathbb{R}_+$ has a greatest lower bound'

This principle can be easily reduced to $(PCM2)$ (provably in $G_2 A^\omega$):
Let $a^{1(0)}$ be such that $\forall n^0(0 \le_\mathbb{R} an)$. Then $(PCM2)(a)$ implies that the decreasing sequence $(\tilde a(n))_n \subset \mathbb{R}_+$ has a limit $\tilde a_0^1$. It is clear that $\tilde a_0$ is the greatest lower bound of $(a(n))_n \subset \mathbb{R}_+$. Thus we have shown

$$G_n A^\omega \vdash \forall a^{1(0)}\big((PCM2)(a) \to (GLB)(a)\big).$$

By this reduction we may replace $(PCM2)(\xi uv)$ by $(GLB)(\xi uv)$ in the assumption of prop.5.1.
There is nothing lost (w.r.t to the rate of growth) in this reduction since in the other direction we have

$$G_n A^\omega + AC^{0,0}\text{-qf} \vdash \forall a^{1(0)}\big((GLB)(a) \to (PCM2)(a)\big):$$

Let $a^{1(0)}$ be as above and a_0 its greatest lower bound. Then $a_0 = \lim_{n\to\infty} \tilde a_n$.
Using $AC^{0,0}$-qf one obtains (see section 4) a modulus of convergence and so a Cauchy modulus for $(\tilde a(n))_n$.

5.3 Π_1^0-CA and Π_1^0-AC

Definition 5.3

1) Π_1^0-$CA(f^{1(0)}) :\equiv \exists g^1 \forall x^0 \big(gx =_0 0 \leftrightarrow \forall y^0 (fxy =_0 0) \big).$

2) Define $A_0^C(f^{1(0)}, x^0, y^0, z^0) :\equiv \forall \tilde{x} \leq_0 x \exists \tilde{y} \leq_0 y \forall \tilde{z} \leq_0 z \big(f\tilde{x}\tilde{y} \neq_0 0 \vee f\tilde{x}\tilde{z} =_0 0 \big).$

A_0^C *can be expressed as a quantifier-free formula in* $G_n A^\omega$ *(see [15]).*

(Note that iteration of $\forall f^{1(0)}(\Pi_1^0$-$CA(f))$ yields CA_{ar}).
In [16] we proved (using cor.2.5)

Proposition 5.4 *Let* $n \geq 1$ *and* $B_0(u^1, v^\tau, w^\gamma) \in \mathcal{L}(G_n A^\omega)$ *be a quantifier-free formula which contains only* u^1, v^τ, w^γ *free, where* $\gamma \leq 2$. *Furthermore let* $\xi, t \in G_n R^\omega$ *and* Δ *be as in thm.2.4. Then the following rule holds*

$$
\begin{cases}
G_n A^\omega + \Delta + AC\text{-}qf \vdash \forall u^1 \forall v \leq_\tau tu \big(\Pi_1^0\text{-}CA(\xi uv) \to \exists w^\gamma B_0(u, v, w) \big) \\
\Rightarrow \exists (eff.) \chi \in G_n R^\omega \text{ such that} \\
G_n A_i^\omega + \Delta + b\text{-}AC \vdash \forall u^1 \forall v \leq_\tau tu \forall \Psi^* \\
\quad ((\Psi^* \text{ satisfies the mon. funct.interpr. of } \forall x^0, h^1 \exists y^0 A_0^C(\xi uv, x, y, hy)) \\
\quad\quad\quad\quad\quad\quad\quad \to \exists w \leq_\gamma \chi u \Psi^* B_0(u, v, w)) \\
\text{and in particular} \\
PRA_i^\omega + \Delta + b\text{-}AC \vdash \forall u^1 \forall v \leq_\tau tu \exists w \leq_\gamma \chi u \Psi B_0(u, v, w),
\end{cases}
$$

where $\Psi := \lambda x^0, h^1 . \max_{i < x+1} \big(\Phi_{it} i 0 h \big) \big(= \lambda x^0, h^1 . \max_{i < x+1} (h^i 0) \big).$

In the conclusion, $\Delta + b\text{-}AC$ *can be replaced by* $\tilde{\Delta}$, *where* $\tilde{\Delta}$ *is defined as in thm.2.4. If* $\Delta = \emptyset$, *then* $b\text{-}AC$ *can be omitted from the proof of the conclusion. If* $\tau \leq 1$ *and the types of the* $\exists$-*quantifiers in* Δ *are* ≤ 1, *then* $G_n A^\omega + \Delta + AC\text{-}qf$ *may be replaced by* $E\text{-}G_n A^\omega + \Delta + AC^{\alpha,\beta}\text{-}qf$, *where* α, β *are as in cor.2.5.*

A similar result holds for Π_1^0-$AC(\xi uv)$, where

$$
\Pi_1^0\text{-}AC(f^{1(0)(0)(0)}) :\equiv
$$
$$
\forall l^0 \big(\forall x^0 \exists y^0 \forall z^0 (flxyz =_0 0) \to \exists g^1 \forall x^0, z^0 (flx(gx)z =_0 0) \big).
$$

5.4 *The Bolzano–Weierstraß property for bounded sequences in* $\mathbb{R}^d$ *(for every fixed d)*

We now consider the Bolzano–Weierstraß principle for sequences in $[-1, 1]^d \subset \mathbb{R}^d$. The restriction to the special bound 1 is convenient but not essential: If $(x_n) \subset [-C, C]^d$ with $C > 0$, we define $x_n' := \frac{1}{C} \cdot x_n$ and apply the

Bolzano–Weierstraß principle to this sequence. For simplicity we formulate the Bolzano– Weierstraß principle w.r.t. the maximum norm $\|\cdot\|_{\max}$. This of course implies the principle for the Euclidean norm $\|\cdot\|_E$ since $\|\cdot\|_E \le \sqrt{d}\cdot\|\cdot\|_{\max}$.

We start with the investigation of the following formulation of the Bolzano–Weierstraß principle:

$$BW : \forall(x_n) \subset [-1,1]^d \exists x \in [-1,1]^d \forall k^0, m^0 \exists n >_0 m\big(\|x - x_n\|_{\max} \le \frac{1}{k+1}\big)$$

i.e. (x_n) possesses a limit point x.

Later on we discuss a second formulation which (relatively to $G_n A^\omega$) is slightly stronger than BW:

$$BW^+ : \left\{ \begin{array}{r} \forall(x_n) \subset [-1,1]^d \exists x \in [-1,1]^d \exists f^1 \big(\forall n^0(fn <_0 f(n+1)) \\ \wedge \forall k^0\big(\|x - x_{fk}\|_{\max} \le \frac{1}{k+1}\big)\big), \end{array} \right.$$

i.e. (x_n) has a subsequence (x_{fn}) which converges (to x) with the modulus $\frac{1}{k+1}$.

Using our representation of $[-1,1]$ from section 3, the principle BW has the following form

$$\forall x_1^{1(0)}, \ldots, x_d^{1(0)}$$

$$\underbrace{\exists a_1, \ldots, a_d \le_1 M \forall k^0, m^0 \exists n >_0 m \bigwedge_{i=1}^{d} \big(|\tilde{a}_i -_{\mathbb{R}} \widetilde{x_i n}| \le_{\mathbb{R}} \frac{1}{k+1}\big),}_{BW(\underline{x}^{1(0)}):\equiv}$$

where M and $y^1 \mapsto \tilde{y}$ are the constructions from our representation of $[-1,1]$ in section 3. We now prove

$(*)\ G_2 A^\omega + AC^{1,0}\text{-qf} \vdash F^- \to \forall x_1^{1(0)}, \ldots, x_d^{1(0)}\big(\Pi_1^0\text{-CA}(\chi\underline{x}) \to BW(\underline{x})\big),$

for a suitable $\chi \in G_2 R^\omega$:

$BW(\underline{x})$ is equivalent to

$$(1)\ \exists a_1, \ldots, a_d \le_1 M \forall k^0 \exists n >_0 k \bigwedge_{i=1}^{d} \big(|\tilde{a}_i -_{\mathbb{R}} \widetilde{x_i n}| \le_{\mathbb{R}} \frac{1}{k+1}\big)$$

which in turn is equivalent to

$$(2)\ \exists a_1, \ldots, a_d \le_1 M \forall k^0 \exists n >_0 k \bigwedge_{i=1}^{d} \big(|\tilde{a}_i k -_{\mathbb{Q}} (\widetilde{x_i n})(k)| \le_{\mathbb{Q}} \frac{3}{k+1}\big).$$

Assume $\neg(2)$, i.e.

$$(3)\ \forall a_1, \ldots, a_d \le_1 M \exists k^0 \forall n >_0 k \bigvee_{i=1}^{d} \big(|\tilde{a}_i k -_{\mathbb{Q}} (\widetilde{x_i n})(k)| >_{\mathbb{Q}} \frac{3}{k+1}\big).$$

Let $\chi \in G_2R^\omega$ be such that

$$G_2A^\omega \vdash \forall x_1^{1(0)}, \ldots, x_d^{1(0)} \forall l^0, n^0 \big(\chi \underline{x} ln =_0 0 \leftrightarrow$$
$$\big[n >_0 \nu_{d+1}^{d+1}(l) \rightarrow \bigvee_{i=1}^{d} |\nu_i^{d+1}(l) -_{\mathbb{Q}} (\widetilde{x_i n})(\nu_{d+1}^{d+1}(l))| >_{\mathbb{Q}} \tfrac{3}{\nu_{d+1}^{d+1}(l)+1}\big]\big).$$

Π_1^0–CA($\chi \underline{x}$) yields the existence of a function h such that

$$(4)\forall l_1^0, \ldots, l_d^0, k^0 \big(hl_1 \ldots l_d k =_0 0 \leftrightarrow \forall n >_0 k \bigvee_{i=1}^{d} \big(|l_i -_{\mathbb{Q}} (\widetilde{x_i n})(k)| >_{\mathbb{Q}} \tfrac{3}{k+1}\big).$$

Using h, (3) has the form

$$(5)\ \forall a_1, \ldots, a_d \leq_1 M \exists k^0 \big(h(\tilde{a}_1 k, \ldots, \tilde{a}_d k, k) =_0 0\big).$$

By Σ_1^0–UB$^-$ (which follows from AC^{1,0}–qf and F^- by [15] (prop. 4.20)) we obtain

$$(6)\ \exists k_0 \forall a_1, \ldots, a_d \leq_1 M \forall m^0 \exists k \leq_0 k_0 \forall n >_0 k \bigvee_{i=1}^{d}$$
$$\big(|(\widetilde{\overline{a_i, m}})(k) -_{\mathbb{Q}} (\widetilde{x_i n})(k)| >_{\mathbb{Q}} \tfrac{3}{k+1}\big)$$

and therefore

$$(7)\ \exists k_0 \forall a_1, \ldots, a_d \leq_1 M \forall m^0 \forall n >_0 k_0 \bigvee_{i=1}^{d} \big(|(\widetilde{\overline{a_i, m}}) -_{\mathbb{R}} \widetilde{x_i n}| >_{\mathbb{R}} \tfrac{1}{k_0+1}\big).$$

Since $|\widetilde{\overline{a_i, 3(m+1)}} -_{\mathbb{R}} \tilde{a}_i| <_{\mathbb{R}} \tfrac{2}{m+1}$ (see the definition of $y \mapsto \tilde{y}$ from section 3) it follows

$$(8)\ \exists k_0 \forall a_1, \ldots, a_d \leq_1 M \forall n >_0 k_0 \bigvee_{i=1}^{d} \big(|\tilde{a}_i -_{\mathbb{R}} \widetilde{x_i n}| >_{\mathbb{R}} \tfrac{1}{2(k_0+1)}\big),\ \text{i.e.}$$

$$(9)\ \exists k_0 \forall (a_1, \ldots, a_d) \in [-1,1]^d \forall n >_0 k_0 \big(\|\underline{a} - \underline{x} n\|_{\max} > \tfrac{1}{2(k_0+1)}\big).$$

By applying this to $\underline{a} := \underline{x}(k_0+1)$ yields the contradiction
$\|\underline{x}(k_0+1) - \underline{x}(k_0+1)\|_{\max} > \tfrac{1}{2(k_0+1)}$, which concludes the proof of $(*)$.

Remark 5.2 In the proof of $(*)$ we used a combination of Π_1^0–CA(ξg) and Σ_1^0–UB$^-$ to obtain a restricted form Π_1^0–UB$^-$↾ of the extension of Σ_1^0–UB$^-$ to Π_1^0-formulas:

$$\Pi_1^0\text{–UB}^-\!\!\restriction\ :\ \begin{cases} \forall f \leq_1 s \exists n^0 \forall k^0 A_0(t^0[f], n, k) \rightarrow \\ \exists n_0 \forall f \leq_1 s \forall m^0 \exists n \leq_0 n_0 \forall k^0 A_0(t[\overline{f, m}], n, k), \end{cases}$$

where k does not occur in $t[f]$ and f does not occur in $A_0(0,0,0)$ and g^1 is the only free variable in $A_0(0,0,0)$.

Π_1^0-UB$^-\!\upharpoonright$ follows by applying Π_1^0-CA to $\lambda n, k.t_{A_0}(a^0, n^0, k^0)$, where t_{A_0} is such that $t_{A_0}(a^0, n^0, k^0) =_0 0 \leftrightarrow A_0(a^0, n^0, k^0)$, and subsequent application of Σ_1^0-UB$^-$. Π_1^0-CA and Σ_1^0-UB$^-$ do not imply the unrestricted form Π_1^0-UB$^-$ of Π_1^0-UB$^-\!\upharpoonright$:

$$\Pi_1^0\text{-UB}^- : \begin{cases} \forall f \leq_1 s \exists n^0 \forall k^0 A_0(f, n, k) \to \\ \exists n_0 \forall f \leq_1 s \forall m^0 \exists n \leq_0 n_0 \forall k^0 A_0((\overline{f, m}), n, k) \end{cases}$$

since a reduction of Π_1^0-UB$^-$ to Σ_1^0-UB$^-$ would require a comprehension functional in f:

$$(+) \quad \exists \Phi \forall f^1, n^0(\Phi f n =_0 0 \leftrightarrow \forall k^0 A_0(f, n, k)).$$

In fact Π_1^0-UB$^-$ can easily be refuted by applying it to $\forall f \leq_1 \lambda x.1 \exists n^0 \forall k^0$ $(fk = 0 \to fn = 0)$, which leads to a contradiction. This reflects the fact that we had to use F^- to derive Σ_1^0-UB$^-$, which is incompatible with $(+)$ since $\Phi + AC^{1,0}$-qf produces (see above) a non–majorizable functional, namely

$$\Psi f^1 := \begin{cases} \min n[fn = 0], & \text{if existent} \\ 0^0, & \text{otherwise,} \end{cases}$$

whereas F^- is true only in the model $\mathcal{M}^\omega$ of all strongly majorizable functionals introduced in [2] (see [15] for details).

Next we prove

$$(\ast\ast) \ G_2 A^\omega + AC^{0,0}\text{-qf} \vdash \forall x_1^{1(0)}, \ldots, x_d^{1(0)} \big(\Sigma_1^0\text{-IA}(\chi\underline{x}) \wedge BW(\underline{x}) \to BW^+(\underline{x})\big)$$

for a suitable term $\chi \in G_2 R^\omega$, where

$$\Sigma_1^0\text{-IA}(f) :\equiv \begin{cases} \forall l^0\big(\exists y^0(fl0y =_0 0) \wedge \forall x^0(\exists y(flxy = 0) \to \exists y(flx'y = 0)) \\ \qquad\qquad\qquad\qquad\qquad \to \forall x \exists y(flxy = 0)\big). \end{cases}$$

$BW(\underline{x})$ implies the existence of $a_1, \ldots, a_d \leq_1 M$ such that

$$(10) \begin{cases} \forall k, m \exists n > m \bigwedge_{i=1}^d \\ \big(|\tilde{a}_i(2(k+1)(k+2)) -_{\mathbb{Q}} (\overline{x_i n})(2(k+1)(k+2))| \leq_{\mathbb{Q}} \frac{1}{k+1}\big). \end{cases}$$

Define (for $x_1^{1(0)}, \ldots, x_d^{1(0)}, l_1^0, \ldots, l_d^0$)

$$F(\underline{x}, \underline{l}, k, m, n) :\equiv \big(\underline{x}n \text{ is the } (m+1)\text{-th element in } (\underline{x}(l))_l \text{ such that} \\ \bigwedge_{i=1}^d \big(|l_i -_{\mathbb{Q}} (\overline{x_i n})(2(k+1)(k+2))| \leq_{\mathbb{Q}} \frac{1}{k+1}\big)\big).$$

One easily verifies that $F(\underline{x}, \underline{l}, k, m, n)$ can be expressed in the form

$$\exists a^0 F_0(\underline{x}, \underline{l}, k, m, n, a) \ ,$$

where F_0 is a quantifier–free formula in $\mathcal{L}(G_2A^\omega)$, which contains only $\underline{x}, \underline{l}, k, m, n, a$ as free variables. Let $\tilde{\chi} \in G_2 R^\omega$ such that

$$\tilde{\chi}(\underline{x}, \underline{l}, k, m, n, a) =_0 0 \leftrightarrow F_0(\underline{x}, \underline{l}, k, m, n, a)$$

and define $\chi(\underline{x}, q, m, p) := \tilde{\chi}(\underline{x}, \nu_1^{d+1}(q), \ldots, \nu_{d+1}^{d+1}(q), m, j_1(p), j_2(p))$. Σ_1^0–IA$(\chi\underline{x})$ yields

$$(11) \quad \begin{cases} \forall l_1, \ldots, l_d, k\big(\exists n\, F(\underline{x}, \underline{l}, k, 0, n) \wedge \forall m\big(\exists nF(\underline{x}, \underline{l}, k, m, n) \\ \qquad \rightarrow \exists nF(\underline{x}, \underline{l}, k, m', n)\big) \rightarrow \forall m\exists nF(\underline{x}, \underline{l}, k, m, n)\big). \end{cases}$$

(10) and (11) imply

$$(12) \quad \begin{cases} \forall k, m\exists n\big(\underline{x}n \text{ is the } (m+1)\text{–th element of } (\underline{x}(l))_l \text{ such that} \\ \bigwedge_{i=1}^{d} \big(|\tilde{a}_i(2(k+1)(k+2)) -_{\mathbb{Q}} (\widetilde{x_in})(2(k+1)(k+2))| \leq_{\mathbb{Q}} \frac{1}{k+1}\big)\big). \end{cases}$$

and therefore

$$(13) \quad \begin{cases} \forall k\exists n\big(\underline{x}n \text{ is the } (k+1)\text{–th element of } (\underline{x}(l))_l \text{ such that} \\ \bigwedge_{i=1}^{d} \big(|\tilde{a}_i(2(k+1)(k+2)) -_{\mathbb{Q}} (\widetilde{x_in})(2(k+1)(k+2))| \leq_{\mathbb{Q}} \frac{1}{k+1}\big)\big). \end{cases}$$

By AC^{0,0}–qf we obtain a function g^1 such that

$$(14) \quad \begin{cases} \forall k\big(\underline{x}(gk) \text{ is the } (k+1)\text{–th element of } (\underline{x}(l))_l \text{ such that} \\ \bigwedge_{i=1}^{d} \big(|\tilde{a}_i(2(k+1)(k+2)) -_{\mathbb{Q}} (\widetilde{x_i(gk)})(2(k+1)(k+2))| \leq_{\mathbb{Q}} \frac{1}{k+1}\big). \end{cases}$$

We show (15) $\forall k(gk < g(k+1))$: Define

$$A_0(\underline{x}l, k) :\equiv \bigwedge_{i=1}^{d} \big(|\tilde{a}_i(2(k+1)(k+2)) -_{\mathbb{Q}} (\widetilde{x_il})(2(k+1)(k+2))| \leq_{\mathbb{Q}} \frac{1}{k+1}\big).$$

Let l be such that $A_0(\underline{x}l, k+1)$. Because of

$$|\tilde{a}_i(2(k+1)(k+2)) -_{\mathbb{Q}} (\widetilde{x_il})(2(k+1)(k+2))| \leq$$
$$|\tilde{a}_i(2(k+2)(k+3)) -_{\mathbb{Q}} (\widetilde{x_il})(2(k+2)(k+3))| + \frac{2}{2(k+1)(k+2)} \overset{A_0(\underline{x}l, k+1)}{\leq}$$
$$\frac{1}{k+2} + \frac{2}{2(k+1)(k+2)} = \frac{1}{k+1},$$

this yields $A_0(\underline{x}l, k)$. Thus the $(k+2)$–th element $\underline{x}l$ such that $A_0(\underline{x}l, k+1)$ is at least the $(k+2)$–th element such that $A_0(\underline{x}l, k)$ and therefore occurs

later in the sequence than the $(k+1)$-th element such that $A_0(\underline{x}l,k)$, i.e. $gk < g(k+1)$.
It remains to show

$$(16)\ \forall k \bigwedge_{i=1}^{d} \left(|\tilde{a}_i -_{\mathbb{R}} \widetilde{x_i(fk)}| \leq_{\mathbb{R}} \frac{1}{k+1}\right),\ \text{where } fk := g(2(k+1)):$$

This follows since
$$\bigwedge_{i=1}^{d} \left(|\tilde{a}_i(2(k+1)(k+2)) -_{\mathbb{Q}} \widetilde{(x_i(gk))}(2(k+1)(k+2))| \leq_{\mathbb{Q}} \frac{1}{k+1}\right)\ \text{implies}$$
$$\bigwedge_{i=1}^{d} \left(|\tilde{a}_i -_{\mathbb{R}} \widetilde{x_i(gk)}| \leq_{\mathbb{R}} \frac{1}{k+1} + \frac{2}{2(k+1)(k+2)+1} \leq \frac{2}{k+1}\right).$$
(15) and (16) imply $BW^+(\underline{x})$ which concludes the proof of $(**)$.

Remark 5.3 One might ask why we did not use the following obvious proof of $BW^+(\underline{x})$ from $BW(\underline{x})$: Let $\underline{a}$ be such that
$$\forall k \exists n > k \bigwedge_{i=1}^{d} \left(|\tilde{a}_i -_{\mathbb{R}} \widetilde{x_i n}| <_{\mathbb{R}} \frac{1}{k+1}\right).\ \text{AC}^{0,0}\text{-qf yields the existence of a}$$
function g such that $\forall k(gk > k \wedge \bigwedge_{i=1}^{d} (|\tilde{a}_i -_{\mathbb{R}} \widetilde{x_i(gk)}| <_{\mathbb{R}} \frac{1}{k+1}))$. Now define
$fk := g^{(k+1)}(0)$. It is clear that f fulfils $BW^+(\underline{x})$.
The problem with this proof is that we cannot use our results from section 2 in the presence of the iteration functional Φ_{it} (see [16] for more information in this point) which is needed to define f as a functional in g. To introduce the graph of Φ_{it} by Σ_1^0-IA and AC-qf does not help since this would require an application of Σ_1^0-IA which involves (besides $\underline{x}$) also g as a genuine function parameter. In contrast to this situation, our proof of $BW(\underline{x}) \to BW^+(\underline{x})$ uses Σ_1^0-IA only for a formula with (besides $\underline{x}$) only $k, \underline{a}k$ as parameters. Since k (as a parameter) remains fixed throughout the induction, $\underline{a}$ only occurs as the **number parameter** $\underline{a}k$ but **not** as **genuine function parameter**. This is the reason why we are able to construct a term χ such that Σ_1^0-IA$(\chi\underline{x}) \wedge BW(\underline{x}) \to BW^+(\underline{x})$.

Using $(*)$ and $(**)$ we are now able to prove

Proposition 5.5 Let $n \geq 2$ and $B_0(u^1, v^\tau, w^\gamma) \in \mathcal{L}(G_n A^\omega)$ be a quantifier-free formula which contains only u^1, v^τ, w^γ free, where $\gamma \leq 2$. Furthermore let $\underline{\xi}, t \in G_n R^\omega$ and Δ be as in thm.2.4. Then for a suitable $\xi' \in G_n R^\omega$ the *following rule holds*

$$
\begin{cases}
G_n A^\omega + \Delta + AC\text{-}qf \vdash \forall u^1 \forall v \leq_\tau tu\big(BW^+(\underline{\xi}uv) \to \exists w^\gamma B_0(u,v,w)\big) \\
\Rightarrow \exists(eff.)\chi \in G_n R^\omega \text{ such that} \\
G_{\max(n,3)} A_i^\omega + \Delta + b\text{-}AC \vdash \\
\quad \forall u^1 \forall v \leq_\tau tu \forall \Psi^* \big((\Psi^* \text{ satisfies the mon. funct. interpr. of} \\
\quad\quad \forall x^0, h^1 \exists y^0 A_0^C(\xi' uv, x, y, hy)) \to \exists w \leq_\gamma \chi u \Psi^* B_0(u,v,w)\big) \\
\text{and in particular} \\
PRA_i^\omega + \Delta + b\text{-}AC \vdash \forall u^1 \forall v \leq_\tau tu \exists w \leq_\gamma \chi u \Psi \, B_0(u,v,w),
\end{cases}
$$

where $\Psi := \lambda x^0, h^1 . \max_{i < x+1} \big(\Phi_{it} i 0 h\big)\big(= \lambda x^0, h^1 . \max_{i<x+1}(h^i 0)\big)$.

In the conclusion, $\Delta + b\text{-}AC$ can be replaced by $\tilde\Delta$, where $\tilde\Delta$ is defined as in thm. 2.4. If $\Delta = \emptyset$, then $b\text{-}AC$ can be omitted from the proof of the conclusion. If $\tau \leq 1$ and the types of the $\exists$-quantifiers in Δ are ≤ 1, then $G_n A^\omega + \Delta + AC\text{-}qf$ may be replaced by $E\text{-}G_n A^\omega + \Delta + AC^{\alpha,\beta}\text{-}qf$, where α, β are as in cor. 2.5.

This results also holds (for a suitable ξ'' instead of ξ') if instead of the single instance $BW^+(\underline{\xi}uv)$, a sequence $\forall l^0 BW^+(\underline{\xi}uvl)$ of instances is used in the proof.

Proof: By $(*), (**)$ and the proof of prop. 3.11 from [16] there are functionals $\varphi_1, \varphi_2 \in G_2 R^\omega$ such that

$$G_2 A^\omega + AC^{1,0}\text{-}qf \vdash F^- \to \forall \underline{x}\big(\Pi_1^0\text{-}CA(\varphi_1 \underline{x}) \wedge \Pi_1^0\text{-}CA(\varphi_2 \underline{x}) \to BW^+(\underline{x})\big).$$

Furthermore $G_2 A^\omega \vdash \Pi_1^0\text{-}CA(\psi f_1 f_2) \to \Pi_1^0\text{-}CA(f_1) \wedge \Pi_1^0\text{-}CA(f_2)$, where

$$
\psi f_1 f_2 x^0 y^0 =_0
\begin{cases}
f_1(j_2 x, y), & \text{if } j_1 x = 0 \\
f_2(j_2 x, y), & \text{otherwise.}
\end{cases}
$$

Hence $G_2 A^\omega + AC^{1,0}\text{-}qf \vdash F^- \to \forall \underline{x}\big(\Pi_1^0\text{-}CA(\varphi_3 \underline{x}) \to BW^+(\underline{x})\big)$, for a suitable $\varphi_3 \in G_2 R^\omega$ and thus

$$G_n A^\omega + \Delta + AC\text{-}qf \vdash F^- \to \forall u^1 \forall v \leq_\tau tu\big(\Pi_1^0\text{-}CA(\varphi_3(\underline{\xi}uv)) \to \exists w B_0\big).$$

By the proof of theorem 4.21 from [15] we obtain

$$G_n A^\omega + \tilde\Delta + (*) + AC\text{-}qf \vdash \forall u^1 \forall v \leq_\tau tu\big(\Pi_1^0\text{-}CA(\varphi_3(\underline{\xi}uv)) \to \exists w B_0\big),$$

where $\tilde\Delta := \{\exists Y \leq_{\rho\delta} s \forall x^\delta, z^\eta A_0(x, Yx, z) : \forall x \exists y \leq sx \forall z^\eta A_0 \in \Delta\}$,

$$
(*) :\equiv \forall n_0 \exists Y \leq \lambda \Phi^{2(0)}, y^{1(0)} . y \forall \Phi, \tilde y^{1(0)}, k^0, \tilde z^1 \forall n \leq_0 n_0
$$
$$
\Big(\bigwedge_{i<n} (\tilde z i \leq \tilde y k i) \to \Phi k(\overline{\tilde z, n}) \leq \Phi k(Y \Phi \tilde y k)\Big).
$$

Prop.5.4 (with $\Delta' := \widetilde{\Delta} \cup \{(*)\}$) yields the conclusion of our proposition in $G_n A_i^\omega + \Delta + (*) + $ b–AC and so (since, again by the proof of theorem 4.21 from [15], $G_3 A_i^\omega \vdash (*)$ and even $G_3 A_i^\omega \vdash (\tilde{*})$) in $G_{\max(3,n)} A_i^\omega + \Delta + $b–AC.

This proof also extends to sequences $\forall l^0 BW^+(\underline{\xi} uvl)$ of instances of BW^+ since by the reasoning above such a sequence reduces to a suitable sequence $\forall l^0 \Pi_1^0$–CA(φuvl) of instances of Π_1^0–CA which can be reduced in turn to a single instance using coding (see [16] for this).

5.5 The Arzelà–Ascoli lemma

Under the name 'Arzelà–Ascoli lemma' we understand (as in the literature on 'reverse mathematics') the following proposition:
Let $(f_l) \subset C[0,1]$ be a sequence of functions[12] which are equicontinuous and have a common bound, i.e. there exists a common modulus of uniform continuity ω for all f_l and a bound $C \in \mathbb{N}$ such that $\|f_l\|_\infty \le C$. Then

(i) (f_l) possesses a limit point w.r.t. $\|\cdot\|_\infty$ which also has the modulus ω, i.e.

$$\exists f \in C[0,1]\big(\forall k^0 \forall m \exists n >_0 m\big(\|f - f_n\|_\infty \le \frac{1}{k+1}\big) \wedge f \text{ has modulus } \omega\big);$$

(ii) there is a subsequence (f_{gl}) of (f_l) which converges with modulus $\frac{1}{k+1}$.

As in the case of the Bolzano–Weierstraß principle we deal first with (i). The slightly stronger assertion (ii) can then be obtained from (i) using Σ_1^0–IA(f) and AC^{0,0}-qf analogously to our proof of $BW^+(\underline{x})$ from $BW(\underline{x})$. For notational simplicity we may assume that $C = 1$. When formalized in $G_n A^\omega$, the version (i) of the Arzelà–Ascoli lemma has the form[13]

$$A\text{-}A(f_{(\cdot)}^{1(0)(0)}, \omega^1) :\equiv \Big(f_{(\cdot)} \le_{1(0)(0)} \lambda l^0, n^0.M \wedge$$

$$\overbrace{\forall l^0, m^0, u^0, v^0\big(|qu -_\mathbb{Q} qv| \le_\mathbb{Q} \frac{1}{\omega(m)+1} \to |\widetilde{f_l}u -_\mathbb{R} \widetilde{f_l}v| \le_\mathbb{R} \frac{1}{m+1}\big)}^{\Pi_1^0 \ni F(f_l,m,u,v) :\equiv \forall a^0 F_0(f_l,m,u,v,a):\equiv}$$

$$\to \exists g \le_{1(0)} \lambda n.M\big(\forall m, u, vF(g,m,u,v) \wedge$$

$$\forall k \exists n >_0 k(\|\lambda x^1.g(x)_\mathbb{R} - \lambda x^1.f_n(x)_\mathbb{R}\|_\infty \le \frac{1}{k+1})\big)\Big).$$

Here M, q and $y^1 \mapsto \tilde{y}$ are the constructions from our representation of $[0,1], [-1,1]$ in section 3. For notational simplicity we omit in the following

---

[12]The restriction to the unit interval $[0,1]$ is convenient for the following proofs but not essential.

[13]$g(x)_\mathbb{R}$ denotes the continuation of $g : [0,1] \cap \mathbb{Q} \to [-1,1]$ to $[0,1]$ which is definable in g and its modulus ω.

$(\tilde{\ })$.

A–A(f, ω) is equivalent to[14]

$$f_{(\cdot)} \leq l^0, n^0.M \wedge \forall l^0, m^0, u^0, v^0 F(f_l, m, u, v) \to$$

$$\exists g \leq_{1(0)} \lambda n.M\big(\forall m, u, v F(g, m, u, v) \wedge$$

$$\forall k \exists n >_0 k \bigwedge_{i=0}^{\omega(k)+1} \big(|g(\tfrac{i}{\omega(k)+1})_{\mathbb{R}}(k) -_{\mathbb{Q}} f_n(\tfrac{i}{\omega(k)+1})_{\mathbb{R}}(k)| \leq_{\mathbb{Q}} \tfrac{5}{k+1}\big)\big).$$

Assume $\neg$A–A(f, ω), i.e. $f_{(\cdot)} \leq \lambda l^0, n^0 M \wedge \forall l, m, u, v F(f_l, m, u, v)$ and

$$(1) \begin{cases} \forall g \leq_{1(0)} \lambda n.M\big(\forall m, u, v\, F(g, m, u, v) \to \exists k \forall n\big(n >_0 k \to \\ \displaystyle\bigvee_{i=0}^{\omega(k)+1} \big(|g(\tfrac{i}{\omega(k)+1})_{\mathbb{R}}(k) -_{\mathbb{Q}} f_n(\tfrac{i}{\omega(k)+1})_{\mathbb{R}}(k)| >_{\mathbb{Q}} \tfrac{5}{k+1}\big)\big)\big). \end{cases}$$

Let α be such that

$$\forall l, k, n\big(\alpha(l^0, k^0, n^0) =_0 0 \leftrightarrow \big[n > k \to$$

$$\bigvee_{i=0}^{\omega(k)+1} \big(|(l)_i -_{\mathbb{Q}} f_n(\tfrac{i}{\omega(k)+1})_{\mathbb{R}}(k)| >_{\mathbb{Q}} \tfrac{5}{k+1}\big)\big]\big).$$

Π_1^0–CA(α') (where $\alpha' in := \alpha(j_1 i, j_2 i, n)$) yields the existence of a function h such that

$$\forall l, k\big(hlk =_0 0 \leftrightarrow \forall n\big(\alpha(l, k, n) = 0\big)\big).$$

Hence

$$(2) \begin{cases} \forall g, k\Big(h\big(\overline{\lambda i.g(\tfrac{i}{\omega(k)+1})_{\mathbb{R}}(k)}(\omega(k)+2), k\big) =_0 0 \leftrightarrow \\ \displaystyle\forall n >_0 k \bigvee_{i=0}^{\omega(k)+1} \big(|g(\tfrac{i}{\omega(k)+1})_{\mathbb{R}}(k) -_{\mathbb{Q}} f_n(\tfrac{i}{\omega(k)+1})_{\mathbb{R}}(k)| >_{\mathbb{Q}} \tfrac{5}{k+1}\big)\Big). \end{cases}$$

(1),(2) and Σ_1^0–UB$^-$ yield (using the fact that g can be coded into a type–1–object by $g'x^0 := g(j_1 x, j_2 x)$)

$$(3) \begin{cases} \exists k_0 \forall g' \leq_1 \lambda x.M(j_2 x)\forall l^0 \\ \big(\forall m, u, v, a \leq k_0 F_0(\lambda x, y.\overline{(g', l)}(j(x, y)), m, u, v, a) \to \exists k \leq k_0 \forall n > k_0 \\ \displaystyle\bigvee_{i=0}^{\omega(k)+1} \big(|(\lambda x, y.\overline{(g', l)}(j(x, y)))(\tfrac{i}{\omega(k)+1})_{\mathbb{R}}(k) -_{\mathbb{Q}} f_n(\tfrac{i}{\omega(k)+1})_{\mathbb{R}}(k)| \\ \qquad\qquad\qquad\qquad\qquad\qquad\qquad\qquad >_{\mathbb{Q}} \tfrac{5}{k+1}\big)\big), \end{cases}$$

---

[14] For better readability we write $\tfrac{i}{\omega(k)+1}$ instead of its code.

and therefore using

$$g_l mn := \begin{cases} gmn, & \text{if } m, n \leq l \\ 0^0, & \text{otherwise, and } g_l =_{1(0)} \lambda x, y.(\overline{(g_l)',r})(j(x,y)) \text{for } r > j(l,l) \end{cases}$$

$$(4) \begin{cases} \exists k_0 \forall g \leq_{1(0)} \lambda n. M \forall l^0 \Big(\forall m, u, v, a \leq k_0 F_0(g_l, m, u, v, a) \rightarrow \\ \exists k \leq k_0 \forall n > k_0 \bigvee_{i=0}^{\omega(k)+1} \big(|g_l(\frac{i}{\omega(k)+1})_{\mathbb{R}}(k) -_{\mathbb{Q}} f_n(\frac{i}{\omega(k)+1})_{\mathbb{R}}(k)| \\ >_{\mathbb{Q}} \frac{5}{k+1} \big) \Big), \end{cases}$$

By putting $g := f_{k_0+1}$ and $l^0 := 3(c+1)$, where c is the maximum of $k_0 + 1$ and the codes of all $\frac{i}{\omega(k)+1}$ for $i \leq \omega(k) + 1$ and $k \leq k_0$, (4) yields the contradiction

$$\exists k \leq k_0 \bigvee_{i=0}^{\omega(k)+1} \big(|f_{k_0+1}(\frac{i}{\omega(k)+1})(k) -_{\mathbb{Q}} f_{k_0+1}(\frac{i}{\omega(k)+1})(k)| >_{\mathbb{Q}} \frac{5}{k+1} \big).$$

α' can be defined as a functional ξ in $f_{(\cdot)}, \omega$, where $\xi \in G_2 R^\omega$. Since the proof above can be carried out in $G_3 A^\omega + AC^{1,0} - qf^{15}$ (under the assumption of F^- and $\Pi_1^0 - CA(\xi(f,\omega))$ using prop. 4.20 from [15]) we have shown that

$$G_3 A^\omega + AC^{1,0} - qf \vdash F^- \rightarrow \forall f^{1(0)(0)}, \omega^1 \big(\Pi_1^0 - CA(\xi(f,\omega)) \rightarrow A-A(f,\omega) \big).$$

Analogously to BW^+ one defines a formalization $A-A^+(f,\omega)$ of the version (ii) of the Arzelà–Ascoli lemma. Similarly to the proof of $BW(\underline{x}) \rightarrow BW^+(\underline{x})$ one shows (using $\Sigma_1^0 - IA(\chi(f,\omega))$ for a suitable $\chi \in G_2 R^\omega$ and $AC^{0,0} - qf$) that $A-A(f,\omega) \rightarrow A-A^+(f,\omega)$. Analogously to prop.5.5 one so obtains

Proposition 5.6 *For $n \geq 3$ proposition 5.5 holds with $BW^+(\xi uv)$ (resp. $\forall l^0 BW^+(\xi uvl)$) replaced by $A-A(\xi uv)$ or $A-A^+(\xi uv)$ (resp. $\forall l^0 A-A(\xi uvl)$ or $\forall l^0 A-A^+(\xi uvl)$).*

5.6 The existence of lim sup and lim inf for bounded sequences in $\mathbb{R}$

Definition 5.7 $a \in \mathbb{R}$ *is the* lim sup *of* $(x_n) \subset \mathbb{R}$ *iff*

$$(*) \quad \forall k^0 \big(\forall m \exists n >_0 m(|a - x_n| \leq \frac{1}{k+1}) \wedge \exists l \forall j >_0 l(x_j \leq a + \frac{1}{k+1}) \big).$$

---

[15] We have to work in $G_3 A^\omega$ instead of $G_2 A^\omega$ since we have used the functional $\Phi_{()} fx = \overline{f}x$.

Remark 5.4 *This definition of* lim sup *is equivalent to the following one:*
(∗∗) *a is the greatest limit point of* (x_n).
The implication (∗) → (∗∗) *is trivial and can be proved e.g. in* $G_2 A^\omega$. *The implication* (∗∗) → (∗) *uses the Bolzano–Weierstraß principle.*
In the following we determine the rate of growth caused by the assertion of the existence of lim sup *(for bounded sequences) in the sense of* (∗) *and thus a fortiori in the sense of* (∗∗).

We may restrict ourselves to sequences of rational numbers: Let $x^{1(0)}$ represent a sequence of real numbers with $\forall n(|x_n| \leq_{\mathbb{R}} C)$. Then $y_n := \widehat{x_n}(n)$ represents a sequence of rational numbers which is bounded by $C + 1$. Let a^1 be the lim sup of (y_n), then a also is the lim sup of x. Hence the existence of lim sup x_n follows from the existence of lim sup y_n. Furthermore we may assume that $C = 1$.

The existence of lim sup for a sequence of rational numbers $\in [-1, 1]$ is formalized in $G_n A^\omega$ (for $n \geq 2$) as follows:

$$\exists \lim\sup(x^1) :\equiv \exists a^1 \forall k^0 \left(\forall m \exists n >_0 m(|a -_{\mathbb{R}} \breve{x}(n)| \leq_{\mathbb{R}} \frac{1}{k+1}) \wedge \right.$$

$$\left. \exists l \forall j >_0 l(\breve{x}(j) \leq_{\mathbb{R}} a + \frac{1}{k+1})\right),$$

where $\breve{x}(n) := \max_{\mathbb{Q}}(-1, \min_{\mathbb{Q}}(xn, 1))$. In the following we use the usual notation $\breve{x}_n$ instead of $\breve{x}(n)$.

We now show that $\exists \lim\sup(x^1)$ can be reduced to a purely **arithmetical** assertion $L(x^1)$ on x^1 in proofs of $\forall u^1 \forall v \leq_\tau tu \exists w^\gamma A_0$–sentences:

$$L(x^1) :\equiv \forall k \exists l >_0 k \forall K \geq_0 l \exists j \forall q, r \geq_0 j$$

$$\underbrace{\forall m, n(K \geq_0 m, n \geq_0 l \to |x_q^m -_{\mathbb{Q}} x_r^n| \leq_{\mathbb{Q}} \frac{1}{k+1})}_{L_0(x,k,l,K,q,r):\equiv},$$

where $x_q^m := \max_{\mathbb{Q}}(\breve{x}_m, \ldots, \breve{x}_{m+q})$ (Note that L_0 can be expressed as a quantifier–free formula in $G_n A^\omega$).

Lemma 5.8

1) $G_2 A^\omega \vdash Mon(\exists k \forall l \exists K \forall j \exists q, r(l > k \to K \geq l \wedge q, r \geq j \wedge \neg L_0))$.

2) $G_2 A^\omega \vdash \forall x^1 \left(\exists \lim\sup(x) \to L(x)\right)$.

3) $G_2 A^\omega \vdash \forall x^1 \left((L(x)^s \to \exists \lim\sup(x)\right)$.
   (The facts 1)–3) combined with the results of section 2 imply that $\exists \lim\sup(\xi uv)$ *can be reduced to* $L(\xi uv)$ *in proofs of sentences* $\forall u^1 \forall v \leq_\tau tu \exists w^\gamma A_0$, *see prop. 5.9 below).*

4) $G_3 A^\omega + \Sigma_2^0\text{-}IA \vdash \forall x^1 L(x)$.

Proof: 1) is obvious.

2) By $\exists \lim\sup(x^1)$ there exists an a^1 such that

(1) $\forall k^0 \forall m \exists n >_0 m(|a -_{\mathbb{R}} \breve{x}_n| \leq_{\mathbb{R}} \frac{1}{k+1})$ and

(2) $\forall k^0 \exists l \forall j >_0 l(\breve{x}_j \leq_{\mathbb{R}} a + \frac{1}{k+1})$. Assume $\neg L(x)$, i.e. there exists a k_0 such that

(3) $\forall l > k_0 \exists K \geq l \forall j \exists q, r \geq j \exists m, n(K \geq m, n \geq l \wedge |x_q^m -_{\mathbb{Q}} x_r^n| > \frac{1}{k_0+1})$.

Applying (2) to $2k_0 + 1$ yields an u_0 such that (4) $\forall j \geq u_0(\breve{x}_j \leq_{\mathbb{R}} a + \frac{1}{2(k_0+1)})$. (3) applied to $l := \max_0(k_0, u_0) + 1$ provides a K_0 with

(5) $K_0 \geq u_0 \wedge \forall j \exists q, r \geq j \exists m, n(K_0 \geq m, n \geq u_0 \wedge |x_q^m -_{\mathbb{Q}} x_r^n| > \frac{1}{k_0+1})$.

(1) applied to $k := 2k_0 + 1$ and $m := K_0$ yields a d_0 such that

(6) $d_0 > K_0 \wedge (|a - \breve{x}_{d_0}| \leq \frac{1}{2(k_0+1)})$.

By (5) applied to $j := d_0$ we obtain

(7) $\begin{cases} K_0 \geq u_0 \wedge d_0 > K_0 \wedge (|a -_{\mathbb{R}} \breve{x}_{d_0}| \leq \frac{1}{2(k_0+1)}) \wedge \\ \exists q, r \geq d_0 \exists m, n(K_0 \geq m, n \geq u_0 \wedge |x_q^m -_{\mathbb{Q}} x_r^n| > \frac{1}{k_0+1}). \end{cases}$

Let q, r, m, n be such that

(8) $q, r \geq d_0 \wedge K_0 \geq m, n \geq u_0 \wedge |x_q^m -_{\mathbb{Q}} x_r^n| > \frac{1}{k_0+1}$.

Then $x_q^m \overset{(6)}{\geq} \breve{x}_{d_0} \geq a - \frac{1}{2(k_0+1)}$ since $m \leq K_0 \leq d_0 \leq m + q$. Analogously: $x_r^n \geq a - \frac{1}{2(k_0+1)}$. On the other hand, (4) implies $x_q^m, x_r^n \leq a + \frac{1}{2(k_0+1)}$. Thus $|x_q^m -_{\mathbb{Q}} x_r^n| \leq \frac{1}{k_0+1}$ which contradicts (8).

3) Let f, g be such that L^s is fulfilled, i.e.

(*) $\begin{cases} \forall k(fk > k \wedge \forall K \geq fk \forall q, r \geq gkK \\ \forall m, n(K \geq m, n \geq fk \rightarrow |x_q^m -_{\mathbb{Q}} x_r^n| \leq_{\mathbb{Q}} \frac{1}{k+1})). \end{cases}$

We may assume that f, g are monotone for otherwise we could define $f^M k := \max_0(f0, \ldots, fk), g^M kK := \max_0 \{gxy : x \leq_0 k \wedge y \leq_0 K\}$
(f^M, g^M can be defined in $G_1 R^\omega$ using Φ_1 and λ–abstraction). If f, g satisfy (*), then f^M, g^M also satisfy (*).
Define

$h(k) :=_0 \begin{cases} \min i[f(k) \leq_0 i \leq_0 f(k) + gk(fk) \wedge \breve{x}_i =_{\mathbb{Q}} x_{gk(fk)}^{fk}], & \text{if existent} \\ 0^0, & \text{otherwise.} \end{cases}$

h can be defined in G_2A^ω as a functional in f, g. The case 'otherwise' does not occur since

$$\forall m, q \exists i (m \leq_0 i \leq_0 m + q \wedge \breve{x}_i =_\mathbb{Q} \max\nolimits_\mathbb{Q}(\breve{x}_m, \ldots, \breve{x}_{m+q})).$$

By the definition of h we have $(+)$ $\breve{x}_{hk} =_\mathbb{Q} x^{fk}_{gk(fk)}$ for all k. Assume that $m \geq k$. By the monotonicity of f, g we obtain $fm \geq_0 fk \wedge gm(fm) \geq_0 gk(fm) \geq_0 gk(fk)$. Hence $(*)$ implies

$$(1)\ |x^{fk}_{gk(fm)} -_\mathbb{Q} x^{fm}_{gm(fm)}| \leq \frac{1}{k+1} \text{ and } (2)\ |x^{fk}_{gk(fk)} -_\mathbb{Q} x^{fk}_{gk(fm)}| \leq \frac{1}{k+1}$$

and therefore $(3)\ |x^{fk}_{gk(fk)} -_\mathbb{Q} x^{fm}_{gm(fm)}| \leq \frac{2}{k+1}$. Thus for $m, \tilde{m} \geq k$ we obtain

$$(4)\ |x^{fm}_{gm(fm)} -_\mathbb{Q} x^{f\tilde{m}}_{g\tilde{m}(f\tilde{m})}| \leq \frac{4}{k+1}.$$

For $\tilde{h}(k) := h(4(k+1))$ this yields $(5)\ \forall k \forall m, \tilde{m} \geq k (|\breve{x}_{\tilde{h}m} -_\mathbb{Q} \breve{x}_{\tilde{h}\tilde{m}}| \leq \frac{1}{k+1})$. Hence for $a :=_1 \lambda m^0 . \breve{x}_{\tilde{h}m}$ we have $\hat{a} =_1 a$, i.e. a represents the limit of the Cauchy sequence $(\breve{x}_{\tilde{h}m})$.

Since $\tilde{h}(k) = h(4(k+1)) \geq f(4(k+1)) \overset{(*)}{\geq} 4(k+1) > k$, we obtain

$$(6)\ \forall k (\tilde{h}(k) > k \wedge |\breve{x}_{\tilde{h}k} -_\mathbb{R} a| \leq_\mathbb{R} \frac{1}{k+1}),$$

i.e. a is a limit point of x. It remains to show that
$(7)\ \forall k \exists l \forall j >_0 l (\breve{x}_j \leq_\mathbb{R} a + \frac{1}{k+1}):$
 Define $c(k) := g(4(k+1), f(4(k+1)))$. Then by $(*)$

$$\forall q, r \geq c(k) (|x^{f(4(k+1))}_q -_\mathbb{Q} x^{f(4(k+1))}_r| \leq \frac{1}{4(k+1)})$$

and by $(+)$ $a(k) =_\mathbb{Q} x^{f(4(k+1))}_{g(4(k+1), f(4(k+1)))}$ and therefore

$$\forall j \geq c(k) (|x^{f(4(k+1))}_j -_\mathbb{Q} a(k)| \leq \frac{1}{4(k+1)}).$$

Hence $\forall j \geq c(k) (\breve{x}_{f(4(k+1))+j} \leq_\mathbb{Q} a(k) + \frac{1}{4(k+1)})$ which implies

$$\forall j \geq c(k) + f(4(k+1)) (\breve{x}_j \leq_\mathbb{R} a + \frac{1}{4(k+1)} + \frac{1}{k+1}).$$

Thus (7) is satisfied by $l := c(2(k+1)) + f(4(2k+1)) + 1$.

4) Assume $\neg L(x)$, i.e. there exists a k_0 such that

$$(+)\ \forall \tilde{l} > k_0 \exists K \geq \tilde{l} \forall j \exists q, r \geq j \exists m, n (K \geq m, n \geq \tilde{l} \wedge |x^m_q -_\mathbb{Q} x^n_r| > \frac{1}{k_0+1}).$$

We show (using Σ_1^0–IA on l^0): $(++) :\equiv$

$$\forall l \geq_0 1 \exists i^0 \underbrace{\left(\begin{array}{l} lth(i) = l \wedge \forall j < l \dot{-} 1\big((i)_j < (i)_{j+1} \big) \\ \wedge \forall j, j' \leq l \dot{-} 1 (j \neq j' \to |\breve{x}_{(i)_j} -_{\mathbb{Q}} \breve{x}_{(i)_{j'}}| > \frac{1}{k_0+1}) \end{array} \right)}_{A_0(i,l):\equiv}.$$

$l = 1$: Obvious. $l \mapsto l + 1$: By the induction hypothesis their exists an i which satisfies $A_0(i, l)$.

Case 1: $\forall j \leq l \dot{-} 1 \exists a \forall b > a\big(|\breve{x}_b -_{\mathbb{Q}} \breve{x}_{(i)_j}| > \frac{1}{k_0+1} \big).$

Then by the collection principle for Π_1^0-formulas Π_1^0–CP there exists an a_0 such that

$$\forall j \leq l \dot{-} 1 \forall b > a_0 \big(|\breve{x}_b -_{\mathbb{Q}} \breve{x}_{(i)_j}| > \frac{1}{k_0 + 1} \big).$$

Hence $i' := i * \langle \max_0(a_0, (i)_{l \dot{-} 1}) + 1 \rangle$ satisfies $A_0(i', l+1)$.

Case 2: $\neg$ Case 1. Let us assume that $\breve{x}_{(i)_0} < \ldots < \breve{x}_{(i)_{l \dot{-} 1}}$ (If not we use a permutation of $(i)_0, \ldots, (i)_{l \dot{-} 1}$). Let $j_0 \leq_0 l \dot{-} 1$ be maximal such that

$$(1) \quad \forall \tilde{m} \exists n \geq_0 \tilde{m} \big(|\breve{x}_n -_{\mathbb{Q}} \breve{x}_{(i)_{j_0}}| \leq \frac{1}{k_0 + 1} \big).$$

(The existence of j_0 follows from the least number principle for Π_2^0-formulas Π_2^0–LNP: Let j_1 be the least number such that $(l \dot{-} 1) \dot{-} j_1$ satisfies (1). Then $j_0 = (l \dot{-} 1) \dot{-} j_1$).

The definition of j_0 implies $\forall j \leq l \dot{-} 1 \big(j > j_0 \to \exists a \forall b > a(|\breve{x}_b -_{\mathbb{Q}} \breve{x}_{(i)_j}| > \frac{1}{k_0+1}) \big)$. Hence (again by Π_1^0–CP)

$$(2) \quad \exists a_1 > j_0 \forall j \leq l \dot{-} 1 \big(j > j_0 \to \forall b > a_1(|\breve{x}_b -_{\mathbb{Q}} \breve{x}_{(i)_j}| > \frac{1}{k_0 + 1}) \big).$$

Let $c \in \mathbb{N}$ be arbitrary. By $(+)$ (applied to $\tilde{l} := \max_0(k_0, c) + 1$) there exists a K_1 such that

$$(3) \quad \forall j \exists q, r \geq j \exists m, n \big(K_1 \geq m, n \geq c, k_0 \wedge |x_q^m -_{\mathbb{Q}} x_r^n| > \frac{1}{k_0 + 1} \big).$$

By (1) applied to $\tilde{m} := K_1$ there exists a $u \geq K_1$ such that $|\breve{x}_u -_{\mathbb{Q}} \breve{x}_{(i)_{j_0}}| \leq \frac{1}{k_0+1}$.

(3) applied to $j := u$ yields q, r, m, n such that

$$(5) \quad q, r \geq u \wedge K_1 \geq m, n \geq c, k_0 \wedge |x_q^m -_{\mathbb{Q}} x_r^n| > \frac{1}{k_0 + 1} \wedge$$
$$x_q^m, x_r^n \geq_{\mathbb{Q}} \breve{x}_{(i)_{j_0}} - \frac{1}{k_0 + 1}$$

(since $m, n \leq u \leq m + q, n + r$).

Because of $m, n \geq c, k_0$ this implies the existence of an $\alpha \geq c, k_0$ such that $\breve{x}_\alpha > \breve{x}_{(i)_{j_0}}$. Thus we have shown

$$(6) \quad \forall c \exists \alpha \geq_0 c, k_0(\breve{x}_\alpha > \breve{x}_{(i)_{j_0}}).$$

For $c := \max_0(a_1, (i)_{l \doteq 1}) + 1$ this yields the existence of an $\alpha_1 > a_1, (i)_{l \doteq 1}$, k_0 such that $\breve{x}_{\alpha_1} > \breve{x}_{(i)_{j_0}}$. Let K_{α_1} be (by (+)) such that

$$(7) \ \forall j \exists q, r \geq j \exists m, n \big(K_{\alpha_1} \geq m, n \geq \alpha_1(\geq a_1, k_0) \wedge |x_q^m -_{\mathbb{Q}} x_r^n| > \frac{1}{k_0 + 1} \big).$$

(6) applies to $c := K_{\alpha_1}$ provides an $\alpha_2 \geq K_{\alpha_1}$ such that $\breve{x}_{\alpha_2} > \breve{x}_{(i)_{j_0}}$. Hence (7) applied to $j := \alpha_2$ yields q, r, m, n with

$$(8) \ q, r \geq \alpha_2 \wedge K_{\alpha_1} \geq m, n \geq \alpha_1 \wedge |x_q^m -_{\mathbb{Q}} x_r^n| > \frac{1}{k_0 + 1} \wedge x_q^m, x_r^n \geq_{\mathbb{Q}} \breve{x}_{\alpha_2}.$$

Since $m, n \geq \alpha_1 > a_1, (i)_{l \doteq 1}$, (8) implies the existence of an $\alpha_3 > (i)_{l \doteq 1}, a_1$ such that

$$(9) \ \breve{x}_{\alpha_3} >_{\mathbb{Q}} \breve{x}_{(i)_{j_0}} + \frac{1}{k_0 + 1}.$$

Since $\breve{x}_{(i)_j} \leq \breve{x}_{(i)_{j_0}}$ for $j \leq j_0$, this implies
$(10) \ \forall j \leq j_0 \big(\breve{x}_{\alpha_3} >_{\mathbb{Q}} \breve{x}_{(i)_j} + \frac{1}{k_0+1} \big).$
Let $j \leq l \doteq 1$ be $> j_0$. Then by (2) and $\alpha_3 > a_1$: $|\breve{x}_{\alpha_3} -_{\mathbb{Q}} \breve{x}_{(i)_j}| > \frac{1}{k_0+1}$.
Put together we have shown

$$(11) \ \alpha_3 > (i)_{l \doteq 1} \wedge \forall j \leq l \doteq 1 \big(|\breve{x}_{\alpha_3} -_{\mathbb{Q}} \breve{x}_{(i)_j}| > \frac{1}{k_0 + 1} \big).$$

Define $i' := i * \langle \alpha_3 \rangle$. Then $A_0(i, l)$ implies $A_0(i', l+1)$, which concludes the proof of (++).
(++) applied to $l := 2(k_0 + 1) + 1$ yields the existence of indices
$i_0 < \ldots < i_{2(k_0+1)}$ such that $|\breve{x}_{(i)_j} -_{\mathbb{Q}} \breve{x}_{(i)_{j'}}| > \frac{1}{k_0+1}$ for
$j, j' \leq 2(k_0 + 1) \wedge j \neq j'$, which contradicts $\forall j^0 (-1 \leq_{\mathbb{Q}} \breve{x}_j \leq_{\mathbb{Q}} 1)$. Hence we have proved $L(x)$. This proof has used Σ_1^0-IA, Π_1^0-CP and Π_2^0-LNP. Since Π_2^0-LNP is equivalent to Σ_2^0-IA (see [20]), and Π_1^0-CP follows from Σ_2^0-IA by [19] (where CP is denoted by M), the proof above can be carried out in $G_3A^\omega + \Sigma_2^0$-IA (these results from [19],[20] are proved there in a purely first–order context but immediately generalize to the case where function parameters are present).

Proposition 5.9 *Let $n \geq 2$ and $B_0(u^1, v^\tau, w^\gamma) \in \mathcal{L}(G_n A^\omega)$ be a quantifier-free formula which contains only u^1, v^τ, w^γ free, where $\gamma \leq 2$. Furthermore* *let* $\xi, t \in G_n R^\omega$ *and* Δ *be as in thm.2.4. Then the following rule holds*

$$
\begin{cases}
G_n A^\omega + \Delta + AC\text{-}qf \vdash \forall u^1 \forall v \leq_\tau tu\big(\exists \limsup(\xi uv) \to \exists w^\gamma B_0(u,v,w)\big) \\
\Rightarrow \exists (eff.)\chi \in G_n R^\omega \text{ such that} \\
G_n A_i^\omega + \Delta + b\text{-}AC \vdash \forall u^1 \forall v \leq_\tau tu \forall \underline{\Psi}^* \\
((\underline{\Psi}^* \text{ satisfies the mon. funct.interpr. of the negative transl. } L(\xi uv)' \\
\text{of } L(\xi uv) \ \to \ \exists w \leq_\gamma \chi u \underline{\Psi}^* B_0(u,v,w)) \\
\text{and in particular } \exists \Psi \in T_1 \text{ such that} \\
PA_i^\omega + \Delta + b\text{-}AC \vdash \forall u^1 \forall v \leq_\tau tu \exists w \leq_\gamma \Psi u \, B_0(u,v,w).
\end{cases}
$$

where T_1 *is the restriction of Gödel's* T *which contains only the recursor* R_ρ
for $\rho \leq 1$. *The Ackermann function (but no functions having an essentially greater order of growth) can be defined in* T_1.
In the conclusion, $\Delta + b\text{-}AC$ *can be replaced by* $\tilde{\Delta}$, *where* $\tilde{\Delta}$ *is defined as in thm.2.4. If* $\Delta = \emptyset$, *then* $b\text{-}AC$ *can be omitted from the proof of the conclusion. If* $\tau \leq 1$ *and the types of the* $\exists$–*quantifiers in* Δ *are* ≤ 1, *then* $G_n A^\omega + \Delta + AC\text{-}qf$ *may be replaced by* $E\text{-}G_n A^\omega + \Delta + AC^{\alpha,\beta}$–*qf, where* α, β *are as in cor.2.5.*

Proof: Prenexation of $\forall u^1 \forall v \leq_\tau tu\big(L(\xi uv) \to \exists w^\gamma B_0(u,v,w)\big)$ yields

$$
G :\equiv \forall u^1 \forall v \leq_\tau tu \exists k \forall l \exists K \forall j \exists q, r, w\big[(l > k \wedge (K \geq l \wedge q, r \geq j \to L_0)) \\
\to B_0(u,v,w)\big].
$$

Lemma 5.8.1) implies

$$(1) \quad G_2 A^\omega \vdash Mon(G).$$

The assumption of the proposition combined with lemma 5.8.3) implies

$$(2) \quad G_n A^\omega + \Delta + AC\text{–qf} \vdash \forall u^1 \forall v \leq_\tau tu\big(L(\xi uv)^S \to \exists w^\gamma B_0(u,v,w)\big)$$

and therefore

$$(3) \quad G_n A^\omega + \Delta + AC\text{–qf} \vdash G^H.$$

Theorem 2.4 applied to (1) and (3) provides the extractability of a tuple $\underline{\varphi} \in G_n R^\omega$ such that

$$(4) \quad G_n A_i^\omega + \Delta + b\text{-AC} \vdash$$
$$\big(\underline{\varphi} \text{ satisfies the monotone functional interpretation of } G'\big).$$

G' intuitionistically implies

$$(5) \quad \forall u^1 \forall v \leq_\tau tu\big(L(\xi uv)' \to \neg\neg\exists w^\gamma B_0(u,v,w)\big).$$

Hence from φ one obtains a term $\tilde{\varphi} \in G_n R^\omega$ such that (provably in $G_n A_i^\omega + \Delta + \text{b-AC}$)

(6) $\exists \psi \big(\tilde{\varphi}$ s–maj $\psi \wedge \forall u^1 \forall v \leq_\tau tu \forall \underline{a} \big(\forall \underline{b}(L(\xi uv)')_D \to B_0(u, v, \psi uv\underline{a}) \big) \big)$,

where $\exists \underline{a} \forall \underline{b} \big(L(\xi uv)' \big)_D$ is the usual functional interpretation of $L(\xi uv)'$.
Let $\underline{\Psi}^*$ satisfy the monotone functional interpretation of $L(\xi uv)'$ then

(7) $\exists \underline{a} \big(\underline{\Psi}^*$ s–maj $\underline{a} \wedge \forall \underline{b}(L(\xi uv)')_D \big)$.

Hence for such a tuple $\underline{a}$ we have

(8) $\lambda u^1 . \tilde{\varphi} u (t^* u) \underline{\Psi}^*$ s–maj $\psi uv\underline{a}$ for $v \leq tu$

(Use lemma 2.2.11 from [15]. t^* in $G_n R^\omega$ is a majorant for t).
Since $\gamma \leq 2$ this yields a $\geq_2$ bound $\chi u \underline{\Psi}^*$ for $\psi uv\underline{a}$ (lemma 2.2.11 from [15]).
The second part of the proposition follows from lemma 5.8.4) and the fact that $G_n A^\omega + \Sigma_2^0$–IA has (via negative translation) a monotone functional interpretation in PA_i^ω by terms $\in T_1$ (By [20] Σ_2^0–IA has a functional interpretation in T_1. Since every term in T_1 has a majorant in T_1, also the monotone functional interpretation can be satisfied in T_1).

Remark 5.5   *By the theorem above the use of the analytical axiom* $\exists \lim \sup(\xi uv)$ *in a given proof of* $\forall u^1 \forall v \leq_\tau tu \exists w^\gamma B_0$ *can be reduced to the use of the arithmetical principle* $L(\xi uv)$*. By lemma 5.8.2) this reduction is optimal (relatively to* $G_2 A^\omega$*).*

6   REFERENCES

[1] Beeson, M.J., Foundations of Constructive Mathematics. Springer Ergebnisse der Mathematik und ihrer Grenzgebiete 3.Folge, Bd.6., Berlin Heidelberg New York Tokyo 1985.

[2] Bezem, M.A., Strongly majorizable functional of finite type: a model for bar recursion containing discontinuous functionals. J. Symb. Logic **50**, pp. 652–660 (1985).

[3] Bishop, E., Foundations of constructive analysis. McGraw–Hill, New–York (1967).

[4] Bishop, E.– Bridges, D. , Constructive analysis. Springer Grundlehren der mathematischen Wissenschaften vol.279, Berlin 1985.

[5] Feferman, S., Theories of finite type related to mathematical practice. In: Barwise, J. (ed.), Handbook of Mathematical Logic, North–Holland, Amsterdam, pp. 913–972 (1977).

[6] Friedman, H., Systems of second order arithmetic with restricted induction (abstract), J. Symbolic Logic **41**, pp. 558-559 (1976).

[7] Gödel, K., Zur intuitionistischen Arithmetik und Zahlentheorie. Ergebnisse eines Mathematischen Kolloquiums, vol. 4 pp. 34–38 (1933).

[8] Gödel, K., Über eine bisher noch nicht benutzte Erweiterung des finiten Standpunktes. Dialectica **12**, pp. 280–287 (1958).

[9] Kleene, S.C., Introduction to Metamathematics. North-Holland (Amsterdam), Noordhoff (Groningen), New–York (Van Nostrand) (1952).

[10] Kleene, S.C., Recursive functionals and quantifiers of finite types, I. Trans. A.M.S. **91**, pp.1–52 (1959).

[11] Ko, K.–I., Complexity theory of real functions. Birkhäuser; Boston, Basel, Berlin (1991).

[12] Kohlenbach, U., Effective bounds from ineffective proofs in analysis: an application of functional interpretation and majorization. J. Symbolic Logic **57**, pp. 1239–1273 (1992).

[13] Kohlenbach, U., Analysing proofs in analysis. In: W. Hodges, M. Hyland, C. Steinhorn, J. Truss, editors, *Logic: from Foundations to Applications. European Logic Colloquium* (Keele, 1993), pp. 225–260, Oxford University Press (1996).

[14] Kohlenbach, U., Real growth in standard parts of analysis. Habilitationsschrift, xv+166 p., Frankfurt (1995).

[15] Kohlenbach, U., Mathematically strong subsystems of analysis with low rate of provably recursive functionals. Arch. Math. Logic **36**, pp. 31–71 (1996).

[16] Kohlenbach, U., Elimination of Skolem functions for monotone formulas. To appear in: Archive for Mathematical Logic.

[17] Kohlenbach, U., The use of a logical principle of uniform boundedness in analysis. To appear in: Proc. 'Logic in Florence 1995'. Kluwer academic publisher.

[18] Luckhardt, H., Extensional Gödel functional interpretation. A consistency proof of classical analysis. Springer Lecture Notes in Mathematics **306** (1973).

[19] Parsons, C., On a number theoretic choice schema and its relation to induction. In: Intuitionism and proof theory, pp. 459-473. North-Holland, Amsterdam (1970).

[20] Parsons, C., On n-quantifier induction. J. Symbolic Logic **37**, pp. 466–482 (1972).

[21] Sieg, W., Fragments of arithmetic. Ann. Pure Appl. Logic **28**, pp. 33-71 (1985).

[22] Simpson, S.G., Reverse Mathematics. Proc. Symposia Pure Math. **42**, pp. 461–471, AMS, Providence (1985).

[23] Spector, C., Provably recursive functionals of analysis: a consistency proof of analysis by an extension of principles formulated in current intuitionistic mathematics. In: Recursive function theory, Proceedings of Symposia in Pure Mathematics, vol. 5 (J.C.E. Dekker (ed.)), AMS, Providence, R.I., pp. 1–27 (1962).

Author Address

BRICS, Department of Computer Science,
University of Aarhus, Ny Munkegade 540,
DK-8000 Aarhus C, Denmark.
email: kohlenb@brics.dk

Satisfaction classes and automorphisms of models of PA

Roman Kossak

1 Introduction

In recent years we have learned a lot about countable recursively saturated models of PA and their automorphisms. We know that there are continuum many nonisomorphic automorphism groups of such models, and we know that each of them contains a copy of the automorphism group of the order preserving permutations of the rationals. We can classify the closed normal subgroups of a given automorphism group, and we have a great deal of information about open maximal subgroups. We know much about the model theory of the arithmetically saturated countable models of PA; in particular we know that the automorphism group of a countable arithmetically saturated model of PA has the small index property. But still many questions remain open: Can we classify all normal and all maximal subgroups of the automorphism group a given model? Do non arithmetically saturated models have the small index property? Is the automorphism group of a countable recursively saturated model decidable?

Many other results and questions could be mentioned here. However, the purpose of this paper is not to give a complete survey; this has been done recently by Kotlarski in [17]. Instead, I will concentrate on a specific feature of countable recursively saturated models of PA — inductive satisfaction classes and their use.

My goal is twofold. Often satisfaction classes allow one to give easy answers to questions that otherwise seem difficult. A list of examples is presented in section 4. I hope those who work in the model theory of recursively saturated models of PA will find this list useful. The other goal is to propose the following problem. Much of the model theory of countable recursively saturated models of PA is based on specific techniques (resplendency arguments, special 'back-and-forth' constructions), but a significant number of results can be obtained as corollaries of classical results concerning models of PA* applied to structures of the form (M, S), where M is a recursively

---

[1]Received September 5, 1996; revised version Marsh 4, 1997.

saturated model of PA and S is an inductive satisfaction class for M. A short survey such results constitutes sections 2 and 3. The advantage this approach is the availability of all the techniques developed for models of PA*, in particular the powerful machinery of definable types of Gaifman [1]. (PA* is Peano Arithmetic formulated in a language extending the language of PA.) The problem, I propose, is to determine what part of the model theory of countable recursively saturated models of PA can be obtained in this way. Of course, the answer might depend on the definition of model theory, and, perhaps, the biggest challenge is to give a precise formulation of the problem.

My presentation of the material will be a bit informal, as it is intended for an easy reading by the reader not necessarily interested in technical details. I assume that the reader is familiar with the notions of recursive saturation and of the standard system of a model of PA. I will also use the notion of arithmetic saturation. A model M of PA is *arithmetically saturated* if it is recursively saturated and its standard system satisfies the arithmetic comprehension schema. Definitions of all other concepts discussed here can be found in [3].

I will only discuss those problems concerning recursively saturated models of PA that can be formulated without direct references to satisfaction classes. There are many interesting results on the theory of satisfaction classes, most of them due to Kotlarski, Ratajczyk, and Smith. The interested reader should consult [17] and [20].

2   Recursive saturation and satisfaction classes

The notion of a satisfaction class requires an arithmetization of the language of arithmetic. The specifics of the arithmetization are not important, as long as it is can be formalized within PA. In what follows we will identify formulas and sentences of PA with their Gödel numbers.

Definition 2.1 *A set $S \subset M \models PA$ is an inductive satisfaction class for M if*
  i) $\mathrm{Th}(M,a)_{a \in M} \subset S$;
  ii) S satisfies Tarski's inductive definition of a satisfaction relation;
  iii) $(M,S) \models PA^*$.

Tarski's argument shows that no inductive satisfaction class can be defined in a model of PA. The same argument can be applied to prove that if S extends $\mathrm{Th}(M,a)_{a \in M}$, and S does not contain sentences false in M, then S is undefinable in M. Definability here, and elswhere in the paper, means definability with parameters.

Here is the basic fact relating inductive satisfaction classes to recursive saturation.

Proposition 2.2 *Let M be a model of PA. i) If M has an inductive satisfaction class, then M is recursively saturated. ii) If M is countable and recursively saturated, then M has an inductive satisfaction class.*

The easy proof consists of an overspill argument in *i*, and a resplendency argument in *ii*.

It should be mentioned that the second part of proposition 2.2 is false for uncountable models. There are uncountable (even ω_1-like ones) recursively saturated models of PA without satisfaction classes. A rather classless model of Kaufmann [2] has this property. Direct constructions of such models can be found in [12] and [14].

If S is an inductive satisfaction class for a model M, then, by overspill S decides all Σ_e sentences, with parameters, in the sense of M, for some nonstandard e. Also, either S decides all sentences in the sense of M, in this case we say that S is *full*, or there is the largest e such that S decides all Σ_e sentences of M, in such a case we will say that S is a Σ_e-inductive satisfaction class for M.

To illustrate how inductive satisfaction classes are used, I will now prove four well-known results, whose first proofs did not use satisfaction classes explicitly. These results form a basis for the model theory of recursively saturated models of PA. The original proofs are not very difficult, so the point here is not that the proofs involving satisfaction classes offer significant simplification (although the new proofs are shorter). What is interesting about the proofs I want to present, is that no reference to recursive saturation is made directly, and that they only use standard results concerning models of PA*.

One of the results, which turns out to be applicable in many situations, and which the reader might not be familiar with, is the following version of Gaifman's theorem on cofinal extensions, due to Kotlarski [16] and Schmerl [18].

Lemma 2.1 *If (M, X) is a model of PA* and N is a cofinal extension of N, then there exists (exactly one) $Y \subset N$ such that $(M, X) \prec (N, Y)$.*

Model theory of recursively saturated models of arithmetic started with Smoryński's paper [21]. One of the important results of that paper was the following analog of the MacDowell-Specker theorem.

Theorem 2.3 *Every countable recursively saturated model of PA has a proper countable recursively saturated elementary end extension.*

Proof: Let M be a countable recursively saturated model of PA. Let S be an inductive satisfaction class for M. By the MacDowell-Specker theorem (M, S) has a countable elementary end extension (N, T). Since T is an inductive satisfaction class for N, N is recursively saturated and the result follows.   □

The next result that contributed much to further developments is known as the Smoryński-Stavi theorem [22].

Theorem 2.4 *Every cofinal extension of a recursively saturated model is recursively saturated.*

Proof: We will consider the countable case first. Let M be a countable recursively saturated model of PA, and let N be its cofinal extension. Let S be an inductive satisfaction class for M. By lemma 2.1 there is $T \subset N$ such that $(M, S) \prec (N, T)$, hence N is recursively saturated.

If M is not countable then M is a union of a tower of elementary submodels each of which has a cofinal countable recursively saturated cofinal submodel, and the result follows, by the previous argument. □

The two theorem quoted above seem to indicate that the model theory of recursively saturated models of PA does not differ much from the model theory of PA. The next proposition, first noted in [5], shows that it is not so.

Recall that if $M \prec N$, then a subset X of M is *coded* in N if $X = M \cap Y$ for some set Y which is definable in N. We say that N is a *conservative extension* of M if the only subsets of M coded in N are the definable subsets of M, or, equivalently, the type of every element of $N \setminus M$ over M is definable. If N is a conservative extension of M, then N must be an end extension. Also, by a suitable version of the MacDowell-Specker theorem, every model of PA has a conservative elementary end extension.

Proposition 2.5 *If N is a recursively saturated model of PA and $M \prec N$, then N is not a conservative extension of M.*

Proof: W.l.o.g. we can assume that N has a cofinal countable elementary submodel; hence, by proposition 2.2 and lemma 2.1, N has an inductive satisfaction class S. Let $T = M \cap S$. Since $(M, T) \models \text{PA}^*$, it is easy to show that T is coded in M. T might not be an inductive satisfaction class for M (and indeed, it is not an inductive satisfaction class for M if M is not recursively saturated), but still by Tarski's argument T is undefinable in M and the result follows. □

According to proposition 2.5, recursively saturated elementary end extension of a model M always codes a nondefinable subset of M, but still we have a weak version of conservativeness for recursively saturated elementary end extensions: if M is countable, then every nondefinable subset of M can be omitted in a recursively saturated elementary end extension of M. This lemma was proved first by Kaufmann [2] using different methods, another short proof using satisfaction classes can be found in [18].

Lemma 2.2 *Let X be a nondefinable subset of a countable recursively saturated model M of PA, then there is a countable recursively saturated elementary end extension N of M in which X is not coded.*

Proof: Let us first suppose that (M, X) is not recursively saturated. Let S be an inductive satisfaction class for M such that (M, S) is recursively saturated. Such satisfaction classes exist by chronic resplendency of countable recursively saturated models (see [3]). Let (N, K) be a conservative elementary end extension of (M, S). Then, if Y is a subset of M coded in N, then Y must be definable in (M, S); hence (M, Y) is recursively saturated. Thus X is not coded in N.

Suppose now that (M, X) is recursively saturated. Consider the theory Σ saying that S is an inductive satisfaction class and X is undefinable in (M, S). Each finite fragment of Σ is modeled by (M, Tr_n), for n large enough, where Tr_n is the universal Σ_n relation (n is standard here). Hence M can be expanded to a model of Σ. If (M, S) is such an expansion, then we again take (N, T) to be a conservative extension of (M, S), and the result follows. $\qquad\square$

3   Minimal satisfaction classes and s-ultrapowers

In the last proof of the previous section we have, tacitly, used definable ultrapowers of structures of the form (N, S). A *definable ultrapower* of a structure $\mathcal{A}$ is an ultrapower of $\mathcal{A}$ built using the set of definable functions of $\mathcal{A}$ and an ultrafilter on the definable subsets of $\mathcal{A}$. For brevity, we will call a recursively saturated model N an *s-ultrapower* if there is a recursively saturated model M and an inductive satisfaction class S for M such that (N, T) is a definable ultrapower of (M, S), for some $T \subset N$.

It is shown in [6] that every countable recursively saturated model of PA can be expanded by adding an inductive satisfaction class in continuum many elementary inequivalent ways. Also, for every such expansion (M, S) there are continuum many nonisomorphic expansions (M, T) which are all elementarily equivalent to (M, S). Using this variety of satisfaction classes it is not difficult to show that for every countable recursively saturated model M, there are continuum many nonisomorphic pairs of the form (M, K) where $K \prec_{end} M$ and M is an s-ultrapower of K. Thus, the question to consider is: Suppose M and N are countable recursively saturated models of PA and $M \prec_{end} N$, is N an s-ultrapower of M? The answer to such a general question is negative. For a model M and $a \in M$, let $M[a]$ denote the largest elementary submodel of M not containing a (if there is such a model). If M is a recursively saturated model of PA, then, for every $a \in M$, M is not a s-ultrapower of $M[a]$, an outline of the argument proving this is given in the discussion of question 4.1 in section 4. We also have the following proposition.

Proposition 3.1 *If M is countable recursively saturated but not arithmetically saturated, model of PA then there is a recursively saturated $K \prec_{end} M$ such that none of the inductive satisfaction classes of K is coded in M.*

For a proof see proposition 4.6 of [7]. Clearly, if K is as in the above proposition, then M is not an s-ultrapower of K.

Problem 3.2 *For a given countable recursively saturated model M of PA, classify all recursively saturated $K \prec_{end} M$ such that M is an s-ultrapower of K.*

An important class of s-ultrapowers is obtained by using minimal satisfaction classes.

For a structure $\mathcal{A}$, $\mathrm{DEF}(\mathcal{A})$ will denote the family of sets definable in $\mathcal{A}$ (with parameters), and $\mathrm{Def}(\mathcal{A})$ will be the set of points definable without parameters in $\mathcal{A}$.

Definition 3.3 *An inductive satisfaction class S for a model M is minimal, if $\mathrm{Def}(M, S) = M$.*

Theorem 3.4 *Every countable recursively saturated model of PA has a minimal inductive satisfaction class. Moreover, if M has a Σ_e inductive satisfaction class, for some $e \in M$, then M has continuum many pairwise elementarily inequivalent structures of the form (M, S), where S is a minimal Σ_e inductive satisfaction class for M.*

Theorem 3.4 was proved in a slightly weaker form in [6], the present formulation is taken from [14]. Here is the outline of the proof. Start with the theory $T_0 = \mathrm{Th}(M)$ plus the set of sentences saying that S is a Σ_e inductive satisfaction class. T_0 is coded in the standard system of M. The crucial point now is that one can find a complete consistent theory T extending T_0, and such that the family of sets of natural numbers represented in T is exactly the standard system of M. If (M_0, S_0) is the minimal model of T, then S_0 is a minimal inductive satisfaction class for M_0. M_0 is elementarily equivalent to M, both models are recursively saturated, and they have the same standard systems, hence M_0 is isomorphic to M, and the result follows.

To obtain the second part of the theorem it is enough to notice that the completion T in the above proof can be obtained in continuum many different ways.

If S is a minimal inductive satisfaction class for a model M, then the srtucture (M, S) is rigid (i.e. has no nontrivial automorphisms). But we can prove more.

Proposition 3.5 *If S is a minimal inductive satisfaction class for a model M, then the structure $(M, \mathrm{DEF}(M, S), \in)$ is rigid.*

The key to the proof of 3.5 is the following lemma:

Lemma 3.1 *If S is a Σ_e inductive satisfaction class for a model M, then there exists a Σ_a inductive satisfaction class that is definable in (M, S) iff $a < e + n$ for some standard n.*

Proofs of 3.5 and 3.1 are not difficult, they can be found in [14].

If K and M are countable recursively saturated models of PA, and M is an elementary end extension of K, then it is important to know which automorphisms of K can be extended to M.

Let $\text{Coded}_M(K)$ denote the family of those subsets of K which are coded in M. If $f \in \text{Aut}(K)$ extends to M, then, f must be an automorphism of $(K, \text{Coded}_M(K), \in)$. In most cases the converse is also true [10]:

Lemma 3.2 *Let M be a countable recursively saturated model of PA, then for all but countably many recursively saturated $K \prec_{end} M$, an automorphism f of K can be extended to an automorphism of M iff f is an automorphism of $(K, \text{Coded}_M(K), \in)$.*

See [11] for the discussion of the special cases in which lemma 3.2 fails.

The interesting problem is: What are the groups of automorphisms of structures of the form $(K, \text{Coded}_M(K), \in)$?. It turns out that they can be almost anything, and the key to this result is an application of minimal satisfaction classes. First an extreme example:

Theorem 3.6 *If M is an s-ultrapower of a recursively saturated model K, built with a minimal inductive satisfaction class for K, then no nontrivial automorphism of K can be extended to M.*

Theorem 3.6 is an easy consequence of 3.5. The proof is given in [14].

Let M be a given countable recursively saturated model of PA, and let G be the automorphism group of M. For $X \subset M$ let $G_{\{X\}}$ be the setwise stabilizer of X and let $G_{(X)}$ be the pointwise stabilizer of X. According to lemma 3.2, for most elementary initial segments $K \prec_{end} M$ the group $G_{\{K\}}/G_{(K)}$ is isomorphic to $\text{Aut}(K, \text{Coded}_M(K), \in)$, in particular, theorem 3.6 can be reformulated as follows. There is $K \prec_{end} M$ such that the quotient group $G_{\{K\}}/G_{(K)} = \text{Aut}(K, \text{Coded}_M(K), \in)$ is trivial. Much more can be done in this direction. Using minimal satisfaction classes and a result of Gaifman [1] we have shown in [9] that:

Theorem 3.7 *For every countable recursively saturated model M and for every countable linearly ordered set $(I, <)$ there is $K \prec_{end} M$ such that M is an s-ultrapower of K, and $G_{\{K\}}/G_{(K)} \cong \text{Aut}(I, <)$.*

Since the models K obtained in the proof of 3.7 satisfy the assumptions of lemma 3.2 (i.e. they are not among the special cases), theorem 3.7 also provides information about the groups $\text{Aut}(K, \text{Coded}_M(K), \in)$.

One of the consequences of theorem 3.7 is that the group G is not divisible; hence it is not elementarily equivalent to the group of order preserving permutations of the rationals.

Recently Schmerl [19] has shown that in the formulation of theorem 3.7 "*linearly ordered set $(I, <)$*" can be replaced by "*linearly ordered structure $(I, <, \ldots)$.*" Schmerl's proof uses s-ultrapowers.

4 Examples and counterexamples

Throughout this section, let M be a countable recursively saturated model of PA, and let $G =$ Aut(M).

Question 4.1 *Can every recursively saturated $K \prec M$ be represented as* Def(M, S) *for some inductive satisfaction class S for M?*

Answer: In general the answer is negative. If $K =$ Def(M, S) for some inductive satisfaction class S, then $(K, T) \prec (M, S)$, where $T = S \cap K$ is an inductive satisfaction class for K coded in M. Thus, the answer is provided by proposition 3.1. Also, since in (M, S) one can define functions majorizing all definable functions of M, if $(K, T) \prec (M, S)$, and $K \prec_{end} M$, then for every $a \in M \setminus K$ there is $b \in M \setminus K$ such that $t(b) < a$, for every function t definable in M without parameters. Hence models $M[a]$ cannot be represented as Def(M, S).

However, if K is a cofinal submodel of M, then the answer is positive. In this case, let T be a minimal inductive satisfaction class for K. By lemma 2.1, there is $S \subset M$ such that $(K, T) \prec (M, S)$. Thus, Def(M, S) $= K$. $\square$

Question 4.2 *Let f be an automorphism of M. Is there an inductive satisfaction class such that $fS = S$?*

Answer: If M is arithmetically saturated, then the answer is negative. By fix(f) we will denote the set of points fixed by f. We will consider fix(f) as a submodel of M. Clearly, fix(f) $\prec M$. If $fS = S$ for some inductive satisfaction class S for M, fix(f) $\cap S$, is an inductive satisfaction class for fix(f). Thus, if fix(f) is nonstandard it is recursively saturated. It is well-known that if M is arithmetically saturated then there are $f \in G$ such that fix(f) is nonstandard and not recursively saturated (cf.[15], [4]), and the result follows.

If M is not arithmetically saturated, then, for every $f \in G$, fix(f) $\cong M$, hence the previous argument cannot be applied. The problem is still open in this case. $\square$

Question 4.3 *Suppose M is arithmetically saturated. Is every $K \prec M$ of the form* fix(f) *for some $f \in G$?*

Answer: The answer is negative. It is shown in [15] that if $K =$ Def(M, S) for an inductive satisfaction class S such that (M, S) is recursively saturated, then K has continuum many elementary substructures which are not of the form fix(f). $\square$

Question 4.3 is related to the following important open problem posed in [4].

Problem 4.4 *Characterize elementary submodels of M which are of the form* fix(f), *for $f \in G$.*

Question 4.5 *Suppose that M is a model of true arithmetic. Is there an $f \in G$ such that $f(a) > a$ for all nonstandard $a \in M$?*

Answer: In [15] a special 'back-and-forth' argument was designed to show that the answer to the question is positive iff M is arithmetically saturated. Here is a much shorter proof using satisfaction classes. The proof is due to Jim Schmerl.

Let M be a countable arithmetically saturated model of true arithmetic. Let $A_0, A_1, \ldots$ be an enumeration of the standard system M. By a result of Gaifman [1], we can find an elementary extension $\mathcal{N}$ of $(\mathbf{N}, A_0, A_1, \ldots)$, which is generated by a set of indiscernibles of the order type of the integers. Hence, $\mathcal{N}$ has an automorphism f such that $f(a) > a$ for all nonstandard a. Let N be the reduct of $\mathcal{N}$ to the language of PA. To finish the proof we have to make two observations. Since M is a recursively saturated model of true arithmetic, the list $A_0, A_1, \ldots$ contains the standard satisfaction class for $\mathbf{N}$, hence N is recursively saturated. Also, since $\mathcal{N}$ is generated by elements realizing a minimal (hence definable) type over $(\mathbf{N}, A_0, A_1, \ldots)$, the standard system of $\mathcal{N}$ (which is the same as the standard system of N) is $\{A_0, A_1, \ldots\}$. Here we are using the fact that $\{A_0, A_1, \ldots\}$ is closed under arithmetic comprehension. Thus N is isomorphic to M and the result follows.   □

Inductive satisfaction classes can be also applied to problems concerning isomorphism types of structures of the form (M, K), where $K \prec_{end} M$. Two such applications are given in [8]. They both involve the following concept.

Definition 4.6 *For $K \prec_{end} M$, let $S(K)$ be the set of those $e \in M$ for which there exists a Σ_e inductive satisfaction class for K, which is coded in M.*

The cut $S(K)$ is definable in (M, K), hence it can be used, in a way similar to that in which the cofinality of a cut is used, to construct structures (M, K) of different isomorphism types. Arguments using the cofinality of a cut apply to non-semiregular cuts only; arguments using $S(K)$ apply to a much larger family of cuts.

Problem 4.7 *Characterize cuts of the form $S(K)$ for a given countable recursively saturated model of PA.*

5 REFERENCES

[1] H. Gaifman, On models and types of Peano's arithmetic, *Annals of Pure and Applied Logic*, vol. 9 (1976), pp. 223-306.

[2] M. Kaufmann, A rather classless model, *Proceedings of the American Mathematical Society*, vol. 62 (1977), pp. 330-333.

[3] R. Kaye, Models of Peano Arithmetic, *Oxford Logic Guides*, Oxford University Press, 1991

[4] R. Kaye, R. Kossak, and H. Kotlarski, Automorphisms of recursively saturated models of arithmetic, *Annals of Pure and Applied Logic*, vol. 55 (1991), pp. 67-99.

[5] R. Kossak, A certain class of models of Peano Arithmetic, *Journal of Symbolic Logic*, vol. 48 (1983), pp. 311-320.

[6] R. Kossak, A note on satisfaction classes, *Notre Dame Journal of Formal Logic*, vol. 26 (1985), pp.1-8.

[7] R. Kossak, Models with the ω-property, *Journal of Symbolic Logic*, vol. 54 (1989), pp. 177-189.

[8] R. Kossak, Four problems concerning recursively saturated models of arithmetic, *Notre Dame Journal of Formal Logic* vol. 36 (1995), pp. 519-530.

[9] R. Kossak, N. Bamber, On two questions concerning the automorphism groups of countable recursively saturated models of PA, *Archive for Mathematical Logic*, vol. 36 (1996) pp. 73-79.

[10] R. Kossak, H. Kotlarski, Results on automorphisms of recursively saturated models of arithmetic, *Fundamenta Mathematicae*, vol. 129 (1988), pp. 9-15.

[11] R. Kossak, H. Kotlarski, On extending automorphisms of models of Peano Arithmetic, *Fundamenta Mathematicae*, vol. 149 (1996), pp. 245-263.

[12] R. Kossak, H. Kotlarski, Game approximations of satisfaction classes and the problem of rather classless models, *Zeitschr. f. math. Logik und Grundlagen d. Math.*, vol. 38 (1992), pp. 21-26.

[13] R. Kossak, H. Kotlarski, and J. H. Schmerl, On maximal subgroups of the automorphism group of a countable recursively saturated model of PA, *Annals of Pure and Applied Logic*, vol. 65 (1993), pp. 125-148.

[14] R. Kossak and J. H. Schmerl, Minimal satisfaction classes with an application to rigid models of Peano Arithmetic, *Notre Dame Journal of Formal Logic*, vol. 32 (1991), pp. 392-398.

[15] R. Kossak, J. H. Schmerl, Arithmetically saturated models of arithmetic, *Notre Dame Journal of Formal Logic* vol. 36 (1995), pp. 531-546.

[16] H. Kotlarski, On cofinal extensions of models of arithmetic, *Journal of Symbolic Logic*, vol. 48 (1983), pp. 311-319.

[17] H. Kotlarski, Automorphisms of countable recursively saturated models of PA: a survey, *Notre Dame Journal of Formal Logic*, vol. 30 (1995), pp. 505-518.

[18] J. H. Schmerl, Recursively saturated rather classless models of Peano Arithmetic, in *Logic Year* ed., M. Lerman, J. Schmerl, R. Soare, *Lecture Notes in Mathematics*, vol. 859, Springer-Verlag, Berlin, 1981.

[19] J. H. Schmerl, Automorphism Groups of Models of Peano Arithmetic, to appear.

[20] S. T. Smith, Nonstandard characterization of recursive saturation and resplendency, *Journal of Symbolic Logic*, vol. 52 (1987), pp. 842-863.

[21] C. Smoryński, Recursively saturated nonstandard models of arithmetic, Journal of Symboloc Logic, vol. 46 (1981), pp. 259-286.

[22] C. Smoryński, J. Stavi, Cofinal extension preserves recursive saturation, in *Model Theory of Algebra and Arithmetic*, eds. L. Pacholski et al., *Lecture Notes in Mathematics*, vol. 843, Springer-Verlag, Berlin 1980.

Author address

Department of Mathematics and Computer Science
BCC/City University of New York
University Ave. & W 181 Street
Bronx, NY 10453, USA
email: rkobb@cunyvm.cuny.edu

Free monoid completeness of the Lambek calculus allowing empty premises

Mati Pentus[2]

ABSTRACT We prove that the Lambek syntactic calculus allowing empty premises is complete with respect to the class of all free monoid models (i. e., the class of all string models, allowing the empty string).

Introduction

Lambek syntactic calculus (introduced in [7]) is one of the logical calculi used in the paradigm of categorial grammar for deriving reduction laws of syntactic types in natural and formal languages. The intended models for these calculi are free semigroup models (also called language models or string models), where each syntactic category is interpreted as a set of non-empty strings over some alphabet of symbols. Models for Lambek calculus were studied in [2], [3], [4], [5], [6], etc. Completeness of the Lambek calculus with respect to string models was proved in [9], [10], and [11]. Closely related is the result about completeness with respect to relational semantics [8].

There is a natural modification of the original Lambek calculus, which we call the Lambek calculus allowing empty premises (cf. [2, p. 44]). This calculus appears to be a fragment of the noncommutative linear logic. The natural class of string models for the Lambek calculus allowing empty premises is the class of all free monoid models, where also empty string is allowed.

In this paper we prove that the Lambek calculus allowing empty premises is complete with respect to these models.

---

[1]Received September 1996; revised version February 1997.

[2]The research described in this publication was made possible in part by the Russian Foundation for Basic Research (project 96-01-01395).

1 Lambek calculus allowing empty premises

We consider the *Lambek calculus allowing empty premises* (cf. [2, p. 44]) and denote it by L^*. This calculus is a modification of the syntactic calculus introduced in [7].

The types of L^* are built of primitive types $p_1, p_2, \ldots$, and three binary connectives $\bullet, \backslash, /$. We shall denote the set of all types by Tp. The set of finite sequences of types (resp. finite non-empty sequences of types) is denoted by Tp^* (resp. Tp^+). The symbol Λ will stand for the empty sequence of types.

Sometimes we shall write $A_1 \bullet \ldots \bullet A_n$ instead of $(\ldots (A_1 \bullet A_2) \bullet \ldots) \bullet A_n$.

Capital letters A, B, ... range over types. Capital Greek letters range over finite (possibly empty) sequences of types.

Sequents of L^* are of the form $\Gamma \to A$. Note that Γ can be the empty sequence.

Axioms: $A \to A$

Rules:

$$\frac{A\Pi \to B}{\Pi \to A \backslash B} \ (\to\backslash) \qquad\qquad \frac{\Phi \to A \quad \Gamma B \Delta \to C}{\Gamma \Phi (A \backslash B) \Delta \to C} \ (\backslash\to)$$

$$\frac{\Pi A \to B}{\Pi \to B/A} \ (\to/) \qquad\qquad \frac{\Phi \to A \quad \Gamma B \Delta \to C}{\Gamma (B/A) \Phi \Delta \to C} \ (/\to)$$

$$\frac{\Gamma \to A \quad \Delta \to B}{\Gamma \Delta \to A \bullet B} \ (\to\bullet) \qquad\qquad \frac{\Gamma A B \Delta \to C}{\Gamma (A \bullet B) \Delta \to C} \ (\bullet\to)$$

$$\frac{\Phi \to B \quad \Gamma B \Delta \to A}{\Gamma \Phi \Delta \to A} \ (CUT)$$

It is known that the cut-elimination theorem holds for this calculus (cf. [2]). We write $L^* \vdash \Gamma \to A$ if the sequent $\Gamma \to A$ is derivable in L^*.

There is an obvious duality phenomenon inherent in L^*.

Definition. The function dual: Tp $\to$ Tp is defined as follows.

$$\begin{aligned}
\text{dual}(p_i) &\rightleftharpoons p_i \\
\text{dual}(A \bullet B) &\rightleftharpoons \text{dual}(B) \bullet \text{dual}(A) \\
\text{dual}(A \backslash B) &\rightleftharpoons \text{dual}(B) / \text{dual}(A) \\
\text{dual}(A/B) &\rightleftharpoons \text{dual}(B) \backslash \text{dual}(A)
\end{aligned}$$

The extension to sequences of types dual: $\text{Tp}^* \to \text{Tp}^*$ is defined as

$$\text{dual}(A_1 \ldots A_n) \rightleftharpoons \text{dual}(A_n) \ldots \text{dual}(A_1).$$

Lemma 1.1 *If $L^* \vdash \Gamma \to A$, then $L^* \vdash \text{dual}(\Gamma) \to \text{dual}(A)$.*

PROOF. Straightforward induction on the derivation of $\Gamma \to A$. ∎

2   Free monoid models

We use the following notation. Let $\mathcal{V}$ be any alphabet, i.e., any set, the elements of which are called symbols. We denote by $\mathcal{V}^+$ the set of all non-empty words over the alphabet $\mathcal{V}$. By $\mathcal{V}^*$ we denote the set of all words over the alphabet $\mathcal{V}$, including the *empty word* ε. Let $\circ$ denote concatenation. Evidently $\mathcal{V}^*$ is a free monoid w.r.t. $\circ$. The unit of the free monoid is ε. Throughout the paper calligraphic letters $\mathcal{U}$, $\mathcal{V}$, $\mathcal{W}$ will denote alphabets.

If α is a word, then $|\alpha|$ (the *length* of α) is the number of symbols in α.

We shall use the following shorthand notation. For any sets $\mathcal{R} \subseteq \mathcal{V}^*$ and $\mathcal{T} \subseteq \mathcal{V}^*$ we write

$$\mathcal{R} \circ \mathcal{T} \rightleftharpoons \{\gamma \in \mathcal{V}^* \mid \text{ there are } \alpha \in \mathcal{R} \text{ and } \beta \in \mathcal{T} \text{ such that } \alpha \circ \beta = \gamma\};$$

$$\mathcal{R} \circ \beta \rightleftharpoons \mathcal{R} \circ \{\beta\}; \qquad \alpha \circ \mathcal{T} \rightleftharpoons \{\alpha\} \circ \mathcal{T}.$$

Since this operation on sets is associative, we omit parentheses in expressions like $\mathcal{R}_1 \circ \mathcal{R}_2 \circ \ldots \circ \mathcal{R}_m$. In the case of $m = 0$ we assume that this expression stands for the set $\{\varepsilon\}$. By $\mathcal{R}^m$ we denote the set $\underbrace{\mathcal{R} \circ \ldots \circ \mathcal{R}}_{m \text{ times}}$.

We shall denote the set of all subsets of a set $\mathcal{S}$ by $\mathbf{P}(\mathcal{S})$.

Definition. A free monoid model $\langle \mathcal{W}^*, w \rangle$ consists of the free monoid $\langle \mathcal{W}^*, \circ, \varepsilon \rangle$ and a valuation $w \colon \mathrm{Tp} \to \mathbf{P}(\mathcal{W}^*)$ associating with each type of L^* a subset of $\mathcal{W}^*$ and satisfying for any types A and B the following conditions.

(1) $w(A\bullet B) = w(A) \circ w(B)$

(2) $w(A\backslash B) = \{\gamma \in \mathcal{W}^* \mid w(A) \circ \gamma \subseteq w(B)\}$

(3) $w(B/A) = \{\gamma \in \mathcal{W}^* \mid \gamma \circ w(A) \subseteq w(B)\}$

For any function $w \colon \mathrm{Tp} \to \mathbf{P}(\mathcal{W}^*)$ and for any types $A_1, \ldots, A_n$, we write $\vec{w}(A_1 \ldots A_n)$ as a shorthand for $w(A_1) \circ \ldots \circ w(A_n)$. Note that $\vec{w}(\Lambda) = \{\varepsilon\}$.

Definition. A sequent $\Gamma{\to}B$ is *true* in a model $\langle \mathcal{W}^*, w \rangle$ iff $\vec{w}(\Gamma) \subseteq w(B)$. A sequent is *false* in a model iff it is not true in the model.

The following well-known soundness theorem holds.

Theorem 2.1 *If a sequent is derivable in the calculus L^*, then the sequent is true in every free monoid model.*

The rest of the paper is devoted to the proof of the corresponding completeness theorem. In view of the following two lemmas it is sufficient to consider only sequents with empty antecedent.

Lemma 2.2 *For any types A_1, ..., A_n, B, the sequent $A_1 \ldots A_n \to B$ is derivable in L^* if and only if the sequent $\Lambda \to (A_1 \bullet \ldots \bullet A_n) \backslash B$ is derivable in L^*.*

Lemma 2.3 *For any free monoid model $\langle \mathcal{W}^*, w \rangle$ and for any types A_1, ..., A_n, B, the sequent $A_1 \ldots A_n \to B$ is true in $\langle \mathcal{W}^*, w \rangle$ if and only if the sequent $\Lambda \to (A_1 \bullet \ldots \bullet A_n) \backslash B$ is true in $\langle \mathcal{W}^*, w \rangle$.*

3   Quasimodels

In this section we introduce the notion of $\mathrm{Tp}(m)$-quasimodels and describe an algorithm of constructing a free monoid model as the limit of an infinite sequence of $\mathrm{Tp}(m)$-quasimodels, which are conservative extensions of each other.

Definition. The *length* of a type is defined as the total number of primitive type occurrences in the type.

$$\|p_i\| \rightleftharpoons 1 \qquad \|A \bullet B\| \rightleftharpoons \|A\| + \|B\|$$

$$\|A \backslash B\| \rightleftharpoons \|A\| + \|B\| \qquad \|A/B\| \rightleftharpoons \|A\| + \|B\|$$

Similarly, for sequences of types we put $\|A_1 \ldots A_n\| \rightleftharpoons \|A_1\| + \ldots + \|A_n\|$.

Definition. The set of primitive types *occurring* in a type is defined as follows.

$$\mathrm{Var}(p_i) \rightleftharpoons \{p_i\} \qquad \mathrm{Var}(A \bullet B) \rightleftharpoons \mathrm{Var}(A) \cup \mathrm{Var}(B)$$

$$\mathrm{Var}(A \backslash B) \rightleftharpoons \mathrm{Var}(A) \cup \mathrm{Var}(B) \qquad \mathrm{Var}(A/B) \rightleftharpoons \mathrm{Var}(A) \cup \mathrm{Var}(B)$$

Definition. For any integer m, we write $\mathrm{Tp}(m)$ for the finite set of types

$$\mathrm{Tp}(m) \rightleftharpoons \{A \in \mathrm{Tp} \mid \mathrm{Var}(A) \subseteq \{p_1, p_2, \ldots, p_m\} \text{ and } \|A\| \leq m\}.$$

By $\mathrm{Tp}(m)^*$ we denote the set of all finite sequences of types from $\mathrm{Tp}(m)$.

Definition. A $\mathrm{Tp}(m)$-*quasimodel* $\langle \mathcal{W}^*, w \rangle$ is a valuation $w \colon \mathrm{Tp} \to \mathbf{P}(\mathcal{W}^*)$ over a free monoid $\langle \mathcal{W}^*, \circ, \varepsilon \rangle$ such that

(1) for any $A \in \mathrm{Tp}$ and $B \in \mathrm{Tp}$, if $A \bullet B \in \mathrm{Tp}(m)$, then $w(A \bullet B) \subseteq w(A) \circ w(B)$;

(2) for any $\Gamma \in \mathrm{Tp}(m)^*$ and $A \in \mathrm{Tp}(m)$, if $L^* \vdash \Gamma \to A$, then $\vec{w}(\Gamma) \subseteq w(A)$;

(3) for any $A \in \mathrm{Tp}(m)$, if $\varepsilon \in w(A)$, then $L^* \vdash \Lambda \to A$.

Lemma 3.1 *Let $\langle \mathcal{W}^*, w \rangle$ be a $\mathrm{Tp}(m)$-quasimodel. Then the following statements hold.*

(i) *If $A \cdot B \in \mathrm{Tp}(m)$, then $w(A \cdot B) = w(A) \circ w(B)$.*

(ii) *If $A\backslash B \in \mathrm{Tp}(m)$, then $w(A\backslash B) \subseteq \{\gamma \in W^* \mid w(A) \circ \gamma \subseteq w(B)\}$.*

(iii) *If $B/A \in \mathrm{Tp}(m)$, then $w(B/A) \subseteq \{\gamma \in W^* \mid \gamma \circ w(A) \subseteq w(B)\}$.*

PROOF. It is sufficient to note that $L^* \vdash AB{\to}A \cdot B$, $L^* \vdash A(A\backslash B){\to}B$, $L^* \vdash B(B/A){\to}B$, and use (2) from the definition of a $\mathrm{Tp}(m)$-quasimodel. ∎

Definition. A sequent $\Gamma{\to}A$ is *true* in a $\mathrm{Tp}(m)$-quasimodel $\langle W^*, w \rangle$ iff $\vec{w}(\Gamma) \subseteq w(A)$.

Definition. A $\mathrm{Tp}(m)$-quasimodel $\langle W^*, w \rangle$ is a *conservative extension* of another $\mathrm{Tp}(m)$-quasimodel $\langle V^*, v \rangle$ iff

(1) $V \subseteq W$;

(2) $w(A) \cap V^* = v(A)$ for any type A.

Evidently, if $\langle W^*, w \rangle$ is a conservative extension of $\langle V^*, v \rangle$, then for any type A we have $v(A) \subseteq w(A)$.

Lemma 3.2 *If $\langle W_2^*, w_2 \rangle$ is a conservative extension of $\langle W_1^*, w_1 \rangle$ and $\langle W_3^*, w_3 \rangle$ is a conservative extension of $\langle W_2^*, w_2 \rangle$, then $\langle W_3^*, w_3 \rangle$ is a conservative extension of $\langle W_1^*, w_1 \rangle$.*

PROOF. In view of $W_1^* \subseteq W_2^*$ we have $w_3(A) \cap W_1^* = w_3(A) \cap (W_2^* \cap W_1^*) = (w_3(A) \cap W_2^*) \cap W_1^*$. Further, $(w_3(A) \cap W_2^*) \cap W_1^* = w_2(A) \cap W_1^* = w_1(A)$. ∎

We shall denote by **Z** the set of all integers and by **N** the set of all natural numbers, including zero.

Definition. We say that a sequence of $\mathrm{Tp}(m)$-quasimodels $\langle W_i^*, w_i \rangle$ ($i \in$ **N**) is *conservative* iff, for every $i \in$ **N**, $\langle W_{i+1}^*, w_{i+1} \rangle$ is a conservative extension of $\langle W_i^*, w_i \rangle$. (Here m is constant.)

Definition. The *limit* of a conservative sequence $\langle W_i^*, w_i \rangle$ ($i \in$ **N**) is the $\mathrm{Tp}(m)$-quasimodel $\langle W_\infty^*, w_\infty \rangle$ defined as follows.

(1) $W_\infty \rightleftharpoons \bigcup_{i \in \mathbf{N}} W_i$

(2) $w_\infty(A) \rightleftharpoons \bigcup_{i \in \mathbf{N}} w_i(A)$

Lemma 3.3 *The definition of the limit is correct, i.e., $\langle W_\infty^*, w_\infty \rangle$ is really a $\mathrm{Tp}(m)$-quasimodel.*

PROOF.

(1) Let $A{\bullet}B \in \mathrm{Tp}(m)$ and $\gamma \in w_\infty(A{\bullet}B)$. Then for some n we have $\gamma \in w_n(A{\bullet}B) \subseteq w_n(A) \circ w_n(B)$. Thus $\gamma = \alpha \circ \beta$, where $\alpha \in w_n(A)$ and $\beta \in w_n(B)$. Evidently $\alpha \in w_\infty(A)$ and $\beta \in w_\infty(B)$, whence $\gamma = \alpha \circ \beta \in w_\infty(A) \circ w_\infty(B)$.

(2) Let $L^* \vdash A_1 \ldots A_l {\to} B$, where $A_1 \in \mathrm{Tp}(m)$, ..., $A_l \in \mathrm{Tp}(m)$, and $B \in \mathrm{Tp}(m)$. Assume that $\gamma \in \vec{w}_\infty(A_1 \ldots A_l)$, i.e., $\gamma = \alpha_1 \circ \ldots \circ \alpha_l$, where $\alpha_1 \in w_\infty(A_1)$, ..., $\alpha_l \in w_\infty(A_l)$. Then $\alpha_1 \in w_{i_1}(A_1)$, ..., $\alpha_l \in w_{i_l}(A_l)$ for some $i_1, \ldots, i_l \in \mathbf{N}$. Put $n \rightleftharpoons \max(i_1, \ldots, i_l)$.

According to Lemma 3.2, $\alpha_1 \in w_n(A_1)$, ..., $\alpha_l \in w_n(A_l)$, whence $\gamma = \alpha_1 \circ \ldots \circ \alpha_l \in \vec{w}_n(A_1 \ldots A_l) \subseteq w_n(B) \subseteq w_\infty(B)$.

(3) Obvious. ∎

Lemma 3.4 *The limit of a conservative sequence is a conservative extension of any of the elements of the sequence.*

PROOF. We verify that $w_\infty(A) \cap \mathcal{W}_i^* = w_i(A)$. For any $k \le i$ we have $w_k(A) \subseteq w_i(A)$. Thus $w_\infty(A) = \bigcup_j w_j(A) = \bigcup_{j \ge i} w_j(A)$, whence $w_\infty(A) \cap \mathcal{W}_i^* = (\bigcup_{j \ge i} w_j(A)) \cap \mathcal{W}_i^* = \bigcup_{j \ge i} (w_j(A) \cap \mathcal{W}_i^*)$. Note that $w_j(A) \cap \mathcal{W}_i^* = w_i(A)$ for any $j \ge i$ (according to Lemma 3.2). Now $\bigcup_{j \ge i} (w_j(A) \cap \mathcal{W}_i^*) = \bigcup_{j \ge i} w_i(A) = w_i(A)$. ∎

4   A simple quasimodel

Definition. We define the *non-negative count* $\bar{\#}$ as the following mapping from types to non-negative integers.

$$\bar{\#}p_i \;\rightleftharpoons\; 1$$
$$\bar{\#}(A{\bullet}B) \;\rightleftharpoons\; \bar{\#}A + \bar{\#}B$$
$$\bar{\#}(A\backslash B) \;\rightleftharpoons\; \begin{cases} \max(0, \bar{\#}B - \bar{\#}A), & \text{if } L^* \vdash \Lambda{\to}A\backslash B \\ \max(1, \bar{\#}B - \bar{\#}A), & \text{if } L^* \not\vdash \Lambda{\to}A\backslash B \end{cases}$$
$$\bar{\#}(A/B) \;\rightleftharpoons\; \begin{cases} \max(0, \bar{\#}A - \bar{\#}B), & \text{if } L^* \vdash \Lambda{\to}A/B \\ \max(1, \bar{\#}A - \bar{\#}B), & \text{if } L^* \not\vdash \Lambda{\to}A/B \end{cases}$$

The non-negative count of a sequence of types is defined in the natural way.

$$\bar{\#}(A_1 \ldots A_l) \rightleftharpoons \bar{\#}A_1 + \ldots + \bar{\#}A_l$$

By definition, $\bar{\#}(\Lambda) \rightleftharpoons 0$.

Lemma 4.1 *For any type A its non-negative satisfies the inequalities $0 \le \bar{\#}A \le \|A\|$.*

PROOF. Induction on $\|A\|$. ∎

Lemma 4.2 *If $\bar{\#}A = 0$, then $L^* \vdash \Lambda{\to}A$.*

PROOF. Induction on $\|A\|$. Induction steps for $B\backslash C$ and B/C are easy. In the case of $B{\cdot}C$ we assume that $\bar{\#}(B{\cdot}C) = 0$ and obtain $\bar{\#}B = 0$, $\bar{\#}C = 0$, and

$$\frac{\Lambda{\to}B \quad \Lambda{\to}C}{\Lambda{\to}B{\cdot}C} \; ({\to}{\cdot}).$$

∎

Lemma 4.3 *If $L^* \vdash \Gamma{\to}A$ then $\bar{\#}\Gamma \geq \bar{\#}A$.*

PROOF. Induction on the length of the derivation.
CASE 1: Axiom.
Obvious.
CASE 2: $({\to}\backslash)$ Given $\dfrac{A\ \Pi{\to}B}{\Pi{\to}A\backslash B} \; ({\to}\backslash)$.

By the induction hypothesis $\bar{\#}A + \bar{\#}\Pi \geq \bar{\#}B$, whence $\bar{\#}\Pi \geq \bar{\#}B - \bar{\#}A$. Obviously, if $\bar{\#}\Pi \geq 1$, then $\bar{\#}\Pi \geq \max(1, \bar{\#}B - \bar{\#}A) \geq \bar{\#}(A\backslash B)$. Let now $\bar{\#}\Pi = 0$ and $\Pi = C_1 \ldots C_n$. Then $\bar{\#}C_i = 0$ for each $i \leq n$. According to Lemma 4.2 $L^* \vdash \Lambda{\to}C_i$ for each $i \leq n$. Applying (CUT) n times we derive $L^* \vdash \Lambda{\to}A\backslash B$, whence $\bar{\#}(A\backslash B) = \max(0, \bar{\#}B - \bar{\#}A)$. From $\bar{\#}B - \bar{\#}A \leq \bar{\#}\Pi = 0$ we obtain $\bar{\#}(A\backslash B) = 0$.
CASE 3: $({\to}/)$
Similar.
CASE 4: $(\backslash{\to})$ Given $\dfrac{\Phi{\to}A \quad \Gamma B\Delta{\to}C}{\Gamma\Phi(A\backslash B)\Delta{\to}C} \; (\backslash{\to})$.

By the induction hypothesis $\bar{\#}\Phi \geq \bar{\#}A$ and $\bar{\#}\Gamma + \bar{\#}B + \bar{\#}\Delta \geq \bar{\#}C$.
Note that $\bar{\#}(A\backslash B) \geq \bar{\#}B - \bar{\#}A$.
Hence $\bar{\#}\Gamma + \bar{\#}\Phi + \bar{\#}(A\backslash B) + \bar{\#}\Delta \geq \bar{\#}\Gamma + \bar{\#}A + (\bar{\#}B - \bar{\#}A) + \bar{\#}\Delta \geq \bar{\#}C$.
CASE 5: $(/{\to})$
Similar.
CASE 6: $({\to}{\cdot})$ Given $\dfrac{\Gamma{\to}A \quad \Delta{\to}B}{\Gamma\Delta{\to}A{\cdot}B} \; ({\to}{\cdot})$.

If $\bar{\#}\Gamma \geq \bar{\#}A$ and $\bar{\#}\Delta \geq \bar{\#}B$, then $\bar{\#}\Gamma + \bar{\#}\Delta \geq \bar{\#}A + \bar{\#}B = \bar{\#}(A{\cdot}B)$.
CASE 7: $({\cdot}{\to})$ Given $\dfrac{\Gamma AB\Delta{\to}C}{\Gamma(A{\cdot}B)\Delta{\to}C} \; ({\cdot}{\to})$.

Evidently $\bar{\#}(\Gamma(A{\cdot}B)\Delta) = \bar{\#}(\Gamma AB\Delta)$. ∎

Remark. For any type A, we have $\bar{\#}A = 0$ if and only if $L^* \vdash \Lambda{\to}A$.

Now we define a $\mathrm{Tp}(m)$-quasimodel $\langle \mathcal{W}_0^*, w_0 \rangle$.

$$\mathcal{W}_0 \rightleftharpoons \{a_0\} \qquad w_0(A) \rightleftharpoons \{a_0^k \mid k \geq \bar{\#}A\}$$

Here a_0^k denotes $\underbrace{a_0 \circ \ldots \circ a_0}_{k \text{ times}}$. In particular, $a_0^0 = \varepsilon$.

Lemma 4.4 $\langle \mathcal{W}_0^*, w_0 \rangle$ *is a* Tp(m)-*quasimodel for any natural number* m.

PROOF. (1) We prove that $w_0(A \bullet B) \subseteq w_0(A) \circ w_0(B)$.
Let $a_0^k \in w_0(A \bullet B)$. We put $k_1 \rightleftharpoons \#A$ and $k_2 \rightleftharpoons k - k_1$. In view of $k \geq \#(A \bullet B) = \#A + \#B$ we have $k_2 \geq \#B$. Evidently $a_0^k = a_0^{k_1} \circ a_0^{k_2}$, $a_0^{k_1} \in w_0(A)$ and $a_0^{k_2} \in w_0(B)$.

(2) We verify that if $L^* \vdash C_1 \ldots C_n \rightarrow A$, then $w_0(C_1) \circ \ldots \circ w_0(C_n) \subseteq w_0(A)$.
Let $a_0^{k_i} \in w_0(C_i)$ for every $i \leq n$. Then $\sum k_i \geq \sum \#C_i \geq \#A$ according to Lemma 4.3. Thus $a_0^{k_1} \circ \ldots \circ a_0^{k_n} \in w_0(A)$.

(3) In view of Lemma 4.2, if $\varepsilon \in w_0(A)$, then $L^* \vdash \Lambda \rightarrow A$. ∎

5   Witnesses

Definition. We fix a countable alphabet $\mathcal{U} = \{a_j \mid j \in \mathbf{N}\}$. By $\mathcal{K}^m$ we denote the class of all Tp(m)-quasimodels $\langle \mathcal{V}^*, v \rangle$, such that $\mathcal{V} \subset \mathcal{U}$, $\mathcal{V}$ is finite, and for every $A \in$ Tp(m) there is $\alpha \in v(A)$ satisfying $|\alpha| \leq m$.

Lemma 5.1 *The* Tp(m)-*quasimodel* $\langle \mathcal{W}_0^*, w_0 \rangle$ *from Lemma 4.4 belongs to the class* $\mathcal{K}^m$.

PROOF. Immediate from Lemma 4.1. ∎

Definition. Let $\langle \mathcal{W}^*, w \rangle$ be a Tp(m)-quasimodel. Let $A, B \in$ Tp, $\alpha \in \mathcal{W}^*$, $\gamma \in \mathcal{W}^*$, and $\gamma \notin w(A \backslash B)$. We say that α is a *witness of* $\gamma \notin w(A \backslash B)$ iff $\alpha \in w(A)$ and $\alpha \circ \gamma \notin w(B)$.

Definition. Let $\langle \mathcal{W}^*, w \rangle$ be a Tp(m)-quasimodel. Let $A, B \in$ Tp, $\alpha \in \mathcal{W}^*$, $\gamma \in \mathcal{W}^*$, and $\gamma \notin w(B/A)$. We say that α is a *witness of* $\gamma \notin w(B/A)$ iff $\alpha \in w(A)$, and $\gamma \circ \alpha \notin w(B)$.

Definition. Let $\mathcal{K}$ be a class of Tp(m)-quasimodels. We say that the class $\mathcal{K}$ is *witnessed* iff

(1) for any $\langle \mathcal{V}^*, v \rangle \in \mathcal{K}$, for any type of the form $A \backslash B$ from Tp(m), and for any $\gamma \in \mathcal{V}^*$, if $\gamma \notin v(A \backslash B)$ then there is a conservative extension $\langle \mathcal{W}^*, w \rangle$ of $\langle \mathcal{V}^*, v \rangle$ in $\mathcal{K}$ and $\langle \mathcal{W}^*, w \rangle$ contains a witness of $\gamma \notin w(A \backslash B)$;

(2) for any $\langle \mathcal{V}^*, v \rangle \in \mathcal{K}$, for any type of the form B/A from Tp(m), and for any $\gamma \in \mathcal{V}^*$, if $\gamma \notin v(B/A)$ then there is a conservative extension $\langle \mathcal{W}^*, w \rangle$ of $\langle \mathcal{V}^*, v \rangle$ in $\mathcal{K}$ and $\langle \mathcal{W}^*, w \rangle$ contains a witness of $\gamma \notin w(B/A)$.

Lemma 5.2 *If the class* $\mathcal{K}^m$ *is witnessed, then there is a free monoid model* $\langle \mathcal{V}^*, v \rangle$ *such that*

(i) *for every type $E \in \mathrm{Tp}(m)$, if $L^* \nvdash \Lambda \to E$, then the sequent $\Lambda \to E$ is false in $\langle V^*, v \rangle$;*

(ii) $v(E) \neq \emptyset$ *for every type $E \in \mathrm{Tp}(m)$;*

(iii) $V \subseteq U$.

At the end of this paper it will be proved that the class $\mathcal{K}^m$ is witnessed. Thus Lemma 5.2 (i) provides a proof of completeness of L^* with respect to free monoid models.

PROOF. Evidently there is a function $\sigma: \mathbf{N} \to \mathrm{Tp}(m) \times U^*$ such that for any $C \in \mathrm{Tp}(m)$ and for any $\gamma \in U^*$ there are infinitely many natural numbers i, for which $\sigma(i) = \langle C, \gamma \rangle$.

Starting with the $\mathrm{Tp}(m)$-quasimodel $\langle \mathcal{W}_0^*, w_0 \rangle$ from Lemma 4.4, we define by induction on i a conservative sequence $\langle \mathcal{W}_i^*, w_i \rangle$ ($i \in \mathbf{N}$), consisting of $\mathrm{Tp}(m)$-quasimodels from the class $\mathcal{K}^m$.

Assume that $\langle \mathcal{W}_i^*, w_i \rangle \in \mathcal{K}^m$ has been constructed. We define $\langle \mathcal{W}_i^*, w_i \rangle$ as follows.

CASE 1:
If $\sigma(i) = \langle A \backslash B, \gamma \rangle$, $\gamma \in \mathcal{W}_i^*$, $\gamma \notin w_i(A \backslash B)$, and there are no witnesses of $\gamma \notin w_i(A \backslash B)$ in $\langle \mathcal{W}_i^*, w_i \rangle$, then we take $\langle \mathcal{W}_{i+1}^*, w_{i+1} \rangle$ to be any conservative extension of $\langle \mathcal{W}_i^*, w_i \rangle$ in $\mathcal{K}^m$, containing a witness of $\gamma \notin w_{i+1}(A \backslash B)$. Such a $\mathrm{Tp}(m)$-quasimodel $\langle \mathcal{W}_{i+1}^*, w_{i+1} \rangle$ exists, since $\mathcal{K}^m$ is witnessed.

CASE 2:
If $\sigma(i) = \langle B/A, \gamma \rangle$, $\gamma \in \mathcal{W}_i^*$, $\gamma \notin w_i(B/A)$, and there are no witnesses of $\gamma \notin w_i(B/A)$ in $\langle \mathcal{W}_i^*, w_i \rangle$, then we take $\langle \mathcal{W}_{i+1}^*, w_{i+1} \rangle$ to be any conservative extension of $\langle \mathcal{W}_i^*, w_i \rangle$ in $\mathcal{K}^m$, containing a witness of $\gamma \notin w_{i+1}(B/A)$.

CASE 3:
Otherwise we put $\langle \mathcal{W}_{i+1}^*, w_{i+1} \rangle \rightleftharpoons \langle \mathcal{W}_i^*, w_i \rangle$.

Let $\langle \mathcal{W}_\infty^*, w_\infty \rangle$ be the limit of the conservative sequence $\langle \mathcal{W}_i^*, w_i \rangle$. We put $V \rightleftharpoons \mathcal{W}_\infty$.

Now we define a valuation $v: \mathrm{Tp} \to \mathbf{P}(\mathcal{W}_\infty^*)$ by induction on the complexity of a type.

$$
\begin{aligned}
v(p_i) &\rightleftharpoons w_\infty(p_i) \\
v(A \cdot B) &\rightleftharpoons v(A) \circ v(B) \\
v(A \backslash B) &\rightleftharpoons \{\gamma \in V^* \mid v(A) \circ \gamma \subseteq v(B)\} \\
v(B/A) &\rightleftharpoons \{\gamma \in V^* \mid \gamma \circ v(A) \subseteq v(B)\}
\end{aligned}
$$

Evidently $\langle V^*, v \rangle$ is a free monoid model. Next we verify by induction on the complexity of C that $w_\infty(C) = v(C)$ for every $C \in \mathrm{Tp}(m)$.

Induction step.

CASE 1: $C = A \cdot B$
Obvious from Lemma 3.1 (i).

CASE 2: $C = A\backslash B$

First we prove that if $\gamma \in w_\infty(A\backslash B)$ then $\gamma \in v(A\backslash B)$. Let $\gamma \in w_\infty(A\backslash B)$. Take any $\alpha \in v(A)$. By the induction hypothesis $\alpha \in w_\infty(A)$. Evidently $\alpha \circ \gamma \in w_\infty^{\to}(A(A\backslash B))$. Hence $\alpha \circ \gamma \in w_\infty(B)$ in view of $L^* \vdash A(A\backslash B) \to B$. By the induction hypothesis $\alpha \circ \gamma \in v(B)$. Thus $\gamma \in v(A\backslash B)$.

Now we prove that if $\gamma \in \mathcal{U}^*$ and $\gamma \notin w_\infty(A\backslash B)$ then $\gamma \notin v(A\backslash B)$. If $\gamma \notin \mathcal{W}_\infty^*$, then this is obvious. Let now $\gamma \in \mathcal{W}_\infty^*$. Recall that $\mathcal{W}_\infty = \bigcup\limits_{j \in \mathbf{N}} \mathcal{W}_j$. Thus $\gamma \in \mathcal{W}_j^*$ for some j. Evidently, there exists an integer $i \geq j$ such that $\sigma(i) = \langle A\backslash B, \gamma \rangle$. According to the construction of $\langle \mathcal{W}_{i+1}^*, w_{i+1} \rangle$ there is a witness $\alpha \in \mathcal{W}_{i+1}^*$ of $\gamma \notin w_\infty(A\backslash B)$. That is, $\alpha \in w_{i+1}(A)$ and $\alpha \circ \gamma \notin w_{i+1}(B)$. Since w_∞ is conservative over w_{i+1}, we have $\alpha \in w_\infty(A)$ and $\alpha \circ \gamma \notin w_\infty(B)$. By the induction hypothesis, $\alpha \in v(A)$ and $\alpha \circ \gamma \notin v(B)$. Thus $\gamma \notin v(A\backslash B)$.

CASE 3: $C = B/A$

Similar to the previous case.

Finally, we prove that the free monoid model $\langle \mathcal{V}^*, v \rangle$ has the desired properties (i)–(iii).

(i) Let $E \in \mathrm{Tp}(m)$ and $L^* \nvdash \Lambda \to E$. We must prove that $\varepsilon \notin v(E)$. According to the definition of a $\mathrm{Tp}(m)$-quasimodel $\varepsilon \notin w_0(E)$. In view of Lemma 3.4 $\varepsilon \notin w_\infty(E) = v(E)$.

(ii) If $E \in \mathrm{Tp}(m)$, then $a_0^m \in w_0(E) \subseteq w_\infty(E) = v(E)$.

(iii) Obvious. ∎

6   Noncommutative linear logic

In this paper we consider only the multiplicative fragment of linear logic.

Noncommutative multiplicative linear formulas are defined as follows. We assume that an enumerable set of *variables* Var $= \{p_1, p_2, \ldots\}$ is given. We introduce the set of formal symbols called *atoms*

$$\mathrm{At} \rightleftharpoons \{p^{\perp n} \mid p \in \mathrm{Var}, \ n \in \mathbf{Z}\}.$$

Intuitively, if $n \geq 0$, then $p^{\perp n}$ means 'p with n right negations' and $p^{\perp(-n)}$ means 'p with n left negations'.

Definition. The set of *normal formulas* (or just *formulas* for shortness) is defined to be the smallest set NFm satisfying the following conditions:

1. At $\subset$ NFm;

2. $\perp \in$ NFm;

3. $1 \in$ NFm;

4. if $A \in$ NFm and $B \in$ NFm, then $(A \otimes B) \in$ NFm and $(A \wp B) \in$ NFm.

Here $\otimes$ is the multiplicative conjunction, called 'tensor', and $\wp$ is the multiplicative disjunction, called 'par'. The constants $\perp$ and 1 are multiplicative falsity and multiplicative truth respectively.

By NFm* we denote the set of all finite sequences of normal formulas. The empty sequence is denoted by Λ.

Definition. We define by induction the right negation $(\ )^{\perp}:$ NFm $\to$ NFm and the left negation $^{\perp}(\ ):$ NFm $\to$ NFm.

$$(p^{\perp n})^{\perp} \;\rightleftharpoons\; p^{\perp(n+1)}$$
$$\perp^{\perp} \;\rightleftharpoons\; 1$$
$$1^{\perp} \;\rightleftharpoons\; \perp$$
$$(A \otimes B)^{\perp} \;\rightleftharpoons\; (B^{\perp}) \wp (A^{\perp})$$
$$(A \wp B)^{\perp} \;\rightleftharpoons\; (B^{\perp}) \otimes (A^{\perp})$$

$$^{\perp}(p^{\perp n}) \;\rightleftharpoons\; p^{\perp(n-1)}$$
$$^{\perp}\!\perp \;\rightleftharpoons\; 1$$
$$^{\perp}1 \;\rightleftharpoons\; \perp$$
$$^{\perp}(A \otimes B) \;\rightleftharpoons\; (^{\perp}B) \wp (^{\perp}A)$$
$$^{\perp}(A \wp B) \;\rightleftharpoons\; (^{\perp}B) \otimes (^{\perp}A)$$

The two negations are extended to sequences of normal formulas as follows.

$$^{\perp}(A_1 \ldots A_n) \;\rightleftharpoons\; {}^{\perp}A_n \ldots {}^{\perp}A_1$$
$$(A_1 \ldots A_n)^{\perp} \;\rightleftharpoons\; A_n^{\perp} \ldots A_1^{\perp}$$

Remark. Several other connectives can be defined in this logic. The most popular ones are two linear implications, defined as

$$A \multimap B \rightleftharpoons A^{\perp} \wp B \quad \text{and} \quad B \multimapinv A \rightleftharpoons B \wp {}^{\perp}A.$$

Lemma 6.1 *For any $A \in$ NFm the equalities $^{\perp}(A^{\perp}) = A$ and $(^{\perp}A)^{\perp} = A$ hold true.*

PROOF. Easy induction on the structure of A. ∎

In [1] V. M. Abrusci introduced a sequent calculus PNCL for the pure noncommutative classical linear propositional logic. In the same paper two one-sided sequent calculi SPNCL and SPNCL' were introduced and it was proved that they are equivalent to PNCL.

We shall use a slightly modified (but equivalent) version of the multiplicative fragment of SPNCL'. The sequents of this calculus are of the form $\to \Gamma$, where $\Gamma \in$ NFm*.

The calculus SPNCL$'$ has the following axioms and rules.

$$\frac{}{\to(p^{\perp(n+1)})(p^{\perp n})}\ (id) \qquad\qquad \frac{}{\to\mathbf{1}}\ (1)$$

$$\frac{\to\Gamma\Delta}{\to\Gamma\perp\Delta}\ (\perp)$$

$$\frac{\to\Gamma AB\Delta}{\to\Gamma(A\wp B)\Delta}\ (\wp) \qquad\qquad \frac{\to\Gamma A \quad \to B\Delta}{\to\Gamma(A\otimes B)\Delta}\ (\otimes)$$

$$\frac{\to A\Gamma}{\to\Gamma(^{\perp\perp}A)}\ (^{\perp\perp}(\,\cdot\,)) \qquad\qquad \frac{\to\Gamma A}{\to(A^{\perp\perp})\Gamma}\ ((\,\cdot\,)^{\perp\perp})$$

Here capital letters A, B, ... stand for formulas, capital Greek letters denote finite (possibly empty) sequences of formulas, p ranges over Var, and n ranges over $\mathbf{Z}$.

Remark. The rule (id) can be written as $\to(C^\perp)C$ (or equivalently as $\to C(^\perp C)$), where $C \in$ At. Actually, the restriction $C \in$ At is not essential. It is imposed in this paper only in order to reduce the number of technical details in some proofs.

We define an embedding of L^* into SPNCL$'$.

Definition. The function $\widehat{(\ )}$: Tp $\to$ NFm is defined as follows.

$$\begin{aligned}
\widehat{p_i} &\rightleftharpoons p_i \\
\widehat{A\bullet B} &\rightleftharpoons \widehat{A}\otimes\widehat{B} \\
\widehat{A\backslash B} &\rightleftharpoons \widehat{A}^\perp \wp \widehat{B} \\
\widehat{A/B} &\rightleftharpoons \widehat{A}\wp {}^\perp\widehat{B}
\end{aligned}$$

If $\Gamma = A_1\ldots A_n$, then by $\widehat{\Gamma}$ we denote the sequence $\widehat{A_1}\ldots\widehat{A_n}$.

Lemma 6.2 *For every normal formula $A \in$ NFm there is at most one type $B \in$ Tp such that $\widehat{B} = A$.*

PROOF. We define a function $\natural$: NFm $\to \mathbf{Z}$ by induction as follows.

$$\begin{aligned}
\natural(p^{\perp n}) &\rightleftharpoons \begin{cases} 0, & \text{if } n \text{ is even} \\ 1, & \text{if } n \text{ is odd} \end{cases} \\
\natural\mathbf{1} &\rightleftharpoons 0 \\
\natural\perp &\rightleftharpoons 1 \\
\natural(A\otimes B) &\rightleftharpoons \natural A + \natural B \\
\natural(A\wp B) &\rightleftharpoons \natural A + \natural B - 1
\end{aligned}$$

It is easy to see that $\natural(A^\perp) = \natural(^\perp A) = 1 - \natural A$. Now we can verify that if $A \in \text{Tp}$, then $\natural\widehat{A} = 0$. Thus there are no types $A_1, A_2 \in \text{Tp}$ such that $\widehat{A_1}^\perp = \widehat{A_2}$ (because $\widehat{A_1}^\perp = 1$ and $\widehat{A_2} = 0$).

Given a formula $D = \widehat{C}$ (where $C \in \text{Tp}$), we can automatically decide what is the main connective in C. If $D \in \text{Var}$, then C is primitive. If the main connective of D is $\otimes$, then the main connective of C is $\bullet$. Finally, if $D = D_1 \wp D_2$, then the main connective of C is $\backslash$ or $/$, depending on whether $\natural D_1 = 1$ or $\natural D_1 = 0$. ∎

Lemma 6.3 *Let $\Gamma \in \text{Tp}^*$ and $A \in \text{Tp}$. The sequent $\Gamma \rightarrow A$ is derivable in L^* if and only if the sequent $\rightarrow \widehat{\Gamma}^\perp \widehat{A}$ is derivable in SPNCL'.*

PROOF. Both directions are proved using induction on derivation length. ∎

7 Proof nets

We define proof nets for the multiplicative fragment of the noncommutative classical linear propositional logic. The concept of proof net introduced here (an extension of that from [1]) appears to be mathematical folklore.

We prove that a sequent is derivable if and only if there exists a proof net for this sequent.

Definition. For the purposes of this paper it is convenient to measure the length of a normal formula using the function $\|\cdot\|\colon \text{NFm} \rightarrow \mathbf{N}$ defined in the following way.

$$\|p^{\perp n}\| \rightleftharpoons 2$$
$$\|\perp\| \rightleftharpoons 2$$
$$\|\mathbf{1}\| \rightleftharpoons 2$$
$$\|A \otimes B\| \rightleftharpoons \|A\| + \|B\|$$
$$\|A \wp B\| \rightleftharpoons \|A\| + \|B\|$$

Remark. We are going to define formally a total order on the set of all $\mathbf{1}$, $\perp$, $\otimes$, $\wp$ and atom occurrences in a formula (in fact this order coincides with the natural order from left to right). To make the forthcoming definition easier, we have used 2 instead of 1 in the base case in the definition of $\|\cdot\|$.

The definition of $\|\cdot\|$ is extended to finite sequences of formulas in the natural way.

$$\|A_1 \ldots A_n\| \rightleftharpoons \|A_1\| + \ldots + \|A_n\|$$

We put $\|\Lambda\| \rightleftharpoons 0$.

The number of formulas in a finite sequence Γ is denoted by $|\Gamma|$. Thus $|A_1 \ldots A_n| = n$.

To formalize the notion of *occurrences* of subformulas we introduce the set

$$\mathrm{Occ} \rightleftharpoons \mathrm{NFm} \times \mathbf{Z}.$$

A pair $\langle B, k\rangle \in \mathrm{Occ}$ will be intuitively interpreted as a subformula occurrence B. Here k in a way characterizes the position of B in the whole formula.

Definition. We define the function $c \colon \mathrm{NFm} \to \mathbf{N}$ (evaluating the "distance" of the "main connective" of a formula from its left end) formally as follows.

$$c(p^{\perp n}) \;\rightleftharpoons\; 1$$
$$c(\perp) \;\rightleftharpoons\; 1$$
$$c(1) \;\rightleftharpoons\; 1$$
$$c(A \otimes B) \;\rightleftharpoons\; \|A\|$$
$$c(A \wp B) \;\rightleftharpoons\; \|A\|$$

Definition. We define the binary relation 'α is a subformula of β' on the set Occ formally as the least transitive binary relation $\prec$ satisfying $\langle A, k - \|A\| + c(A)\rangle \prec \langle (A\lambda B), k\rangle$ and $\langle B, k + c(B)\rangle \prec \langle (A\lambda B), k\rangle$ for every $\lambda \in \{\otimes, \wp\}$, $A \in \mathrm{NFm}$, $B \in \mathrm{NFm}$, and $k \in \mathbf{Z}$.

Definition. The binary relation $\preceq$ on the set Occ is defined in the natural way: $\alpha \preceq \beta$ if and only if $\alpha \prec \beta$ or $\alpha = \beta$.

Given a standalone formula $A \in \mathrm{NFm}$, we usually associate it with the pair $\langle A, c(A)\rangle \in \mathrm{Occ}$. Then each subformula occurrence B is associated with a pair $\langle B, k\rangle \in \mathrm{Occ}$ such that $\langle B, k\rangle \preceq \langle A, c(A)\rangle$ and k is (intuitively) the "$\|\cdot\|$-distance" of the "main connective" of B from the left end of A.

Lemma 7.1 *Let $A \in \mathrm{NFm}$. Then*

(i) *the set $\{\alpha \in \mathrm{Occ} \mid \alpha \preceq \langle A, c(A)\rangle\}$ contains $\|A\| - 1$ elements;*

(ii) *for every $k \in \mathbf{Z}$ such that $0 < k < \|A\|$, there is a unique formula $B \in \mathrm{NFm}$ satisfying $\langle B, k\rangle \preceq \langle A, c(A)\rangle\}$.*

Definition. For any sequence of normal formulas $\Gamma = A_1 \ldots A_n$ we construct a finite set

$$\Omega_\Gamma \subset (\mathrm{NFm} \cup \{\diamond\}) \times \mathbf{N},$$

where $\diamond$ is a new formal symbol which does not belong to NFm.

The set Ω_Γ will act as the domain of all proof structures for the sequent $\to \Gamma$.

$$\Omega_\Gamma \;\rightleftharpoons\; \{\langle B, k + \|A_1 \ldots A_{i-1}\|\rangle \mid 1 \le i \le n \text{ and } \langle B, k\rangle \preceq \langle A_i, c(A_i)\rangle\}$$
$$\cup \{\langle \diamond, \|A_1 \ldots A_{i-1}\|, \rangle \mid 1 \le i \le n\}$$

Example 7.2 Let $\Gamma = ((q^{\perp 3} \otimes p^{\perp 8}) \wp p^{\perp 7}) q^{\perp 2}$. Then $\Omega_\Gamma = \{\alpha_0, \ldots, \alpha_7\}$, where

$$
\begin{aligned}
\alpha_0 &\rightleftharpoons \langle \diamond, 0 \rangle, \\
\alpha_1 &\rightleftharpoons \langle q^{\perp 3}, 1 \rangle, \\
\alpha_2 &\rightleftharpoons \langle (q^{\perp 3} \otimes p^{\perp 8}), 2 \rangle, \\
\alpha_3 &\rightleftharpoons \langle p^{\perp 8}, 3 \rangle, \\
\alpha_4 &\rightleftharpoons \langle ((q^{\perp 3} \otimes p^{\perp 8}) \wp p^{\perp 7}), 4 \rangle, \\
\alpha_5 &\rightleftharpoons \langle p^{\perp 7}, 5 \rangle, \\
\alpha_6 &\rightleftharpoons \langle \diamond, 6 \rangle, \\
\alpha_7 &\rightleftharpoons \langle q^{\perp 2}, 7 \rangle.
\end{aligned}
$$

The set Ω_Γ can be considered as consisting of six disjoint parts

$$
\Omega_\Gamma = \Omega_\Gamma^\diamond \cup \Omega_\Gamma^{At} \cup \Omega_\Gamma^\perp \cup \Omega_\Gamma^1 \cup \Omega_\Gamma^\otimes \cup \Omega_\Gamma^\wp,
$$

where

$$
\begin{aligned}
\Omega_\Gamma^\diamond &\rightleftharpoons \{\langle \diamond, k, \Pi \rangle \in \Omega_\Gamma\}; \\
\Omega_\Gamma^{At} &\rightleftharpoons \{\langle p^{\perp n}, k, \Pi \rangle \in \Omega_\Gamma\}; \\
\Omega_\Gamma^\perp &\rightleftharpoons \{\langle \perp, k, \Pi \rangle \in \Omega_\Gamma\}; \\
\Omega_\Gamma^1 &\rightleftharpoons \{\langle 1, k, \Pi \rangle \in \Omega_\Gamma\}; \\
\Omega_\Gamma^\otimes &\rightleftharpoons \{\langle A \otimes B, k, \Pi \rangle \in \Omega_\Gamma\}; \\
\Omega_\Gamma^\wp &\rightleftharpoons \{\langle A \wp B, k, \Pi \rangle \in \Omega_\Gamma\}.
\end{aligned}
$$

We shall often write $\Omega_\Gamma^{\wp\diamond}$ for $\Omega_\Gamma^\diamond \cup \Omega_\Gamma^\wp$.

Lemma 7.3 $|\Omega_\Gamma| = \|\Gamma\|$.

Definition. The invariant $\flat$, associating an integer with Ω_Γ, is defined as

$$
\flat(\Omega_\Gamma) \rightleftharpoons |\Omega_\Gamma^{\wp\diamond}| - |\Omega_\Gamma^\otimes| - |\Omega_\Gamma^\perp| + |\Omega_\Gamma^1|.
$$

Definition. For every subset Θ of Ω_Γ we put

$$
\flat(\Theta) \rightleftharpoons |\Omega_\Gamma^{\wp\diamond} \cap \Theta| - |\Omega_\Gamma^\otimes \cap \Theta| - |\Omega_\Gamma^\perp \cap \Theta| + |\Omega_\Gamma^1 \cap \Theta|.
$$

Remark. $\flat(\Omega_{C^\perp}) = \flat(\Omega_{\perp C}) = 2 - \flat(\Omega_C)$.

Lemma 7.4 *For all Γ*

(i) $|\Omega_\Gamma^{At}| + |\Omega_\Gamma^\perp| + |\Omega_\Gamma^1| = |\Omega_\Gamma^\otimes| + |\Omega_\Gamma^\wp| + |\Omega_\Gamma^\diamond|$;

(ii) *if $\rightarrow\Gamma$ is derivable in* SPNCL$'$, *then* $\flat(\Omega_\Gamma) = 2$.

PROOF. (i) Easy induction on $\|\Gamma\|$.
(ii) Straightforward induction on the length of the derivation in SPNCL'.
■

For each sequent $\to\Gamma$ we define two binary relations on Ω_Γ.

Definition. Let $\alpha \in \Omega_\Gamma$ and $\beta \in \Omega_\Gamma$. Then $\alpha \prec_\Gamma \beta$ if and only if $\alpha \notin \Omega_\Gamma^\circ$, $\beta \notin \Omega_\Gamma^\circ$, and $\alpha \prec \beta$.

Remark. The relation $\prec_\Gamma$ is a strict partial order on Ω_Γ.

Definition. Let $\langle A, k, \Delta \rangle \in \Omega_\Gamma$ and $\langle B, l, \Pi \rangle \in \Omega_\Gamma$. Then $\langle A, k, \Delta \rangle <_\Gamma \langle B, l, \Pi \rangle$ if and only if $k < l$.

Remark. The relation $<_\Gamma$ is an irreflexive linear order on Ω_Γ.

Definition. For any sequent $\to\Gamma$ we denote by $\boldsymbol{\Omega}_\Gamma$ the triple $\langle \Omega_\Gamma, \prec_\Gamma, <_\Gamma \rangle$.

Lemma 7.5 *Let $\alpha, \beta, \gamma \in \Omega_\Gamma$ and $\alpha <_\Gamma \beta <_\Gamma \gamma$.*

(i) *If $\alpha \prec_\Gamma \gamma$, then $\beta \prec_\Gamma \gamma$.*

(ii) *If $\gamma \prec_\Gamma \alpha$, then $\beta \prec_\Gamma \alpha$.*

Definition. Let $\langle \Omega, \mathcal{C} \rangle$ be an undirected graph, where Ω is the set of vertices and $\mathcal{C}$ is the set of edges. Let $<$ be a strict linear order on the set Ω. We say that the graph $\langle \Omega, \mathcal{C} \rangle$ is $<$-*planar* iff for every edge $\{\alpha, \beta\} \in \mathcal{C}$ and every edge $\{\gamma, \delta\} \in \mathcal{C}$, if $\alpha < \gamma < \beta$, then $\alpha < \delta < \beta$ or $\delta = \alpha$ or $\delta = \beta$.

Remark. Intuitively, a graph is $<$-planar if and only if its edges can be drawn without intersections on a semiplane while the vertices of the graph are ordered according to $<$ on the border of the semiplane.

Lemma 7.6 *If $\langle \Omega, \mathcal{C}_1 \rangle$ is $<$-planar and $\mathcal{C}_2 \subseteq \mathcal{C}_1$, then $\langle \Omega, \mathcal{C}_2 \rangle$ is $<$-planar.*

Lemma 7.7 *Let $\langle \Omega, \mathcal{C} \rangle$ be an undirected graph, where $\Omega = \Omega_1 \cup \Omega_2$ and $\Omega_1 \cap \Omega_2 = \emptyset$. Let $<$ and $<'$ be two linear orders on Ω such that*

$$(\forall \alpha \in \Omega_1)(\forall \beta \in \Omega_2)\ \alpha < \beta;$$
$$(\forall \alpha \in \Omega_1)(\forall \beta \in \Omega_2)\ \beta <' \alpha;$$
$$(\forall \alpha \in \Omega_1)(\forall \beta \in \Omega_1)\ \alpha < \beta \text{ iff } \alpha <' \beta;$$
$$(\forall \alpha \in \Omega_2)(\forall \beta \in \Omega_2)\ \alpha < \beta \text{ iff } \alpha <' \beta.$$

Then $\langle \Omega, \mathcal{C} \rangle$ is $<$-planar if and only if $\langle \Omega, \mathcal{C} \rangle$ is $<'$-planar.

Definition. If $\mathcal{C}$ is a set of directed edges, then by $\mathcal{C}^\#$ we denote the associated set of undirected edges.

$$\mathcal{C}^\# \rightleftharpoons \{\{\alpha, \beta\} \mid \langle \alpha, \beta \rangle \in \mathcal{C}\}$$

Definition. A *proof structure* is a quadruple $\langle \boldsymbol{\Omega}_\Gamma, \mathcal{A}, \mathcal{B}, \mathcal{E} \rangle$, where

(A1) $\mathcal{A} \subseteq \Omega_\Gamma^\otimes \times \Omega_\Gamma^{\wp\circ}$;

(A2) $\mathcal{B} \subseteq \Omega_\Gamma^\perp \times (\Omega_\Gamma^{At} \cup \Omega_\Gamma^1)$;

(A3) $\mathcal{E} \subseteq \Omega_\Gamma^{At} \times \Omega_\Gamma^{At}$;

(A4) the relations $\mathcal{A}$, $\mathcal{B}$, and $\mathcal{E}$ are total functions on domains $\Omega_\Gamma^\otimes$, $\Omega_\Gamma^\perp$, and Ω_Γ^{At} respectively.

(A5) if $\langle \alpha, \beta \rangle \in \mathcal{E}$, then $\langle \beta, \alpha \rangle \in \mathcal{E}$;

(A6) if $\langle \alpha, \beta \rangle \in \mathcal{E}$ and $\alpha <_\Gamma \beta$, then there are $p \in \mathrm{Var}$ and $n \in \mathbf{Z}$ such that $\alpha = p^{\perp(n+1)}$ and $\beta = p^{\perp n}$;

(A7) the graph $\langle \Omega_\Gamma, (\mathcal{A} \cup \mathcal{B} \cup \mathcal{E})^\# \rangle$ is $<_\Gamma$-planar.

If $\alpha \in \Omega_\Gamma^\otimes$, then we denote by $\mathcal{A}\alpha$ the only element $\beta \in \Omega_\Gamma$ such that $\langle \alpha, \beta \rangle \in \mathcal{A}$. Similarly for $\mathcal{B}$ and $\mathcal{E}$.

Definition. A *proof net* is a proof structure $\langle \Omega_\Gamma, \mathcal{A}, \mathcal{B}, \mathcal{E} \rangle$ such that

(A8) $\flat(\Omega_\Gamma) = 2$;

(A9) the graph $\langle \Omega_\Gamma, \prec_\Gamma \cup \mathcal{A} \rangle$ is acyclic (i. e., the transitive closure of $\prec_\Gamma \cup \mathcal{A}$ is irreflexive).

Example 7.8 We continue Example 7.2, where

$$\Gamma = ((q^{\perp 3} \otimes p^{\perp 8}) \, \wp \, p^{\perp 7}) q^{\perp 2}.$$

Let $\mathcal{A} = \{\langle \alpha_2, \alpha_6 \rangle\}$, $\mathcal{B} = \emptyset$, and $\mathcal{E} = \{\langle \alpha_1, \alpha_7 \rangle, \langle \alpha_3, \alpha_5 \rangle, \langle \alpha_5, \alpha_3 \rangle, \langle \alpha_7, \alpha_1 \rangle\}$. Then $\langle \Omega_\Gamma, \mathcal{A}, \mathcal{B}, \mathcal{E} \rangle$ is a proof net.

Remark. In the definition of a proof structure one may in addition require that, if $\langle \alpha, \beta \rangle \in \mathcal{B}$ and $\langle \beta, \gamma \rangle \in \mathcal{E}$, then $\beta <_\Gamma \gamma$.

Before establishing that a sequent is derivable if and only if it has a proof net we prove some auxiliary lemmas.

Definition. Let $\Gamma \in \mathrm{NFm}^*$, $\alpha, \beta \in \Omega_\Gamma$, and $\alpha <_\Gamma \beta$. Then by $\Theta_\Gamma^{\alpha,\beta}$ we denote the set $\{\gamma \in \Omega_\Gamma \mid \alpha <_\Gamma \gamma <_\Gamma \beta\}$ and by $\Xi_\Gamma^{\alpha,\beta}$ we denote the set $\{\gamma <_\Gamma \alpha \text{ or } \beta <_\Gamma \gamma\}$.

Lemma 7.9 *Let* $\langle \Omega_\Gamma, \mathcal{A}, \mathcal{B}, \mathcal{E} \rangle$ *be a proof structure,* $\{\alpha, \beta\} \in \mathcal{A}^\#$, *and* $\alpha <_\Gamma \beta$. *Then* $\flat(\Theta_\Gamma^{\alpha,\beta}) \geq 1$ *and* $\flat(\Xi_\Gamma^{\alpha,\beta}) \geq 1$.

PROOF. For shortness we denote $\Theta \rightleftharpoons \Theta_\Gamma^{\alpha,\beta}$ and $\Xi \rightleftharpoons \Xi_\Gamma^{\alpha,\beta}$. We shall verify only $\flat(\Theta) \geq 1$. The proof of $\flat(\Xi) \geq 1$ is analogous.

According to Lemma 7.6 the graph $\langle \Omega_\Gamma, \mathcal{A}^\# \rangle$ is $<_\Gamma$-planar. Thus the set $\mathcal{A}$ is divided into three disjoint subsets

$$\mathcal{A} = \mathcal{A}_0 \cup \mathcal{A}^\Theta \cup \mathcal{A}^\Xi,$$

where $\mathcal{A}_0^\# = \{\{\alpha, \beta\}\}$, $\mathcal{A}^\Theta \subseteq \Theta \times (\Theta \cup \{\alpha, \beta\})$ and $\mathcal{A}^\Xi \subseteq \Xi \times (\Xi \cup \{\alpha, \beta\})$.

Similarly, the graph $\langle \Omega_\Gamma, \{\{\alpha, \beta\}\} \cup \mathcal{B}^\# \rangle$ is $<_\Gamma$-planar and thus $\mathcal{B}$ is divided into two disjoint subsets

$$\mathcal{B} = \mathcal{B}^\Theta \cup \mathcal{B}^\Xi,$$

where $\mathcal{B}^\Theta \subseteq \Theta \times \Theta$ and $\mathcal{B}^\Xi \subseteq \Xi \times \Xi$.

Once again, the graph $\langle \Omega_\Gamma, \{\{\alpha, \beta\}\} \cup \mathcal{E}^\# \rangle$ is $<_\Gamma$-planar and thus $\mathcal{E}$ is divided into two disjoint subsets

$$\mathcal{E} = \mathcal{E}^\Theta \cup \mathcal{E}^\Xi,$$

where $\mathcal{E}^\Theta \subseteq \Theta \times \Theta$ and $\mathcal{E}^\Xi \subseteq \Xi \times \Xi$.

Note that $\langle \Theta \cup \{\alpha, \beta\}, (\mathcal{A}_0 \cup \mathcal{A}^\Theta \cup \mathcal{B}^\Theta \cup \mathcal{E}^\Theta)^\# \rangle$ is an undirected graph. Furthermore, this graph is $<_\Theta$-planar, where $<_\Theta$ is the restriction of $<_\Gamma$ on the set $\Theta \cup \{\alpha, \beta\}$.

Let us draw this $<_\Theta$-planar graph on a semiplane as described after the definition of a $<$-planar graph. We denote the segment of the semiplane border between α and β by $[\alpha, \beta]$. The border segment $[\alpha, \beta]$ and the edge $\{\alpha, \beta\}$ surround a closed area, which contains all edges from the set $(\mathcal{A}^\Theta \cup \mathcal{B}^\Theta \cup \mathcal{E}^\Theta)^\#$. The edges from $(\mathcal{B}^\Theta \cup \mathcal{E}^\Theta)^\#$ divide this area into $|(\mathcal{B}^\Theta \cup \mathcal{E}^\Theta)^\#| + 1$ regions. We are interested in all these regions except the one adjacent to the edge $\{\alpha, \beta\}$.

Consider any of these regions. We claim that it has at least one nontrivial segment of $[\alpha, \beta]$ at its border. (Otherwise every vertex from $\Omega_\Gamma^{\text{At}} \cup \Omega_\Gamma^1 \cup \Omega_\Gamma^\perp$ adjacent to the region considered would belong to two edges from $(\mathcal{B}^\Theta \cup \mathcal{E}^\Theta)^\#$, but this is impossible.)

Any such segment of $[\alpha, \beta]$ contains at least one element of $\Omega_\Gamma^\otimes \cup \Omega_\Gamma^{\wp\circ}$. Thus some elements of $\Omega_\Gamma^\otimes \cup \Omega_\Gamma^{\wp\circ}$ are adjacent to the region considered. It is impossible that all of these would belong to $\Omega_\Gamma^\otimes$ (because $\mathcal{A}$ is a total function).

Thus the number of regions considered does not exceed the cardinality of the set $\Omega_\Gamma^{\wp\circ} \cap \Theta$.

$$|\Omega_\Gamma^{\wp\circ} \cap \Theta| \geq |(\mathcal{B}^\Theta \cup \mathcal{E}^\Theta)^\#|$$

Taking into account that

$$|(\mathcal{B}^\Theta)^\#| = |\mathcal{B}^\Theta| = |\Omega_\Gamma^\perp \cap \Theta|$$

and

$$|(\mathcal{E}^\Theta)^\#| = \frac{1}{2}|\mathcal{E}^\Theta| = \frac{1}{2}|\Omega_\Gamma^{\text{At}} \cap \Theta|$$

we obtain that

$$|\Omega_\Gamma^{\wp\circ} \cap \Theta| \geq |\Omega_\Gamma^\perp \cap \Theta| + \frac{1}{2}|\Omega_\Gamma^{\text{At}} \cap \Theta|.$$

Analogously to Lemma 7.4 (i) we notice that

$$|(\Omega_\Gamma^{\wp\circ} \cup \Omega_\Gamma^\otimes) \cap \Theta| = |(\Omega_\Gamma^\perp \cup \Omega_\Gamma^1 \cup \Omega_\Gamma^{\text{At}}) \cap \Theta| - 1.$$

Subtracting

$$|\Omega_\Gamma^{\wp\circ} \cap \Theta| + |\Omega_\Gamma^\otimes \cap \Theta| = |\Omega_\Gamma^\perp \cap \Theta| + |\Omega_\Gamma^1 \cap \Theta| + |\Omega_\Gamma^{\mathrm{At}} \cap \Theta| - 1$$

from

$$2|\Omega_\Gamma^{\wp\circ} \cap \Theta| \geq 2|\Omega_\Gamma^\perp \cap \Theta| + |\Omega_\Gamma^{\mathrm{At}} \cap \Theta|$$

we obtain the desired inequality

$$|\Omega_\Gamma^{\wp\circ} \cap \Theta| - |\Omega_\Gamma^\otimes \cap \Theta| \geq |\Omega_\Gamma^\perp \cap \Theta| - |\Omega_\Gamma^1 \cap \Theta| + 1.$$

■

Lemma 7.10 *Let* $\langle \Omega_\Gamma, \mathcal{A}, \mathcal{B}, \mathcal{E} \rangle$ *be a proof net,* $\{\alpha, \beta\} \in \mathcal{A}^\#$*, and* $\alpha <_\Gamma \beta$*. Then* $\flat(\Theta_\Gamma^{\alpha,\beta}) = 1$ *and* $\flat(\Xi_\Gamma^{\alpha,\beta}) = 1$*.*

PROOF. Note that $2 = \flat(\Omega_\Gamma) = \flat(\Theta_\Gamma^{\alpha,\beta}) + \flat(\Xi_\Gamma^{\alpha,\beta}) + \flat(\alpha, \beta) = \flat(\Theta_\Gamma^{\alpha,\beta}) + \flat(\Xi_\Gamma^{\alpha,\beta})$, since $\flat(\alpha, \beta) = 0$. It remains to use the previous lemma. ■

Proposition 7.11 *Let* $\langle \Omega_{\Gamma\Delta(A\otimes B)\Pi}, \mathcal{A}, \mathcal{B}, \mathcal{E} \rangle$ *be a proof net and* $\mathcal{A}\langle A\otimes B,$ $\|\Gamma\| + \|\Delta\| + \|A\|\rangle = \langle \diamond, \|\Gamma\|\rangle$*. Then the sequents* $\to\Delta A$ *and* $\to\Gamma B\Pi$ *are derivable in* SPNCL′*.*

PROOF. Proof structures for $\to\Delta A$ and $\to\Gamma B\Pi$ are easily constructed from the given proof net. To verify that they are proof nets we use Lemma 7.10. ■

Theorem 7.12 *A sequent* $\to\Gamma$ *is derivable in* SPNCL′ *if and only if there exists a proof net* $\langle \Omega_\Gamma, \mathcal{A}, \mathcal{B}, \mathcal{E} \rangle$ *for the sequent* $\to\Gamma$*.*

PROOF. Sketch. Proving the 'only if' part is easy. To prove the 'if' part we proceed by induction on the cardinality of the set $\Omega_\Gamma^\otimes \cup \Omega_\Gamma^\wp$.

Induction base. Let $\Omega_\Gamma^\otimes \cup \Omega_\Gamma^\wp = \emptyset$. From $\flat(\Omega_\Gamma) = 2$ we conclude that either $\Gamma = \perp\ldots\perp 1 1 \perp\ldots\perp$ or $\Gamma = \perp\ldots\perp q^{\perp k}\perp\ldots\perp p^{\perp n}\perp\ldots\perp$. In the latter case $q = p$ and $k = n + 1$ in view of (A6). Evidently all sequents $\to\perp\ldots\perp 1 1 \perp\ldots\perp$ and $\to\perp\ldots\perp p^{\perp n+1}\perp\ldots\perp p^{\perp n}\perp\ldots\perp$ are derivable in SPNCL′.

Induction step. Assume now that $\Omega_\Gamma^\otimes \cup \Omega_\Gamma^\wp$ is not empty. We introduce on $\Omega_\Gamma^\otimes \cup \Omega_\Gamma^\wp$ a binary relation $\ll$ stipulating that $\alpha \ll \beta$ if and only if $\alpha \prec_\Gamma \beta$ or $\langle \alpha, \beta \rangle \in \mathcal{A}$. In other words, $\ll$ is the restriction of $\prec_\Gamma \cup \mathcal{A}$ on the set $\Omega_\Gamma^\otimes \cup \Omega_\Gamma^\wp$.

According to (A9) there is an element $\delta_0 \in \Omega_\Gamma^\otimes \cup \Omega_\Gamma^\wp$ maximal with respect to $\ll$. We consider two cases.

CASE 1: $\delta_0 \in \Omega_\Gamma^\wp$

We can use the induction hypothesis and apply the rule $(\wp)$.

CASE 2: $\delta_0 \in \Omega_\Gamma^\otimes$

In view of $\mathcal{A}$ being a function there exists $\beta \in \Omega_\Gamma^{\wp\circ}$ such that $\langle \delta_0, \beta \rangle \in \mathcal{A}$. Since δ_0 is maximal with respect to $\ll$, we have $\beta \in \Omega_\Gamma^\circ$.

We consider two subcases.

CASE 2a: $\beta = \langle \diamond, 0 \rangle$ (i.e., β is the least element of Ω_Γ w. r. t. $<_\Gamma$)
In view of Proposition 7.11 we can use the induction hypothesis and apply the rule $(\otimes)$.

CASE 2b: $\beta \neq \langle \diamond, 0 \rangle$
We use Lemma 7.7 and the rules $({}^{\perp \perp}(\, \cdot \,))$, $((\, \cdot \,)^{\perp \perp})$ to reduce this case to the previous one. ∎

Remark. Analogous result can be easily established also for the multiplicative fragment of cyclic linear logic defined in [12].

8   Properties of proof nets

Lemma 8.1 *Let $\langle \Omega_\Gamma, \mathcal{A}, \mathcal{B}, \mathcal{E} \rangle$ be a proof structure. If the graph $\langle \Omega_\Gamma, \prec_\Gamma \cup \mathcal{A} \rangle$ contains a cycle, then there exists a cycle*

$$(\alpha_1, \beta_1, \alpha_2, \beta_2, \ldots, \alpha_n, \beta_n)$$

such that

(i) *$\alpha_i \in \Omega_\Gamma^\wp$ and $\beta_i \in \Omega_\Gamma^\otimes$ for each $i \leq n$;*

(ii) *$\alpha_i \prec_\Gamma \beta_i$ for each $i \leq n$;*

(iii) *$\langle \beta_i, \alpha_{i+1} \rangle \in \mathcal{A}$ for each $i < n$;*

(iv) *$\langle \beta_n, \alpha_1 \rangle \in \mathcal{A}$;*

(v) *either $\alpha_1 <_\Gamma \beta_1 <_\Gamma \alpha_2 <_\Gamma \cdots <_\Gamma \beta_n$ or $\beta_n <_\Gamma \alpha_n <_\Gamma \beta_{n-1} <_\Gamma \cdots <_\Gamma \alpha_1$.*

Definition. Let $g: \Omega_1 \to \Omega_2$ be a bijection and $\mathcal{R}$ be a binary relation on Ω_1. Then by $\mathcal{R}^g$ we denote the binary relation $\{\langle g(\alpha), g(\beta) \rangle \mid \langle \alpha, \beta \rangle \in \mathcal{R}\}$ on Ω_2.

Proposition 8.2 *Let $\langle \Omega_\Gamma, \mathcal{A}, \mathcal{B}, \mathcal{E} \rangle$ be a proof net. Let Γ' be obtained from Γ by replacing an occurrence of a subformula $(A \otimes (B \otimes C))$ by $((A \otimes B) \otimes C)$ or vice versa. Let g denote the unique isomorphism of $\langle \Omega_\Gamma, <_\Gamma \rangle$ and $\langle \Omega_{\Gamma'}, <_{\Gamma'} \rangle$. Then $\langle \Omega_{\Gamma'}, \mathcal{A}^g, \mathcal{B}^g, \mathcal{E}^g \rangle$ is a proof net.*

PROOF. Sketch. Let Γ' be obtained from Γ by replacing an occurrence of a subformula $(A \otimes (B \otimes C))$ by $((A \otimes B) \otimes C)$. Assume that the graph $\langle \Omega_\Gamma, \prec_\Gamma \cup \mathcal{A} \rangle$ is acyclic, whereas the graph $\langle \Omega_{\Gamma'}, <_{\Gamma'} \cup \mathcal{A}^g \rangle$ is not. Applying Lemma 8.1 we find in $\langle \Omega_{\Gamma'}, <_{\Gamma'} \cup \mathcal{A}^g \rangle$ a cycle of special form $(\alpha_1, \beta_1, \alpha_2, \beta_2, \ldots, \alpha_n, \beta_n)$.

Evidently there is $m \leq n$ such that $\beta_m = \langle (A \otimes B) \otimes C, k + \|A\| + \|B\| \rangle$ and $\alpha_m \preceq_{\Gamma'} \langle A, k + c(A) \rangle$ for some k.

We denote $\gamma \rightleftharpoons \langle A \otimes B, k + \|A\| \rangle$. In view of (A7) and the special form of the cycle there must be $l \leq n$ such that $\alpha_l <_{\Gamma'} \mathcal{A}^g \gamma <_{\Gamma'} \beta_l$ or $\alpha_l = \mathcal{A}^g \gamma$. But then there is another cycle in $\langle \Omega_{\Gamma'}, <_{\Gamma'} \cup \mathcal{A}^g \rangle$ containing the edge $\langle \gamma, \mathcal{A}^g \gamma \rangle$ and not involving the vertex β_m (see Lemma 7.5). This cycle is mapped by g^{-1} to a cycle in $\langle \Omega_\Gamma, \prec_\Gamma \cup \mathcal{A} \rangle$. Contradiction. ∎

Proposition 8.3 *Let* $\langle \Omega_{\Gamma(A \otimes (B \otimes C))\Pi}, \mathcal{A}, \mathcal{B}, \mathcal{E} \rangle$ *be a proof net and*

$$\mathcal{A}\langle A \otimes (B \otimes C), \|\Gamma\| + \|A\| \rangle = \mathcal{A}\langle B \otimes C, \|\Gamma\| + \|A\| + \|B\| \rangle.$$

Then the sequent $\rightarrow B$ *is derivable in* SPNCL'.

PROOF. Sketch. The proof structure for $\rightarrow B$ is copied from the relevant part of the given proof net. To prove (A8) we apply Lemma 7.10 twice. ∎

Proposition 8.4 *Let* $\langle \Omega_{\Gamma \Delta_1 (A_1 \otimes B_1) \Pi_1}, \mathcal{A}_1, \mathcal{B}_1, \mathcal{E}_1 \rangle$ *be a proof net and* $\langle \Omega_{\Gamma \Delta_2 (A_2 \otimes B_2) \Pi_2}, \mathcal{A}_2, \mathcal{B}_2, \mathcal{E}_2 \rangle$ *be another proof net. If* $\mathcal{A}_1 \langle A_1 \otimes B_1, \|\Gamma\| + \|\Delta_1\| + \|A_1\| \rangle = \mathcal{A}_2 \langle A_2 \otimes B_2, \|\Gamma\| + \|\Delta_2\| + \|A_2\| \rangle \in \Omega_\Gamma \cup \{\langle \diamond, \|\Gamma\| \rangle\}$, *then the sequents* $\rightarrow \Gamma \Delta_1 (A_1 \otimes B_2) \Pi_2$ *and* $\rightarrow \Gamma \Delta_2 (A_2 \otimes B_1) \Pi_1$ *are derivable in* SPNCL'.

PROOF. Sketch. Let $\beta_1 \rightleftharpoons \langle A_1 \otimes B_1, \|\Gamma\| + \|\Delta_1\| + \|A_1\| \rangle$, $\beta_2 \rightleftharpoons \langle A_2 \otimes B_2, \|\Gamma\| + \|\Delta_2\| + \|A_2\| \rangle$, and $\alpha \rightleftharpoons \mathcal{A}_1 \beta_1 = \mathcal{A}_2 \beta_2$.

To obtain a proof structure for $\rightarrow \Gamma \Delta_1 (A_1 \otimes B_2) \Pi_2$ we combine the parts of the given proof nets corresponding to $\Theta_{\Gamma \Delta_1 (A_1 \otimes B_1) \Pi_1}^{\alpha \beta_1}$ and $\Xi_{\Gamma \Delta_2 (A_2 \otimes B_2) \Pi_2}^{\alpha \beta_2}$.

Using Lemma 8.1 one can verify that the proof structure is a proof net.

The claim $\rightarrow \Gamma \Delta_2 (A_2 \otimes B_1) \Pi_1$ follows from the other one due to the symmetry of the conditions of the theorem. ∎

Proposition 8.5 *Let* SPNCL' $\vdash \rightarrow B$ *and* $\alpha \in \Omega_\Gamma^{p \diamond}$. *Then there are* $C, D \in$ NFm *and there is a proof net*

$$\langle \Omega_{\Gamma C (\perp \otimes (B \otimes \perp)) D}, \mathcal{A}, \mathcal{B}, \mathcal{E} \rangle$$

such that $\mathcal{A}\langle \perp \otimes (B \otimes \perp), \|\Gamma\| + \|C\| + \|\perp\| \rangle = \mathcal{A}\langle B \otimes \perp, \|\Gamma\| + \|C\| + \|\perp\| + \|B\| \rangle = \alpha$.

PROOF. Sketch. Given a sequence $\Gamma \in$ NFm* and a vertex $\alpha \in \Omega_\Gamma^{p \diamond}$ it is easy to construct two formulas C, D and a proof net $\langle \Omega_{\Gamma C \otimes D}, \mathcal{A}_0, \mathcal{B}_0, \mathcal{E}_0 \rangle$ such that $\mathcal{A}_0 \langle C \otimes D, \|\Gamma\| + \|C\| \rangle = \alpha$ and each edge $\langle \langle E, k \rangle, \langle F, m \rangle \rangle \in \mathcal{A}_0 \cup \mathcal{B}_0 \cup \mathcal{E}_0$ satisfies $k + m = 2\|\Gamma\|$.

On the other hand, there is a proof net for the sequent $\rightarrow B$. It remains to combine these two proof nets. Again, Lemma 8.1 is useful for checking (A9). ∎

Proposition 8.6 *Let* $\langle \Omega_{\Gamma(B \otimes (C \otimes D))\Pi}, \mathcal{A}_1, \mathcal{B}_1, \mathcal{E}_1 \rangle$ *be a proof net. Let the sequent* $\rightarrow C^\perp E$ *be derivable in* SPNCL'. *Then there exists a proof net*

$$\langle \Omega_{\Gamma(B \otimes (E \otimes D))\Pi}, \mathcal{A}, \mathcal{B}, \mathcal{E} \rangle$$

such that $\mathcal{A}\langle B \otimes (E \otimes D), \|\Gamma\| + \|B\|\rangle = \mathcal{A}_1\langle B \otimes (C \otimes D), \|\Gamma\| + \|B\|\rangle$ and
$\mathcal{A}\langle E \otimes D, \|\Gamma\| + \|B\| + \|E\|\rangle = \mathcal{A}_1\langle C \otimes D, \|\Gamma\| + \|B\| + \|C\|\rangle$.

PROOF. Sketch. According to Theorem t-complete there is a proof net

$$\langle \Omega_{C^\perp E}, \mathcal{A}_2, \mathcal{B}_2, \mathcal{E}_2 \rangle.$$

There is a natural one-to-one mapping (an anti-isomorphism of linear orders) between the part of $\Omega_{\Gamma(B \otimes (C \otimes D))\Pi}$ corresponding to C and the part of $\Omega_{C^\perp E}$ corresponding to $C^\perp$. We denote the graph of this mapping by $\mathcal{G}$ and the graph of its inverse by $\mathcal{G}^{-1}$.

We define $\mathcal{H}$ as the transitive closure of $\mathcal{A}_1 \cup \mathcal{B}_1 \cup \mathcal{E}_1 \cup \mathcal{A}_2 \cup \mathcal{B}_2 \cup \mathcal{E}_2 \cup \mathcal{G} \cup \mathcal{G}^{-1}$ on the disjoint union of $\Omega_{\Gamma(B \otimes (C \otimes D))\Pi}$ and $\Omega_{C^\perp E}$.

Finally, $\mathcal{A}$, $\mathcal{B}$, and $\mathcal{E}$ are chosen so that $\mathcal{A} \cup \mathcal{B} \cup \mathcal{E}$ coincides with the restriction of $\mathcal{H}$ to the domain excluding C and $C^\perp$. ∎

9   Tp(m)-maps

The aim of this section is to introduce Tp(m)-maps $\langle \mathbf{V}_n, v_n \rangle$ and $\langle \mathbf{V}'_n, v'_n \rangle$, which will later be used in the proof of Lemma 10.20, where we construct a Tp(m)-quasimodel containing a witness for a given word $\delta \notin v(E \backslash F)$ (resp. $\delta \notin v(F/E)$).

We need some notation. If $\mathcal{R}$ and $\mathcal{T}$ are two binary relations on a set $\mathbf{D}$, then we define

$$\mathcal{R} \odot \mathcal{T} \rightleftharpoons \{\langle r, t \rangle \in \mathbf{D} \times \mathbf{D} \mid (\exists s \in \mathbf{D})\ \langle r, s \rangle \in \mathcal{R}\ \text{and}\ \langle s, t \rangle \in \mathcal{T}\}.$$

Evidently, $\odot$ is associative.

Given a set $\mathbf{D}$ and a function $w \colon \mathrm{Tp} \to \mathbf{P}(\mathbf{D} \times \mathbf{D})$ we denote by $\vec{w}$ the function from Tp^* to $\mathbf{P}(\mathbf{D} \times \mathbf{D})$ defined as follows:

$$\vec{w}(\Lambda) \rightleftharpoons \{\langle s, s \rangle \mid s \in \mathbf{D}\};$$
$$\vec{w}(\Gamma A) \rightleftharpoons \vec{w}(\Gamma) \odot w(A).$$

Remark. $\vec{w}(\Gamma\Pi) = \vec{w}(\Gamma) \odot \vec{w}(\Pi)$.

Definition. A Tp(m)-*map* $\langle \mathbf{D}, \mathbf{W}, w \rangle$ consists of a finite set $\mathbf{D}$, a reflexive linear order $\mathbf{W} \subseteq \mathbf{D} \times \mathbf{D}$, and a valuation $w \colon \mathrm{Tp} \to \mathbf{P}(\mathbf{W})$ such that

(1) for any $A \in \mathrm{Tp}$, $B \in \mathrm{Tp}$, if $A \bullet B \in \mathrm{Tp}(m)$, then $w(A \bullet B) \subseteq w(A) \odot w(B)$;

(2) for any $\Gamma \in \mathrm{Tp}(m)^*$, $B \in \mathrm{Tp}(m)$, if $L^* \vdash \Gamma \to B$, then $\vec{w}(\Gamma) \subseteq w(B)$;

(3) for any $B \in \mathrm{Tp}$, if $\langle s, s \rangle \in w(B)$ for some $s \in \mathbf{D}$, then $L^* \vdash \Lambda \to B$.

Lemma 9.1 *For any given number $m \in \mathbf{N}$ there is a family of* Tp(m)-*maps $\langle \mathbf{D}_\Gamma, \mathbf{V}_\Gamma, v_\Gamma \rangle$ indexed by sequences of types $\Gamma \in \mathrm{Tp}^*$ and there are elements $\chi_\Gamma \in \mathbf{D}_\Gamma$ such that*

(i) $(\forall \Gamma \in \mathrm{Tp}^*)\ (\forall \Pi \in \mathrm{Tp}^*)\ \mathbf{D}_\Gamma \subseteq \mathbf{D}_{\Pi\Gamma};$

(ii) $(\forall \Gamma \in \mathrm{Tp}^*)\ (\forall \Pi \in \mathrm{Tp}^*)\ \mathbf{V}_\Gamma \subseteq \mathbf{V}_{\Pi\Gamma};$

(iii) $(\forall \Gamma \in \mathrm{Tp}^*)\ (\forall \Pi \in \mathrm{Tp}^*)\ (\forall \langle s,t \rangle \in \mathbf{V}_{\Pi\Gamma})$ if $s \in \mathbf{D}_\Gamma$ then $\langle s,t \rangle \in \mathbf{V}_\Gamma;$

(iv) $(\forall \Gamma \in \mathrm{Tp}^*)\ (\forall \Pi \in \mathrm{Tp}^*)\ (\forall A \in \mathrm{Tp})\ v_\Gamma(A) \subseteq v_{\Pi\Gamma}(A);$

(v) $(\forall \Gamma \in \mathrm{Tp}^*)\ (\forall \Pi \in \mathrm{Tp}^*)\ (\forall A \in \mathrm{Tp})\ (\forall \langle s,t \rangle \in v_{\Pi\Gamma}(A))$ if $s \in$ $\mathbf{D}_\Gamma$ then
$\langle s,t \rangle \in v_\Gamma(A);$

(vi) $(\forall \Gamma \in \mathrm{Tp}^*)\ (\forall \Delta \in \mathrm{Tp}^*)$ if $\chi_\Gamma \in \mathbf{D}_\Delta$ then $(\exists \Pi \in \mathrm{Tp}^*)\ \Delta = \Pi\Gamma;$

(vii) $(\forall \Gamma \in \mathrm{Tp}^*)\ (\forall C \in \mathrm{Tp})\ \langle \chi_\Gamma, \chi_\Lambda \rangle \in v_\Gamma(C) \Leftrightarrow L^* \vdash \Gamma{\rightarrow}C;$

(viii) $(\forall \Gamma \in \mathrm{Tp}^*)\ (\forall B \in \mathrm{Tp})\ \langle \chi_{B\Gamma}, \chi_\Gamma \rangle \in v_{B\Gamma}(B).$

PROOF. The construction of $\mathrm{Tp}(m)$-maps is based on the proof nets introduced in section 7. The domain $\mathbf{D}_\Gamma$ of the $\mathrm{Tp}(m)$-map corresponding to Γ will be a finite subset of $\mathrm{NFm} \times \mathrm{Occ} \times \mathrm{NFm}^*$.

Let $\widehat{\Gamma}^\perp = A_1 \dots A_n$. We define $\mathbf{D}_\Gamma^-$ as the set of 3-tuples $\langle B, k, A_1 \dots A_j \rangle$ such that $\langle B, k \rangle \in \Omega_{\widehat{\Gamma}^\perp}^{\wp\circ}$ and $j \in \mathbf{N}$ is the smallest natural number satisfying $\|A_1 \dots A_j\| \geq k.$

We put $\chi_\Gamma \rightleftharpoons \langle \diamond, \|\widehat{\Gamma}^\perp\|, \widehat{\Gamma}^\perp \rangle$ and $\mathbf{D}_\Gamma \rightleftharpoons \mathbf{D}_\Gamma^- \cup \{\chi_\Gamma\}.$

For any 3-tuple $s = \langle B, k, \Phi \rangle$ we denote by $\widetilde{s}$ the 2-tuple $\langle B, k \rangle$. Evidently the mapping $s \mapsto \widetilde{s}$ establishes a one-to-one correspondence between $\mathbf{D}_\Gamma^-$ and $\Omega_{\widehat{\Gamma}^\perp}^{\wp\circ}$.

The linear order $\mathbf{V}_\Gamma \subseteq \mathbf{D}_\Gamma \times \mathbf{D}_\Gamma$ is defined by stipulating that a pair $\langle \langle B_1, k_1, \Phi_1 \rangle, \langle B_2, k_2, \Phi_2 \rangle \rangle$ belongs to $\mathbf{V}_\Gamma$ if and only if $k_1 \geq k_2.$

Finally, the function $v_\Gamma \colon \mathrm{Tp} \to \mathbf{P}(\mathbf{V}_\Gamma)$ is defined by stating that $\langle s,t \rangle \in v_\Gamma(C)$ if and only if there are $E \in \mathrm{NFm}$, $F \in \mathrm{NFm}$, $\Delta \in \mathrm{NFm}^*$, $\Pi \in \mathrm{NFm}^*$, and there is a proof net $\langle \Omega_{\widehat{\Gamma}^\perp \Delta(E \otimes (\widehat{C} \otimes F))\Pi}, \mathcal{A}, \mathcal{B}, \mathcal{E} \rangle$ such that $\mathcal{A}\langle E \otimes (\widehat{C} \otimes F), \|\widehat{\Gamma}^\perp\| + \|\Delta\| + \|E\| \rangle = \widetilde{s}$ and $\mathcal{A}\langle \widehat{C} \otimes F, \|\widehat{\Gamma}^\perp\| + \|\Delta\| + \|E\| + \|\widehat{C}\| \rangle = \widetilde{t}.$

First we verify that for each $\Gamma \in \mathrm{Tp}^*$ the triple $\langle \mathbf{D}_\Gamma, \mathbf{V}_\Gamma, v_\Gamma \rangle$ is a $\mathrm{Tp}(m)$-map for every $m \in \mathbf{N}$.

(1) Let $\langle s,t \rangle \in v_\Gamma(A{\bullet}B).$
This means that

- there is a proof net $\langle \Omega, \mathcal{A}, \mathcal{B}, \mathcal{E} \rangle$ for a derivable sequent of the form

$$\rightarrow \widehat{\Gamma}^\perp \Delta(E \otimes ((\widehat{A} \otimes \widehat{B}) \otimes F))\Pi;$$

- $\mathcal{A}\langle E \otimes ((\widehat{A} \otimes \widehat{B}) \otimes F), \|\widehat{\Gamma}^\perp\| + \|\Delta\| + \|E\| \rangle = \widetilde{s};$

- $\mathcal{A}\langle (\widehat{A} \otimes \widehat{B}) \otimes F, \|\widehat{\Gamma}^\perp\| + \|\Delta\| + \|E\| + \|\widehat{A}\| + \|\widehat{B}\| \rangle = \widetilde{t}.$

Evidently there is $u \in \mathbf{D}_\Gamma$ such that $\mathcal{A}\langle \widehat{A} \otimes \widehat{B}, \|\widehat{\Gamma}^\perp\| + \|\Delta\| + \|E\| + \|\widehat{A}\|\rangle = \widetilde{u}$. Using Proposition 8.2 it is easy to establish that $\langle s, u \rangle \in v_\Gamma(A)$ and $\langle u, t \rangle \in v_\Gamma(B)$, whence $\langle s, t \rangle \in v_\Gamma(A) \odot v_\Gamma(B)$. Thus we have established that $v_\Gamma(A \bullet B) \subseteq v_\Gamma(A) \odot v_\Gamma(B)$.

(2) Let $L^* \vdash A_1 \ldots A_n \rightarrow B$. We must verify that $v_\Gamma(A_1) \odot \ldots \odot v_\Gamma(A_n) \subseteq v_\Gamma(B)$. If $n = 0$, then we use Proposition 8.5.

Assume now that $n > 0$. Let $\langle s, u \rangle \in v_\Gamma(A_1)$ and $\langle u, t \rangle \in v_\Gamma(A_2)$. According to the definition of v_Γ there are proof nets

$$\langle \Omega_{\widehat{\Gamma}^\perp \; \Delta_1(E_1 \otimes (\widehat{A_1} \otimes F_1))\Pi_1}, \mathcal{A}_1, \mathcal{B}_1, \mathcal{E}_1 \rangle$$

and

$$\langle \Omega_{\widehat{\Gamma}^\perp \; \Delta_2(E_2 \otimes (\widehat{A_2} \otimes F_2))\Pi_2}, \mathcal{A}_2, \mathcal{B}_2, \mathcal{E}_2 \rangle$$

such that $\mathcal{A}_1 \langle E_1 \otimes (\widehat{A_1} \otimes F_1), \|\widehat{\Gamma}^\perp\| + \|\Delta_1\| + \|E_1\|\rangle = \widetilde{s}$, $\mathcal{A}_1 \langle \widehat{A_1} \otimes F_1, \|\widehat{\Gamma}^\perp\| + \|\Delta_1\| + \|E_1\| + \|\widehat{A_1}\|\rangle = \widetilde{u}$, $\mathcal{A}_2 \langle E_2 \otimes (\widehat{A_2} \otimes F_2), \|\widehat{\Gamma}^\perp\| + \|\Delta_2\| + \|E_2\|\rangle = \widetilde{u}$, $\mathcal{A}_2 \langle \widehat{A_2} \otimes F_2, \|\widehat{\Gamma}^\perp\| + \|\Delta_2\| + \|E_2\| + \|\widehat{A_2}\|\rangle = \widetilde{t}$. From Proposition 8.2 and Proposition 8.4 we obtain a proof net

$$\langle \Omega_{\widehat{\Gamma}^\perp \; \Delta_1(E_1 \otimes ((\widehat{A_1} \otimes \widehat{A_2}) \otimes F_2))\Pi_2}, \mathcal{A}, \mathcal{B}, \mathcal{E} \rangle$$

such that

$$\mathcal{A}\langle E_1 \otimes ((\widehat{A_1} \otimes \widehat{A_2}) \otimes F_2), \|\widehat{\Gamma}^\perp\| + \|\Delta_1\| + \|E_1\|\rangle =$$
$$\widetilde{s}, \mathcal{A}\langle (\widehat{A_1} \otimes \widehat{A_2}) \otimes F_2, \|\widehat{\Gamma}^\perp\| + \|\Delta_1\| + \|E_1\| + \|\widehat{A_1}\| + \|\widehat{A_2}\|\rangle = \widetilde{t}.$$

Thus $\langle s, t \rangle \in v_\Gamma(A_1 \bullet A_2)$. We have established that $v_\Gamma(A_1) \odot v_\Gamma(A_2) \subseteq v_\Gamma(A_1 \bullet A_2)$. By induction on n we obtain $\vec{v_\Gamma}(A_1 \ldots A_n) \subseteq v_\Gamma(A_1 \bullet \ldots \bullet A_n)$.

It remains to apply Proposition 8.6.

(3) Let $\langle \Omega_{\widehat{\Gamma}^\perp \; \Delta(E \otimes (\widehat{B} \otimes F))\Pi}, \mathcal{A}, \mathcal{B}, \mathcal{E} \rangle$ be a proof net such that

$$\mathcal{A}\langle E \otimes (\widehat{B} \otimes F), \|\widehat{\Gamma}^\perp\| + \|\Delta\| + \|E\|\rangle = \mathcal{A}\langle \widehat{B} \otimes F, \|\widehat{\Gamma}^\perp\| + \|\Delta\| + \|E\| + \|\widehat{B}\|\rangle.$$

According to Proposition 8.3 the sequent $\rightarrow \widehat{B}$ is derivable in SPNCL$'$.

Now we verify that the elements χ_Γ and $\mathrm{Tp}(m)$-maps $\langle \mathbf{V}_\Gamma, v_\Gamma \rangle$ satisfy (i)–(viii).

(i)

Evident from $\widehat{(\Pi\Gamma)}^\perp = \widehat{\Gamma}^\perp \widehat{\Pi}^\perp$.

(ii)

Similar.

(iii)

Obvious from the fact that if $\langle B_1, k_1, \Phi_1 \rangle \in \mathbf{D}_\Gamma$, $\langle B_2, k_2, \Phi_2 \rangle \in \mathbf{D}_\Gamma$, and $k_1 \geq k_2$, then there is $\Pi \in \mathrm{NFm}^*$ such that $\Phi_1 = \Pi\Phi_2$.

(iv)

We verify that $v_\Gamma(A) \subseteq v_{\Pi\Gamma}(A)$. Let $\langle s, t \rangle \in v_\Gamma(A)$. This means that there is a proof net $\langle \Omega_{\widehat{\Gamma}^\perp \; \Delta(E \otimes (\widehat{A} \otimes F))\Pi}, \mathcal{A}, \mathcal{B}, \mathcal{E}\rangle$ such that $\mathcal{A}\langle E \otimes (\widehat{A} \otimes F), \|\widehat{\Gamma}^\perp\| + \|\Delta\| + \|E\|\rangle = \widetilde{s}$ and $\mathcal{A}\langle \widehat{A} \otimes F, \|\widehat{\Gamma}^\perp\| + \|\Delta\| + \|E\| + \|\widehat{A}\|\rangle = \widetilde{t}$. Let $\Pi = B_1 \ldots B_n$. Then one can easily construct another proof net $\langle \Omega', \mathcal{A}', \mathcal{B}', \mathcal{E}'\rangle$ for the sequent

$$\to \widehat{\Gamma}^\perp \, \widehat{B_n}^\perp \ldots \widehat{B_1}^\perp \, (\widehat{B_1} \otimes \ldots \otimes \widehat{B_n} \otimes \perp)\Delta(E \otimes (\widehat{A} \otimes F))$$

such that $\mathcal{A}'\langle E \otimes (\widehat{A} \otimes F), \|\widehat{\Gamma}^\perp\| + \|\widehat{\Pi}^\perp\| + \|\widehat{\Pi}\| + \|\perp\| + \|\Delta\| + \|E\|\rangle = \widetilde{s}$ and $\mathcal{A}'\langle \widehat{A} \otimes F, \|\widehat{\Gamma}^\perp\| + \|\widehat{\Pi}^\perp\| + \|\widehat{\Pi}\| + \|\perp\| + \|\Delta\| + \|E\| + \|\widehat{A}\|\rangle = \widetilde{t}$. Thus $\langle s, t\rangle \in v_{\Pi\Gamma}(A)$.

(v)
Similar to (iii).

(vi)
Follows from the definition of $\mathbf{D}_\Gamma$.

(vii)
Let $L^* \vdash \Gamma \to C$. According to Lemma 6.3 and Theorem 7.12 there is a proof net for the sequent $\widehat{\Gamma}^\perp \, \widehat{C}$. By an easy modification we obtain a proof net $\langle \Omega_{\widehat{\Gamma}^\perp \; (1 \otimes (\widehat{C} \otimes 1))}, \mathcal{A}, \mathcal{B}, \mathcal{E}\rangle$ such that $\mathcal{A}\langle 1 \otimes (\widehat{C} \otimes 1), \|\widehat{\Gamma}^\perp\| + \|1\|\rangle = \chi_\Gamma$ and $\mathcal{A}\langle \widehat{C} \otimes 1, \|\widehat{\Gamma}^\perp\| + \|1\| + \|\widehat{C}\|\rangle = \chi_\Lambda$.

For the converse assume that $\langle \Omega_{\widehat{\Gamma}^\perp \; \Delta(E \otimes (\widehat{C} \otimes F))\Pi}, \mathcal{A}, \mathcal{B}, \mathcal{E}\rangle$ is a proof net such that $\mathcal{A}\langle E \otimes (\widehat{C} \otimes F), \|\widehat{\Gamma}^\perp\| + \|\Delta\| + \|E\|\rangle = \chi_\Gamma$ and $\mathcal{A}\langle \widehat{C} \otimes F, \|\widehat{\Gamma}^\perp\| + \|\Delta\| + \|E\| + \|\widehat{C}\|\rangle = \chi_\Lambda$. Using Proposition 7.11 twice we can obtain a proof net for $\to \widehat{\Gamma}^\perp \, \widehat{C}$.

(viii)
Let $\Gamma = A_1 \ldots A_n$. It is easy to construct a proof net

$$\langle \Omega_{\widehat{A_n}^\perp \ldots \widehat{A_1}^\perp \, \widehat{B}^\perp \, (1 \otimes (\widehat{B} \otimes (\widehat{A_1} \otimes \ldots \otimes \widehat{A_n})))}, \mathcal{A}, \mathcal{B}, \mathcal{E}\rangle$$

such that $\mathcal{A}\langle 1 \otimes (\widehat{B} \otimes (\widehat{A_1} \otimes \ldots \otimes \widehat{A_n})), \|\widehat{\Gamma}^\perp\| + \|\widehat{B}^\perp\| + \|1\|\rangle = \chi_{BA_1 \ldots A_n}$ and $\mathcal{A}\langle \widehat{B} \otimes (\widehat{A_1} \otimes \ldots \otimes \widehat{A_n}), \|\widehat{\Gamma}^\perp\| + \|\widehat{B}^\perp\| + \|1\| + \|\widehat{B}\|\rangle = \chi_{A_1 \ldots A_n}$. ∎

Definition. For any two integers m and n, we write $\mathrm{LST}_{m,n}$ for the following finite subset of $\mathrm{Tp}(m)^*$.

$$\mathrm{LST}_{m,n} \rightleftharpoons \{A_1 \ldots A_l \mid 1 \le l \le n, \; A_1 \in \mathrm{Tp}(m), \ldots, A_l \in \mathrm{Tp}(m)\}$$

Lemma 9.2 *For any given number $m \in \mathbf{N}$ there is a family of $\mathrm{Tp}(m)$-maps $\langle \mathbf{D}_n, \mathbf{V}_n, v_n\rangle$ indexed by $n \in \mathbf{N}$, there is an element g, and there is a family of elements h_Γ indexed by $\Gamma \in \mathrm{Tp}(m)^*$, such that*

(i) $(\forall n) \; g \in \mathbf{D}_n$;

(ii) $(\forall n) \; (\forall \Gamma \in \mathrm{LST}_{m,n}) \; h_\Gamma \in \mathbf{D}_n$;

(iii) $(\forall n) \; (\forall \Gamma \in \mathrm{LST}_{m,n}) \; (\forall C \in \mathrm{Tp}(m)) \; \langle h_\Gamma, g\rangle \in v_n(C) \Leftrightarrow L^* \vdash \Gamma \to C$;

(iv) $(\forall n)\ (\forall \Gamma \in \text{LST}_{m,n-1})\ (\forall B \in \text{Tp}(m))\ \langle h_{B\Gamma}, h_\Gamma\rangle \in v_n(B).$

PROOF. Take arbitrary $m, n \in \mathbf{N}$. We construct the $\text{Tp}(m)$-map $\langle \mathbf{D}_n, \mathbf{V}_n, v_n\rangle$, using the $\text{Tp}(m)$-maps $\langle \mathbf{D}_\Gamma, \mathbf{V}_\Gamma, v_\Gamma\rangle$ from the previous lemma.

We put $\mathbf{D}_n \rightleftharpoons \bigcup_{\Gamma \in \text{LST}_{m,n}} \mathbf{D}_\Gamma$. Let $\mathbf{V}_n$ be any linear order containing the binary relation $\bigcup_{\Gamma \in \text{LST}_{m,n}} \mathbf{V}_\Gamma$. The valuation v_n is defined by $v_n(C) \rightleftharpoons \bigcup_{\Gamma \in \text{LST}_{m,n}} v_\Gamma(C).$

We put $g \rightleftharpoons \chi_\Lambda$ and $h_\Gamma \rightleftharpoons \chi_\Gamma$.

It remains to check that $\langle \mathbf{D}_n, \mathbf{V}_n, v_n\rangle$ is a $\text{Tp}(m)$-map.

(1) Obvious.

(2) Let $L^* \vdash A_1 \ldots A_k \rightarrow B$, $A_i \in \text{Tp}(m)$, and $B \in \text{Tp}(m)$. Assume that $\langle s_0, s_1\rangle \in v_n(A_1), \ldots, \langle s_{k-1}, s_k\rangle \in v_n(A_k)$. Then there is $\Delta \in \text{LST}_{m,n}$ such that $\langle s_0, s_1\rangle \in v_\Delta(A_1)$.

By induction on $i < k$ it can be proved that $\langle s_i, s_{i+1}\rangle \in v_\Delta(A_{i+1})$ for the same Δ. Thus $\langle s_0, s_k\rangle \in v_\Delta(B) \subseteq v_n(B)$.

(3) Obvious. ∎

We shall also need the dual of Lemma 9.2.

Lemma 9.3 *For any given number $m \in \mathbf{N}$ there is a family of $\text{Tp}(m)$-maps $\langle \mathbf{D}'_n, \mathbf{V}'_n, v'_n\rangle$ indexed by $n \in \mathbf{N}$, there is an element g', and there is a family of elements h'_Γ indexed by $\Gamma \in \text{Tp}(m)^*$, such that*

(i) $(\forall n)\ g' \in \mathbf{D}'_n$;

(ii) $(\forall n)\ (\forall \Gamma \in \text{LST}_{m,n})\ h'_\Gamma \in \mathbf{D}'_n$;

(iii) $(\forall n)\ (\forall \Gamma \in \text{LST}_{m,n})\ (\forall C \in \text{Tp}(m))\ \langle g', h'_\Gamma\rangle \in v'_n(C) \Leftrightarrow L^* \vdash \Gamma \rightarrow C$;

(iv) $(\forall n)\ (\forall \Gamma \in \text{LST}_{m,n-1})\ (\forall B \in \text{Tp}(m))\ \langle h'_\Gamma, h'_{\Gamma B}\rangle \in v'_n(B).$

10 Construction of witnesses

In this section we prove that the class $\mathcal{K}^m$ is witnessed.

We assume being given a number $m \in \mathbf{N}$, a $\text{Tp}(m)$-quasimodel $\langle \mathcal{V}^*, v\rangle \in \mathcal{K}^m$, two types E and F such that $E\backslash F \in \text{Tp}(m)$, and a word $\delta \in \mathcal{V}^*$ such that $\delta \notin v(E\backslash F)$. We fix m, $\mathcal{V}$, v, E, F, and δ until the end of this section. Our aim is to find a $\text{Tp}(m)$-quasimodel $\langle \mathcal{W}^*, w\rangle \in \mathcal{K}^m$ and a word $\zeta \in \mathcal{W}^*$ such that $\zeta \in w(E)$, $\zeta \circ \delta \notin w(F)$, and $\langle \mathcal{W}^*, w\rangle$ is a conservative extension of $\langle \mathcal{V}^*, v\rangle$.

First, we put $n \rightleftharpoons |\delta| + 1$. For the given m and n we take the $\text{Tp}(m)$-map $\langle \mathbf{D}'_n, \mathbf{V}'_n, v'_n\rangle$ from Lemma 9.3. Let $k \rightleftharpoons |\mathbf{D}'_n|$. The reflexive linear order $\mathbf{V}'_n$ is isomorphic to the natural order $\leq$ of the set $[0, k-1] \rightleftharpoons \{i \in \mathbf{Z} \mid 0 \leq i \leq k - 1\}$.

Throughout this section we shall identify $\mathbf{D}_n$ with $[0, k-1]$ and the linear order $\mathbf{V}'_n$ with $\leq$.

Let $x, z, y_1, y_2, \ldots y_k$ be any $k+2$ distinct elements of $\mathcal{U} = \{a_j \mid j \in \mathbf{N}\}$, which do not occur in $\mathcal{V}$. We denote $\mathcal{Y} \rightleftharpoons \{x, z, y_1, y_2, \ldots, y_k\}$ and put $\mathcal{W} \rightleftharpoons \mathcal{V} \cup \mathcal{Y}$.

We shall work with subwords of

$$(x^m \circ y_1 \circ z^m) \circ (x^m \circ y_2 \circ z^m) \circ \ldots \circ (x^m \circ y_k \circ z^m).$$

Here $x^m \rightleftharpoons \underbrace{x \circ \ldots \circ x}_{m \text{ times}}$.

To define the mapping w we need several auxiliary words and sets. For any integers s and t such that $0 \leq s \leq t \leq k$ we define the word $\pi\langle s, t\rangle \in \mathcal{Y}^*$ as follows:

$$\pi\langle s, t\rangle \rightleftharpoons (x^m \circ y_{s+1} \circ z^m) \circ (x^m \circ y_{s+2} \circ z^m) \circ \ldots \circ (x^m \circ y_t \circ z^m).$$

By definition, $\pi\langle s, s\rangle = \varepsilon$ for every s.

Note that if $0 \leq r \leq s \leq t \leq k$, then $\pi\langle r, s\rangle \circ \pi\langle s, t\rangle = \pi\langle r, t\rangle$.

We shall denote by $\mathrm{Subword}(\beta)$ the set of all subwords of β. Formally,

$$\mathrm{Subword}(\beta) \rightleftharpoons \{\alpha \in \mathcal{W}^* \mid \beta = \gamma_1 \circ \alpha \circ \gamma_2 \text{ for some } \gamma_1, \gamma_2 \in \mathcal{W}^*\}.$$

We define the finite set $\mathcal{R}$ as follows.

$$\mathcal{R} \;\rightleftharpoons\; \{\rho \in \mathcal{V}^* \mid \rho \circ \alpha = \delta \text{ for some } \alpha \in \mathcal{V}^*\}$$

We define several functions associating subsets of $\mathcal{W}^*$ with sequences of types from $\mathrm{Tp}(m)$. For any $\Theta \in \mathrm{Tp}(m)^*$ we put

$$u_0(\Theta) \;\rightleftharpoons\; \{\pi\langle s, t\rangle \mid 0 \leq s < t < k \text{ and } \langle s, t\rangle \in \vec{v_n}(\Theta)\};$$

$$u_2(\Theta) \;\rightleftharpoons\; \{\pi\langle s, k\rangle \circ \rho \mid 0 \leq s < k, \; \rho \in \mathcal{R}, \text{ and } \exists \Delta \in \mathrm{Tp}(m)^*,$$
$$|\Delta| \leq |\rho|, \; \rho \in \vec{v}(\Delta), \; \langle s, h'_{E\Delta}\rangle \in \vec{v_n}(\Theta)\};$$

$$u(\Theta) \;\rightleftharpoons\; u_0(\Theta) \cup u_2(\Theta) \cup \vec{v}(\Theta).$$

We define some subsets of $\mathcal{W}^*$.

$$\mathcal{M}_1 \;\rightleftharpoons\; \{\alpha \in \mathcal{W}^* \mid \alpha \notin \mathcal{V}^* \text{ and } \alpha \notin \mathrm{Subword}(\pi\langle 0, k\rangle \circ \delta)\}$$
$$\mathcal{M}_2 \;\rightleftharpoons\; z \circ \mathcal{W}^*$$
$$\mathcal{M}_3 \;\rightleftharpoons\; \mathcal{W}^* \circ x$$
$$\mathcal{M} \;\rightleftharpoons\; \mathcal{M}_1 \cup \mathcal{M}_2 \cup \mathcal{M}_3$$

We define a function $\mathrm{Subst}_\mathcal{M}: \mathcal{W}^* \to \mathbf{P}(\mathcal{W}^*)$ and two valuations $\tilde{v}: \mathrm{Tp}(m) \to \mathbf{P}(\mathcal{W}^*)$ and $w: \mathrm{Tp}(m) \to \mathbf{P}(\mathcal{W}^*)$.

$$\mathrm{Subst}_{\mathcal{M}}(\varepsilon) \;\; \rightleftharpoons \;\; \{\varepsilon\}$$

$\mathrm{Subst}_{\mathcal{M}}(\alpha \circ q) \;\; \rightleftharpoons \;\; \mathrm{Subst}_{\mathcal{M}}(\alpha) \circ (\{q\} \cup \mathcal{M})$ if $\alpha \in \mathcal{W}^*$ and $q \in \mathcal{W}$

Informally, for every word $\beta \in \mathcal{W}^*$, the set $\mathrm{Subst}_{\mathcal{M}}(\beta)$ consists of all words that are obtained replacing some (may be none) of symbol occurrences in β by words from the set $\mathcal{M}$.

$$\tilde{v}(A) \;\; \rightleftharpoons \;\; \bigcup_{\alpha \in v(A)} \mathrm{Subst}_{\mathcal{M}}(\alpha)$$

$$w(A) \;\; \rightleftharpoons \;\; u(A) \cup \tilde{v}(A)$$

Finally, we put $\zeta \rightleftharpoons \pi\langle g', k\rangle$.

Lemma 10.1 $\zeta \in u(E)$.

PROOF. From $L^* \vdash E{\to}E$ and Lemma 9.3 (iii) we obtain $\langle g', h'_E\rangle \in v'_n(E)$. Thus $\pi\langle g', k\rangle \in u_2(E)$. ∎

Lemma 10.2 $\zeta \circ \delta \notin u(F)$.

PROOF. Assume, for the contrary, that $\pi\langle g', k\rangle \circ \delta \in u(F)$. Evidently $\pi\langle g', k\rangle \in \mathcal{Y}^+$ and $\delta \in \mathcal{V}^+$. Thus $\pi\langle g', k\rangle \circ \delta \in u_2(F)$. According to the definition of u_2 there is $\Delta \in \mathrm{Tp}(m)^*$ such that $|\Delta| \leq |\delta|$, $\delta \in \tilde{v}(\Delta)$ and $\langle g', h'_{E\Delta}\rangle \in v'_n(F)$. From $n = |\delta| + 1$ we get $|\Delta| \leq n - 1$, whence $E\Delta \in \mathrm{LST}_{m,n}$. From Lemma 9.3 (iii) we obtain $L^* \vdash E\Delta{\to}F$. Applying the rule $({\to}\backslash)$ we derive $L^* \vdash \Delta{\to}E\backslash F$, whence $\tilde{v}(\Delta) \subseteq v(E\backslash F)$. Thus $\delta \in v(E\backslash F)$. Contradiction. ∎

Lemma 10.3 If $0 \leq r \leq s < k$, $A \in \mathrm{Tp}(m)$, and $\langle r, s\rangle \in v'_n(A)$, then $\pi\langle r, s\rangle \in u(A)$.

PROOF. Let $\langle r, s\rangle \in v'_n(A)$. If $r < s$, then $\pi\langle r, s\rangle \in u_0(A)$. If $r = s$, then $L^* \vdash \Lambda{\to}A$ and thus $\varepsilon \in v(A)$. ∎

Lemma 10.4 If $A{\bullet}B \in \mathrm{Tp}(m)$, then $u(A{\bullet}B) \subseteq u(A) \circ u(B)$.

PROOF. Let $A{\bullet}B \in \mathrm{Tp}(m)$ and $\gamma \in u(A{\bullet}B)$.
CASE 1: $\gamma \in u_0(A{\bullet}B)$
By definition $\gamma = \pi\langle r, t\rangle$, $0 \leq r < t < k$, $\langle r, t\rangle \in \vec{v'_n}(A{\bullet}B) = v'_n(A{\bullet}B)$ for some r and t.

Since $\langle \mathbf{V}'_n, v'_n\rangle$ is a $\mathrm{Tp}(m)$-map, there is $s \in [0, k-1]$ such that $\langle r, s\rangle \in v'_n(A)$ and $\langle s, t\rangle \in v'_n(B)$.

Now $\gamma = \pi\langle r, t\rangle = \pi\langle r, s\rangle \circ \pi\langle s, t\rangle \in u_0(A) \circ u_0(B)$ according to Lemma 10.3.

CASE 2: $\gamma \in u_2(A \bullet B)$

By definition $\gamma = \pi\langle r, k \rangle \circ \rho$, $\rho \in \mathcal{R}$, $\Delta \in \mathrm{Tp}(m)^*$, $|\Delta| \leq |\rho|$, $\rho \in \vec{v}(\Delta)$, $\langle r, h'_{E\Delta} \rangle \in v'_n(A \bullet B)$, $0 \leq r < k$.

Again, there is s such that $\langle r, s \rangle \in v'_n(A)$ and $\langle s, h'_{E\Delta} \rangle \in v'_n(B)$. According to Lemma 10.3, $\pi\langle r, s \rangle \in u_0(A)$. On the other hand $\pi\langle s, k \rangle \circ \rho \in u_2(B)$, whence $\gamma \in u_0(A) \circ u_2(B)$.

CASE 3: $\gamma \in v(A \bullet B)$

Obvious from $v(A \bullet B) \subseteq v(A) \circ v(B)$. ∎

We define
$$T \rightleftharpoons \underbrace{\mathcal{M} \circ \ldots \circ \mathcal{M}}_{m \text{ times}}$$
and establish several properties of $\mathcal{M}$ and T.

Lemma 10.5

(i) *If $\beta \in \mathcal{W}^*$, $\alpha \in \mathrm{Subword}(\beta)$, and $\alpha \in \mathcal{M}_1$, then $\beta \in \mathcal{M}_1$.*

(ii) $\mathcal{M}_2 \circ \mathcal{W}^* \subseteq \mathcal{M}_2$

(iii) $\mathcal{W}^* \circ \mathcal{M}_3 \subseteq \mathcal{M}_3$

PROOF. (i) According to the definition of $\mathcal{M}_1$, if $\beta \in \mathcal{W}^*$ and $\beta \notin \mathcal{M}_1$, then $\beta \in \mathcal{V}^*$ or $\beta \in \mathrm{Subword}(\pi\langle 0, k \rangle \circ \delta)$. But then one has also $\alpha \in \mathcal{V}^*$ or $\alpha \in \mathrm{Subword}(\pi\langle 0, k \rangle \circ \delta)$, whence $\alpha \notin \mathcal{M}_1$. ∎

Lemma 10.6

(i) $\mathcal{M} \circ \mathcal{M} \subseteq \mathcal{M}$

(ii) $T \subseteq \mathcal{M}$

PROOF.

(i)

Let $\alpha \in \mathcal{M}$ and $\beta \in \mathcal{M}$. We verify that $\alpha \circ \beta \in \mathcal{M}$. If $\alpha \in \mathcal{M}_1$ then $\alpha \circ \beta \in \mathcal{M}_1$. If $\alpha \in \mathcal{M}_2$ then $\alpha \circ \beta \in \mathcal{M}_2$. If $\beta \in \mathcal{M}_1$ then $\alpha \circ \beta \in \mathcal{M}_1$. If $\beta \in \mathcal{M}_3$ then $\alpha \circ \beta \in \mathcal{M}_3$.

The only complicated case is $\alpha \in \mathcal{M}_3$ and $\beta \in \mathcal{M}_2$, i.e., $\alpha = \alpha' \circ x$ and $\beta = z \circ \beta'$. Note that then $x \circ z \in \mathrm{Subword}(\alpha \circ \beta)$ and $x \circ z \in \mathcal{M}_1$. Thus $\alpha \circ \beta \in \mathcal{M}_1$.

(ii)

Follows from (i). ∎

We introduce some subsets of $\mathcal{W}^*$.

$$\begin{aligned}
\mathcal{P}_0 &\rightleftharpoons \{\pi\langle s, t \rangle \mid 0 \leq s < t < k\} \\
\mathcal{P}_1 &\rightleftharpoons \{\pi\langle s, k \rangle \mid 0 \leq s < k\} \\
\mathcal{P}_2 &\rightleftharpoons \mathcal{P}_1 \circ \mathcal{R} \\
\mathcal{P} &\rightleftharpoons \mathcal{P}_0 \cup \mathcal{P}_2
\end{aligned}$$

Note that $\varepsilon \in \mathcal{R}$ and thus $\mathcal{P}_1 \subsetneq \mathcal{P}_2$. Note that $u_0(\Theta) \subsetneq \mathcal{P}_0$, $u_2(\Theta) \subset \mathcal{P}_2$, and $u(\Theta) \subseteq \mathcal{P} \cup \mathcal{V}^*$.

Lemma 10.7 *If $A \in \mathrm{Tp}(m)$, then $\mathcal{T} \subseteq w(A)$.*

PROOF. Since $\langle \mathcal{V}^*, v \rangle \in \mathcal{K}^m$, we can choose a word $\alpha \in v(A)$ such that $|\alpha| \leq m$. Evidently $\underbrace{\mathcal{M} \circ \ldots \circ \mathcal{M}}_{|\alpha| \text{ times}} \subseteq \mathrm{Subst}_{\mathcal{M}}(\alpha) \subseteq \tilde{v}(A) \subseteq w(A)$. In view of $\mathcal{M} \circ \mathcal{M} \subseteq \mathcal{M}$ and taking into account that $|\alpha| \leq m$, we have $\underbrace{\mathcal{M} \circ \ldots \circ \mathcal{M}}_{m \text{ times}} \subseteq$ $\underbrace{\mathcal{M} \circ \ldots \circ \mathcal{M}}_{|\alpha| \text{ times}}$. Thus $\mathcal{T} = \underbrace{\mathcal{M} \circ \ldots \circ \mathcal{M}}_{m \text{ times}} \subseteq w(A)$. ■

Lemma 10.8

(i) $\mathcal{P} \cap \mathcal{V}^* = \emptyset$

(ii) $\mathcal{P} \cap \mathcal{M} = \emptyset$

(iii) $\mathcal{V}^* \cap \mathcal{M} = \emptyset$

PROOF.
(i)
Evident.
(ii)
Let $\alpha \in \mathcal{P}$. Then the leftmost symbol of α is x and the rightmost symbol of α belongs to $\mathcal{V} \cup \{z\}$. Thus $\alpha \notin \mathcal{M}_2$ and $\alpha \notin \mathcal{M}_3$. Note that $\mathcal{P} \subseteq \mathrm{Subword}(\pi \langle 0, k \rangle \circ \delta)$. Thus $\alpha \notin \mathcal{M}_1$.
(iii)
Obvious. ■

Lemma 10.9

(i) $v(A) \subseteq u(A)$

(ii) $u(A) \subseteq v(A) \cup \mathcal{P}$

Lemma 10.10

(a) $\mathcal{V}^* \circ \mathcal{M} \subseteq \mathcal{M}$

(b) $\mathcal{M} \circ \mathcal{V}^* \subseteq \mathcal{M}$

(c) $\mathcal{P} \circ \mathcal{M} \subseteq \mathcal{T}$

(d) $\mathcal{M} \circ \mathcal{P} \subseteq \mathcal{T}$

(e) $\mathcal{V}^+ \circ \mathcal{P} \subseteq \mathcal{T}$

(f) $\mathcal{P}_0 \circ \mathcal{V}^+ \subseteq \mathcal{T}$

(g) $\mathcal{P}_1 \circ \{\beta \in \mathcal{W}^+ \mid \beta \notin \mathcal{R}\} \subseteq T$

(h) *If* $0 \le r < s < k$, $0 \le s' < t < k$, *and* $s \ne s'$, *then* $\pi\langle r, s\rangle \circ \pi\langle s', t\rangle \in T$.

PROOF.

(a)
Let $\alpha \in \mathcal{V}^+$ and $\beta \in \mathcal{M}$. We verify that $\alpha \circ \beta \in \mathcal{M}$. If $\beta \in \mathcal{M}_1$ then $\alpha \circ \beta \in \mathcal{M}_1$. If $\beta \in \mathcal{M}_3$ then $\alpha \circ \beta \in \mathcal{M}_3$. The only complicated case is $\beta = z \circ \beta'$. Note that in that case $\alpha \circ z \in \text{Subword}(\alpha \circ \beta)$ and $\alpha \circ z \in \mathcal{M}_1$.

(b)
Let $\alpha \in \mathcal{M}$ and $\beta \in \mathcal{V}^+$. We verify that $\alpha \circ \beta \in \mathcal{M}$. If $\alpha \in \mathcal{M}_1$ then $\alpha \circ \beta \in \mathcal{M}_1$. If $\alpha \in \mathcal{M}_2$ then $\alpha \circ \beta \in \mathcal{M}_2$. The only complicated case is $\alpha = \alpha' \circ x$. In this case $x \circ \beta \in \text{Subword}(\alpha \circ \beta)$ and $x \circ \beta \in \mathcal{M}_1$.

(c)
Let $\gamma = \alpha \circ \beta$, where $\alpha \in \mathcal{P}$ and $\beta \in \mathcal{M}$.
CASE 1: $\alpha \in \mathcal{P}_0 \cup \mathcal{P}_1$
By definition $\gamma = \pi\langle s, t\rangle \circ \beta$, where $0 \le s < t < k$.
   Evidently $\gamma = \underbrace{x \circ \ldots \circ x}_{m-1 \text{ times}} \circ \phi$, where $\phi = x \circ y_{s+1} \circ z^m \circ \pi\langle s+1, t\rangle \circ \beta$. We must verify that $\phi \in \mathcal{M}$.
   If $\beta \in \mathcal{M}_1$ then $\phi \in \mathcal{M}_1$. If $\beta \in \mathcal{M}_3$ then $\phi \in \mathcal{M}_3$. Let now $\beta \in \mathcal{M}_2$, i.e. $\beta = z \circ \beta'$. Evidently $z^{m+1} \in \text{Subword}(y_{s+1} \circ z^m \circ \pi\langle s+1, t\rangle \circ z \circ \beta'$ and $z^{m+1} \in \mathcal{M}_1$.
CASE 2: $\alpha \in \mathcal{P}_2$ and $\alpha \notin \mathcal{P}_1$
By definition $\gamma = \pi\langle s, k\rangle \circ \rho \circ \beta$, $\rho \in \mathcal{R}$, $\rho \ne \varepsilon$.
   Evidently $\gamma = \underbrace{x \circ \ldots \circ x}_{m-1 \text{ times}} \circ \phi$, where $\phi = x \circ y_{s+1} \circ z^m \circ \pi\langle s+1, k\rangle \circ \rho \circ \beta$.
The only complicated case is $\beta \in \mathcal{M}_2$, i.e., $\beta = z \circ \beta'$. Note that $\rho \circ z \in \text{Subword}(\phi)$ and $\rho \circ z \in \mathcal{M}_1$.

(d) and (e)
Let $\alpha \in \mathcal{M} \cup \mathcal{V}^+$ and $\beta \in \mathcal{P}$. We must prove that $\alpha \circ \beta \in \underbrace{\mathcal{M} \circ \ldots \circ \mathcal{M}}_{m \text{ times}}$.

CASE 1: $\beta = \pi\langle s, t\rangle \in \mathcal{P}_0$
Evidently $\alpha \circ \beta = \phi \circ \underbrace{z \circ \ldots \circ z}_{m-1 \text{ times}}$, where $\phi = (\alpha \circ \pi\langle s, t-1\rangle \circ x^m \circ y_t \circ z)$.
Obviously $z \in \mathcal{M}_2$. It remains to verify that $\phi \in \mathcal{M}$.
CASE 1a: $\alpha \in \mathcal{M}_1$
Obvious from Lemma 10.5 (i).
CASE 1b: $\alpha \in \mathcal{M}_2$
Obvious from Lemma 10.5 (ii).
CASE 1c: $\alpha \in \mathcal{M}_3$
   Note that the rightmost symbol of α is x and the first m symbols of the word $\pi\langle s, t-1\rangle \circ x^m \circ y_t \circ z$ are x^m. Thus $x^{m+1} \in \text{Subword}(\phi)$. In view of $x^{m+1} \in \mathcal{M}_1$ we have $\phi \in \mathcal{M}_1$.

CASE 1d: $\alpha \in \mathcal{V}^+$

Evidently $\alpha \circ x \in \text{Subword}(\phi)$. On the other hand, $\alpha \circ x \in \mathcal{V}^+ \circ \mathcal{Y}^+$ and $\mathcal{V}^+ \circ \mathcal{Y}^+ \subseteq \mathcal{M}_1$. According to Lemma 10.5 (i'), $\phi \in \mathcal{M}_1$.

CASE 2: $\beta \in \mathcal{P}_2$

By definition $\beta = \pi\langle s, k\rangle \circ \rho$, $\rho \in \mathcal{R}$.

Now $\alpha \circ \beta = \phi \circ \underbrace{z \circ \ldots \circ z}_{m-2 \text{ times}} \circ (z \circ \rho)$, where ϕ is the same as in the previous case. We have already verified that $z \in \mathcal{M}$ and $\phi \in \mathcal{M}$. Evidently $z \circ \rho \in \mathcal{M}_2$.

(f)

Let $\alpha \in \mathcal{P}_0$ and $\beta \in \mathcal{V}^+$. By definition $\alpha = \pi\langle s, t\rangle$, $0 \le s < t < k$.

Evidently $\alpha \circ \beta = \underbrace{x \circ \ldots \circ x}_{m-1 \text{ times}} \circ \phi$, where $\phi = x \circ y_{s+1} \circ z^m \circ \pi\langle s+1, t\rangle \circ \beta$.

Note that $y_t \circ z^m \circ \beta \in \text{Subword}(\phi)$. On the other hand, $y_t \circ z^m \circ \beta \in \mathcal{M}_1$, since $t \ne k$. Thus $\phi \in \mathcal{M}_1$.

(g)

Let $\alpha \in \mathcal{P}_1$, $\beta \in \mathcal{W}^*$, and $\beta \notin \mathcal{R}$. By definition $\alpha = \pi\langle s, k\rangle$, where $0 \le s < k$.

Evidently $\alpha \circ \beta = \underbrace{x \circ \ldots \circ x}_{m-1 \text{ times}} \circ \phi$, where $\phi = x \circ y_{s+1} \circ z^m \circ \pi\langle s+1, k\rangle \circ \beta$.

Note that $z \circ \beta \in \text{Subword}(\phi)$. On the other hand, $z \circ \beta \in \mathcal{M}_1$, since β is not a left subword of δ (see the definition of $\mathcal{R}$). Thus $\phi \in \mathcal{M}_1$.

(h)

Evidently $\pi\langle r, s\rangle \circ \pi\langle s', t\rangle = \phi \circ \underbrace{z \circ \ldots \circ z}_{m-1 \text{ times}}$, where

$$\phi = (\pi\langle r, s\rangle \circ \pi\langle s', t-1\rangle \circ x^m \circ y_t \circ z).$$

We only need to prove that $\phi \in \mathcal{M}$. Note that $y_s \circ z^m \circ x^m \circ y_{s'+1} \in \text{Subword}(\phi)$. On the other hand $y_s \circ z^m \circ x^m \circ y_{s'+1} \in \mathcal{M}_1$, since $s \ne s'$. According to Lemma 10.5 (i'), $\phi \in \mathcal{M}_1$. ∎

To make the formulation of the following lemmas more readable we introduce two subsets of $\mathcal{W}^*$.

$$\mathcal{Q} \rightleftharpoons \mathcal{P} \cup \mathcal{V}^* \cup \mathcal{M}$$
$$\mathcal{Q}_\infty \rightleftharpoons \{\varepsilon\} \cup \mathcal{Q} \cup (\mathcal{Q} \circ \mathcal{Q}) \cup (\mathcal{Q} \circ \mathcal{Q} \circ \mathcal{Q}) \cup \ldots$$

Lemma 10.11

(i) $\mathcal{T} \circ \mathcal{Q} \subseteq \mathcal{T}$

(ii) $\mathcal{Q} \circ \mathcal{T} \subseteq \mathcal{T}$

(iii) $\mathcal{P} \circ \mathcal{V}^* \subseteq \mathcal{P} \cup \mathcal{T}$

(iv) $\mathcal{P} \circ \mathcal{P} \subseteq \mathcal{P} \cup \mathcal{T}$

(v) $(\mathcal{P} \cup \mathcal{T}) \circ \mathcal{Q} \subseteq (\mathcal{P} \cup \mathcal{T})$

(vi) $\mathcal{Q} \circ (\mathcal{P} \cup \mathcal{T}) \subseteq (\mathcal{P} \cup \mathcal{T})$

PROOF.
(i)
Evident from Lemma 10.10 (d), (b), and Lemma 10.6.
(ii)
Evident from Lemma 10.10 (c), (a), and Lemma 10.6.
(iii)
Let $\alpha \in \mathcal{P}$ and $\beta \in \mathcal{V}^*$. We must prove that $\alpha \circ \beta \in \mathcal{P} \cup \mathcal{T}$. If $\beta = \varepsilon$, then $\alpha \circ \beta = \alpha \in \mathcal{P}$. Let $\beta \in \mathcal{V}^+$.
CASE 1: $\alpha \in \mathcal{P}_0$
According to Lemma 10.10 (f), $\alpha \circ \beta \in \mathcal{T}$.
CASE 2: $\alpha = \pi\langle s, k\rangle \circ \rho \in \mathcal{P}_2$
If $\rho \circ \beta \in \mathcal{R}$, then $\alpha \circ \beta \in \mathcal{P}_2$, else $\alpha \circ \beta \in \mathcal{T}$ in view of Lemma 10.10 (g).
(iv)
Let $\alpha \in \mathcal{P}$ and $\beta \in \mathcal{P}$. We must prove that $\alpha \circ \beta \in \mathcal{P} \cup \mathcal{T}$.
CASE 1: $\alpha \in \mathcal{P}_0 \cup \mathcal{P}_1$
By definition $\alpha = \pi\langle r, s\rangle$, where $0 \le r < s \le k$.
CASE 1a: $\beta \in \mathcal{P}_0 \cup \mathcal{P}_1$
By definition $\beta = \pi\langle s', t\rangle$, where $0 \le s' < t \le k$.
If $s = s'$, then $\alpha \circ \beta = \pi\langle r, s\rangle \circ \pi\langle s, t\rangle = \pi\langle r, t\rangle \in \mathcal{P}$ according to the definition of the function π. If $s \ne s'$, then $\alpha \circ \beta \in \mathcal{T}$ according to Lemma 10.10 (h).
CASE 1b: $\beta \in \mathcal{P}_2$
Evidently $\alpha \circ \beta \in \alpha \circ \mathcal{P}_1 \circ \mathcal{R}$. According to case 1a, $\alpha \circ \beta \in (\mathcal{P} \cup \mathcal{T}) \circ \mathcal{R} \subseteq (\mathcal{P} \cup \mathcal{T}) \circ \mathcal{V}^*$. From (iii) and (i) we obtain $(\mathcal{P} \cup \mathcal{T}) \circ \mathcal{V}^* \subseteq \mathcal{P} \cup \mathcal{T}$.
CASE 2: $\alpha \in \mathcal{P}_2, \alpha \notin \mathcal{P}_1$
By definition $\alpha = \pi\langle s, k\rangle \circ \rho, \rho \in \mathcal{R}, \rho \ne \varepsilon$.
Evidently, $\mathcal{P}_2 \circ \mathcal{P} = \mathcal{P}_1 \circ \mathcal{R} \circ \mathcal{P} \subseteq \mathcal{P}_1 \circ (\mathcal{V}^* \circ \mathcal{P})$.
From (ii) and Lemma 10.10 (e) we get $\mathcal{P}_1 \circ (\mathcal{V}^+ \circ \mathcal{P}) \subseteq \mathcal{P}_1 \circ \mathcal{T} \subseteq \mathcal{T}$.
(v)
Immediate from (i), (iv), (iii), and Lemma 10.10 (c).
(vi)
Immediate from (ii), (iv), and Lemma 10.10 (e), (d). ∎

Lemma 10.12

(i) $\mathcal{Q}_\infty \circ \mathcal{P} \circ \mathcal{Q}_\infty \subseteq \mathcal{P} \cup \mathcal{T}$

(ii) $\mathcal{Q}_\infty \circ \mathcal{P} \circ \mathcal{Q}_\infty \circ \mathcal{M} \circ \mathcal{Q}_\infty \subseteq \mathcal{T}$

(iii) $\mathcal{Q}_\infty \circ \mathcal{M} \circ \mathcal{Q}_\infty \circ \mathcal{P} \circ \mathcal{Q}_\infty \subseteq \mathcal{T}$

PROOF.

(i)

From Lemma 10.11 (vi) we obtain

$$\mathcal{Q}_\infty \circ (\mathcal{P} \cup \mathcal{T}) \subseteq \mathcal{P} \cup \mathcal{T}.$$

From Lemma 10.11 (v) we obtain

$$(\mathcal{P} \cup \mathcal{T}) \circ \mathcal{Q}_\infty \subseteq \mathcal{P} \cup \mathcal{T}.$$

Thus $(\mathcal{Q}_\infty \circ \mathcal{P}) \circ \mathcal{Q}_\infty \subseteq (\mathcal{P} \cup \mathcal{T}) \circ \mathcal{Q}_\infty \subseteq \mathcal{P} \cup \mathcal{T}$.

(ii)

According to (i), Lemma 10.10 (c), and Lemma 10.6 $(\mathcal{Q}_\infty \circ \mathcal{P} \circ \mathcal{Q}_\infty) \circ \mathcal{M} \subseteq \mathcal{T}$. It remains to apply Lemma 10.11 (i).

(iii)

According to (i), Lemma 10.10 (d), and Lemma 10.6 $\mathcal{M} \circ (\mathcal{Q}_\infty \circ \mathcal{P} \circ \mathcal{Q}_\infty) \subseteq \mathcal{T}$. It remains to apply Lemma 10.11 (ii). ■

Lemma 10.13 *Let* $\Delta \in \mathrm{Tp}(m)^*$, $\Pi \in \mathrm{Tp}(m)^*$, *and* $\varepsilon \in u(\Delta)$. *Then* $v_n(\vec{\Delta}\Pi) = v_n(\vec{\Pi})$ *and* $v_n(\vec{\Pi}\Delta) = v_n(\vec{\Pi})$.

PROOF. Evidently $\varepsilon \in \vec{v}(\Delta)$.

Let $\Delta = A_1 \dots A_l$. Take arbitrary $i \le l$. Evidently, $\varepsilon \in A_i$ and thus $L^* \vdash \Lambda \to A_i$. We obtain $\langle s, s \rangle \in v_n(A_i)$ for every $s \in [0, k-1]$. Hence $\langle s, s \rangle \in \vec{v_n}(\Delta)$ for every $s \in [0, k-1]$. ■

Lemma 10.14 *Let* $\Theta \in \mathrm{Tp}(m)^*$ *and* $B \in \mathrm{Tp}(m)$. *Then* $u(\Theta) \circ u(B) \subseteq u(\Theta B) \cup \mathcal{T}$.

PROOF. Let $\gamma \in u(\Theta) \circ u(B)$. Then $\gamma = \alpha \circ \beta$ for some $\alpha \in u(\Theta) = u_0(\Theta) \cup u_2(\Theta) \cup \vec{v}(\Theta)$ and $\beta \in u(B) = u_0(B) \cup u_2(B) \cup v(B)$. We consider the corresponding nine cases and prove that $\alpha \circ \beta \in u_0(\Theta B) \cup u_2(\Theta B) \cup \vec{v}(\Theta B) \cup \mathcal{T}$.

CASE 1: $\alpha \in u_0(\Theta)$

By definition $\alpha = \pi\langle r, s \rangle$, $0 \le r < s < k$, $\langle r, s \rangle \in \vec{v_n}(\Theta)$.

CASE 1a: $\beta \in u_0(B)$

By definition $\beta = \pi\langle s', t \rangle$, $0 \le s' < t < k$, $\langle s', t \rangle \in v'_n(B)$.

If $s \ne s'$, then $\alpha \circ \beta \in \mathcal{T}$ in view of Lemma 10.10 (h).

If $s = s'$, then $\alpha \circ \beta = \pi\langle r, s \rangle \circ \pi\langle s, t \rangle = \pi\langle r, t \rangle \in u_0(\Theta B)$, since $\langle r, t \rangle \in \vec{v_n}(\Theta) \odot v'_n(B) = \vec{v_n}(\Theta B)$.

CASE 1b: $\beta \in u_2(B)$

By definition $\beta = \pi\langle s', k \rangle \circ \rho$, $\rho \in \mathcal{R}$, $\Delta \in \mathrm{Tp}(m)^*$, $|\Delta| \le |\rho|$, $\rho \in \vec{v}(\Delta)$, $\langle s', h'_{E\Delta} \rangle \in v'_n(B)$, $0 \le s' < k$.

If $s \ne s'$, then $\alpha \circ \beta = \pi\langle r, s \rangle \circ \pi\langle s', k \rangle \circ \rho \in \mathcal{T} \circ \rho \subseteq \mathcal{T} \circ \mathcal{V}^* \subseteq \mathcal{T}$ in view of Lemma 10.10 (h) and (i).

If $s = s'$, then $\alpha \circ \beta = \pi\langle r, k \rangle \circ \rho \in u_2(\Theta B)$, since $\langle r, h'_{E\Delta} \rangle \in \vec{v_n}(\Theta) \odot v'_n(B) = \vec{v_n}(\Theta B)$.

CASE 1c: $\beta \in v(B)$
If $\beta \neq \varepsilon$, then $\alpha \circ \beta \in \mathcal{P}_0 \circ \mathcal{V}^+ \subseteq \mathcal{T}$ in view of Lemma 10.10 (f).
If $\beta = \varepsilon$, then $\alpha \circ \beta = \alpha \in u_0(\Theta) \subseteq u_0(\Theta B)$ in view of Lemma 10.13.

CASE 2: $\alpha \in u_2(\Theta)$
$\alpha = \pi\langle r, k\rangle \circ \rho,\ \rho \in \mathcal{R},\ \Delta \in \mathrm{Tp}(m)^*,\ |\Delta| \leq |\rho|,\ \rho \in \vec{v}(\Delta),\ \langle r, h'_{E\Delta}\rangle \in \vec{v_n}(\Theta),$
$0 \leq r < k$
Note that $\alpha \in \mathcal{P}_1 \circ \mathcal{V}^*$.
CASE 2ab: $\beta \in u_0(B) \cup u_2(B) \subseteq \mathcal{P}$
In view of Lemma 10.10 (g) we have $\alpha \circ \beta \in \mathcal{P}_1 \circ (\mathcal{V}^* \circ \mathcal{P}) \subseteq \mathcal{T}$.
CASE 2c: $\beta \in v(B)$
If $\rho \circ \beta \notin \mathcal{R}$, then $\alpha \circ \beta = \pi\langle r, k\rangle \circ (\rho \circ \beta) \in \mathcal{T}$ in view of Lemma 10.10 (g).
  Now we prove that if $\rho \circ \beta \in \mathcal{R}$ then $\alpha \circ \beta \in u_2(\Theta B)$. We take $\Delta' = \Delta B$
and $\rho' = \rho \circ \beta$. Evidently $\rho \circ \beta \in \vec{v}(\Delta) \circ v(B) = \vec{v}(\Delta B)$.
  If $\beta \neq \varepsilon$, then $|\Delta'| = |\Delta| + 1 \leq |\rho| + 1 \leq |\rho| + |\beta| = |\rho'|$. Further,
$|E\Delta| = |\Delta| + 1 \leq |\rho'| \leq |delta| = n - 1$. By Lemma 9.3 (iv) we have
$\langle h'_{E\Delta}, h'_{E\Delta B}\rangle \in v'_n(B)$.
  Thus $\langle r, h'_{E\Delta B}\rangle \in \vec{v_n}(\Theta) \odot v'_n(B) = \vec{v_n}(\Theta B)$.
If $\beta = \varepsilon$, then $\alpha \circ \beta = \alpha \in u_2(\Theta) \subseteq u_2(\Theta B)$ in view of Lemma 10.13.

CASE 3: $\alpha \in \vec{v}(\Theta)$
CASE 3ab: $\beta \in u_0(B) \cup u_2(B) \subseteq \mathcal{P}$
If $\alpha \neq \varepsilon$, then in view of Lemma 10.10 (e) we have $\alpha \circ \beta \in \mathcal{V}^* \circ \mathcal{P} \subseteq \mathcal{T}$.
  If $\alpha = \varepsilon$, then $\alpha \circ \beta = \beta \in u(B) \subseteq u(\Theta B)$ in view of Lemma 10.13.
CASE 3c: $\beta \in v(B)$
Evidently $\alpha \circ \beta \in \vec{v}(\Theta) \circ v(B) = \vec{v}(\Theta B)$. ∎

Lemma 10.15 *Let* $l \geq 0$, $B_1 \in \mathrm{Tp}(m)$, ..., $B_l \in \mathrm{Tp}(m)$. *Then*

$$u(B_1) \circ \ldots \circ u(B_l) \subseteq u(B_1 \ldots B_l) \cup \mathcal{T}.$$

PROOF. Induction on l.
  Induction base. For $l = 0$ we have to prove that $\{\varepsilon\} \subseteq u(\Lambda) \cup \mathcal{T}$. Indeed,
$\varepsilon \in \vec{v}(\Lambda) \subseteq u(\Lambda)$.
  Induction step. We must prove that if $u(B_1) \circ \ldots \circ u(B_l) \subseteq u(B_1 \ldots B_l) \cup$
$\mathcal{T}$ then $u(B_1) \circ \ldots \circ u(B_l) \circ u(B_{l+1}) \subseteq u(B_1 \ldots B_l B_{l+1}) \cup \mathcal{T}$. In view
of the induction hypothesis, it is sufficient to verify that $(u(B_1 \ldots B_l) \cup$
$\mathcal{T}) \circ u(B_{l+1}) \subseteq u(B_1 \ldots B_l B_{l+1}) \cup \mathcal{T}$.
  From Lemma 10.14 we obtain $u(B_1 \ldots B_l) \circ u(B_{l+1}) \subseteq u(B_1 \ldots B_l B_{l+1}) \cup$
$\mathcal{T}$. According to Lemma 10.11 (i), $\mathcal{T} \circ u(B_{l+1}) \subseteq \mathcal{T}$. ∎

Lemma 10.16 *If* $B_1, \ldots, B_l, C \in \mathrm{Tp}(m)$ *and* $L^* \vdash B_1 \ldots B_l {\rightarrow} C$, *then*

$$u(B_1) \circ \ldots \circ u(B_l) \subseteq u(C) \cup \mathcal{T}.$$

PROOF. Let $B_1, \ldots, B_l, C \in \mathrm{Tp}(m)$ and $L^* \vdash B_1 \ldots B_l {\rightarrow} C$. According
to Lemma 10.15, $u(B_1) \circ \ldots \circ u(B_l) \subseteq u(B_1 \ldots B_l) \cup \mathcal{T}$. It remains to

prove that $u(B_1 \ldots B_l) \subseteq u(C)$. This follows from $\vec{v}(B_1 \ldots B_l) \subseteq v(C)$ and $\vec{v_n}(B_1 \ldots B_l) \subseteq v'_n(C)$ (cf. the definition of u at page 197). ∎

Lemma 10.17 *Let* $A \in \mathrm{Tp}(m)$. *Then*

(i) $v(A) \subseteq \tilde{v}(A)$;

(ii) $\tilde{v}(A) \subseteq v(A) \cup \mathcal{M}$.

PROOF. (i) Let $\alpha \in v(A)$. Evidently $\alpha \in \mathrm{Subst}_{\mathcal{M}}(\alpha) \subseteq \tilde{v}(A)$.

(ii) By induction on $|\alpha|$ we prove that $\mathrm{Subst}_{\mathcal{M}}(\alpha) \subseteq \{\alpha\} \cup \mathcal{M}$ for any $\alpha \in \mathcal{V}^*$. For the induction step it is sufficient to verify that $(\{\alpha\} \cup \mathcal{M}) \circ (\{q\} \cup \mathcal{M}) \subseteq \{\alpha \circ q\} \cup \mathcal{M}$ whenever $\alpha \in \mathcal{V}^*$ and $q \in \mathcal{V}$. This follows from Lemma 10.10 (a), (b) and Lemma 10.6 (i). ∎

Lemma 10.18 *If* $A \in \mathrm{Tp}(m)$, *then* $w(A) \cap \mathcal{V}^* = v(A)$.

PROOF. Let $A \in \mathrm{Tp}(m)$. According to Lemma 10.9 (ii) and Lemma 10.8 (i) $u(A) \cap \mathcal{V}^* = v(A)$.

According to Lemma 10.17 (ii) and Lemma 10.8 (iii) $\tilde{v}(A) \cap \mathcal{V}^* = v(A)$. ∎

Lemma 10.19 $\langle \mathcal{W}^*, w \rangle$ *is a* $\mathrm{Tp}(m)$-*quasimodel.*

PROOF. We verify the conditions (1), (2) and (3) from the definition of a $\mathrm{Tp}(m)$-quasimodel at page 174.

(1)

Let $A \bullet B \in \mathrm{Tp}(m)$ and $\gamma \in w(A \bullet B)$. We must prove that $\gamma \in w(A) \circ w(B)$.

CASE 1: $\gamma \in u(A \bullet B)$

Obvious from Lemma 10.4.

CASE 2: $\gamma \in \tilde{v}(A \bullet B)$

By definition, $\gamma \in \mathrm{Subst}_{\mathcal{M}}(\gamma')$ for some $\gamma' \in v(A \bullet B) \subseteq v(A) \circ v(B)$. Thus $\gamma' = \alpha' \circ \beta'$, where $\alpha' \in v(A)$ and $\beta' \in v(B)$. Evidently, $\mathrm{Subst}_{\mathcal{M}}(\gamma') = \mathrm{Subst}_{\mathcal{M}}(\alpha') \circ \mathrm{Subst}_{\mathcal{M}}(\beta') \subseteq \tilde{v}(A) \circ \tilde{v}(B)$.

(2)

Let $A_1, \ldots, A_l, B \in \mathrm{Tp}(m)$, $L^* \vdash A_1 \ldots A_l \to B$, $\alpha_1 \in w(A_1), \ldots \alpha_l \in w(A_l)$. We must prove that $\alpha_1 \circ \ldots \circ \alpha_l \in w(B)$.

CASE 1: $(\forall i \leq l)\, \alpha_i \in u(A_i)$

According to Lemma 10.16, $\alpha_1 \circ \ldots \circ \alpha_l \in u(B) \cup \mathcal{T}$.

In view of Lemma 10.7, $\alpha_1 \circ \ldots \circ \alpha_l \in w(B)$.

CASE 2: $(\forall j \leq l)\, \alpha_j \in \tilde{v}(A_j)$

This means that for every index $j \leq l$ there is a word $\beta_j \in v(A_j)$ such that $\alpha_j \in \mathrm{Subst}_{\mathcal{M}}(\beta_j)$. Evidently, $\alpha_1 \circ \ldots \circ \alpha_l \in \mathrm{Subst}_{\mathcal{M}}(\beta_1 \circ \ldots \circ \beta_l)$. Note that $\beta_1 \circ \ldots \circ \beta_l \in v(A_1) \circ \ldots \circ v(A_l) \subseteq v(B)$, since $\langle \mathcal{V}^*, v \rangle$ is a $\mathrm{Tp}(m)$-quasimodel. Thus $\alpha_1 \circ \ldots \circ \alpha_l \in \tilde{v}(B)$.

CASE 3: $(\exists i \leq l)\, \alpha_i \notin u(A_i)$ and $(\exists j \leq l)\, \alpha_j \notin \tilde{v}(A_j)$

Evidently $\alpha_i \in \tilde{v}(A_i)$. From Lemma 10.17 (ii) we obtain $\alpha_i \in v(A_i) \cup \mathcal{M}$. In view of $v(A_i) \subseteq u(A_i)$ we have $\alpha_i \notin v(A_i)$. Thus $\alpha_i \in \mathcal{M}$.

Evidently $\alpha_j \in u(A_j) \subseteq v(A_j) \cup \mathcal{P}$. On the other hand, from Lemma 10.17 (i) we obtain $\alpha_j \notin v(A_j)$. Thus $\alpha_j \in \mathcal{P}$.

Note that $\alpha_h \in \mathcal{Q}$ for every $h \leq l$. According to Lemma 10.12 (i) and (ii), $\alpha_1 \circ \ldots \circ \alpha_l \in \mathcal{T}$. It remains to apply Lemma 10.7.

(3)

Immediate from Lemma 10.18. ∎

Lemma 10.20

(i) $\langle \mathcal{W}^*, w \rangle \in \mathcal{K}^m$;

(ii) $\langle \mathcal{W}^*, w \rangle$ *is a conservative extension of* $\langle \mathcal{V}^*, v \rangle$;

(iii) $\zeta \in w(E)$;

(iv) $\zeta \circ \delta \notin w(F)$.

PROOF. (i) Obvious.

(ii) Immediate from Lemma 10.18.

(iii) Obvious from Lemma 10.1.

(iv) Follows from Lemma 10.2, Lemma 10.17 (ii), and Lemma 10.8 (i), (ii). ∎

Lemma 10.21 *The class* $\mathcal{K}^m$ *is witnessed.*

PROOF. Immediate from Lemma 10.20 and its dual for F/E. ∎

11   Main result

Theorem 11.1 *Let* $\Gamma \in \mathrm{Tp}^*$ *and* $A \in \mathrm{Tp}$. *Then* $L^* \vdash \Gamma{\to}A$ *if and only if the sequent* $\Gamma{\to}A$ *is true in every free monoid model over a countable alphabet.*

PROOF. The 'only if' part coincides with Theorem 2.1. The 'if' part is immediate from Lemma 10.21, Lemma 5.2, Lemma 2.2, and Lemma 2.3. ∎

Theorem 11.2 *Let* $\Gamma \in \mathrm{Tp}^*$ *and* $A \in \mathrm{Tp}$. *Then* $L^* \vdash \Gamma{\to}A$ *if and only if the sequent* $\Gamma{\to}A$ *is true in every free monoid model over a two symbol alphabet.*

PROOF. Following the proof of Theorem 11.1 we reduce the proof to the case of a sequent $\Lambda{\to}F$ and we find a free monoid model $\langle \mathcal{V}^*, v \rangle$, where $\mathcal{V} \subseteq \{a_j \mid j \in \mathbf{N}\}$ such that $\varepsilon \notin v(F)$ and $v(A) \neq \emptyset$ for every $A \in \mathrm{Tp}(m)$. Here $m = \|F\|$.

We take $\mathcal{W} \rightleftharpoons \{b, c\}$ and define a function $g: \mathcal{V}^* \to \mathcal{W}^*$ as follows.

$$g(a_j) \rightleftharpoons b \circ \underbrace{c \circ \ldots \circ c}_{j \text{ times}} \circ b \qquad g(\alpha \circ \beta) \rightleftharpoons g(\alpha) \circ g(\beta)$$

Note that g is injective.

Now we put $w(p_i) \rightleftharpoons \{g(\gamma) \mid \gamma \in v(p_i)\}$ for every primitive type p_i and define $w(A)$ for complex types by induction according to the definition of a free monoid model (cf. p. 173).

By induction on the complexity of A it is easy to prove that $w(A) = \{g(\gamma) \mid \gamma \in v(A)\}$ for every $A \in \mathrm{Tp}(m)$. In the proof of $\{\gamma \in \mathcal{W}^* \mid w(A) \circ \gamma \subseteq w(B)\} \subseteq w(A \backslash B)$ we use Lemma 5.2 (ii) and the fact that if $\alpha' \in \mathcal{V}^*$, $\beta' \in \mathcal{V}^*$, $\gamma \in \mathcal{W}^*$, and $g(\alpha') \circ \gamma = g(\beta')$ then there is $\gamma' \in \mathcal{V}^*$ such that $\gamma = g(\gamma')$.

Similarly for the dual case $\{\gamma \in \mathcal{W}^* \mid \gamma \circ w(A) \subseteq w(B)\} = w(B/A)$. Other cases of the induction step are trivial. ∎

Acknowledgements

I would like to thank S. Artemov, L. Beklemishev, and M. Kanovich for constructive discussions of topics related to this paper. I am grateful to the anonymous referee for a number of suggestions that improved the exposition.

12   References

[1] M. V. Abrusci. Phase semantics and sequent calculus for pure noncommutative classical linear propositional logic, *The Journal of Symbolic Logic*, 56(4):1403–1451, 1991.

[2] J. van Benthem. *Language in Action: Categories, Lambdas and Dynamic Logic*. North-Holland, Amsterdam, (Studies in Logic 130), 1991.

[3] W. Buszkowski. Compatibility of categorial grammar with an associated category system. *Zeitschrift für Mathematische Logik und Grundlagen der Mathematik*, 28:229–238, 1982.

[4] W. Buszkowski. Completeness results for Lambek syntactic calculus. *Zeitschrift für Mathematische Logik und Grundlagen der Mathematik*, 32:13–28, 1986.

[5] K. Došen. A completeness theorem for the Lambek calculus of syntactic categories. *Zeitschrift für Mathematische Logik und Grundlagen der Mathematik*, 31:235–241, 1985.

[6] K. Došen. A brief survey of frames for the Lambek calculus. *Zeitschrift für Mathematische Logik und Grundlagen der Mathematik*, 38:179–188, 1992.

[7] J. Lambek. The mathematics of sentence structure. *American Mathematical Monthly*, 65(3):154–170, 1958.

[8] Sz. Mikulás. *Taming Logics*. ILLC Dissertation series 1995–12, Institute for Logic, Language and Computation, University of Amsterdam, 1995.

[9] M. Pentus. Lambek calculus is L-complete. ILLC Prepublication series LP–93–14, Institute for Logic, Language and Computation, University of Amsterdam, 1993.

[10] M. Pentus. Language completeness of the Lambek calculus. *Proc. 9th Annual IEEE Symp. on Logic in Computer Science*, Paris, France, 487–496, 1994.

[11] M. Pentus. Models for the Lambek calculus. *Annals of Pure and Applied Logic*, 75:179–213, 1995.

[12] D. N. Yetter. Quantales and noncommutative linear logic. *Journal of Symbolic Logic*, 55(1):41–64, 1990.

Author address

Department of Mathematical Logic
Faculty of Mechanics and Mathematics
Moscow State University
Moscow, RUSSIA, 119899
e-mail: pentus@lpcs.math.msu.ru

Simple groups definable in O-minimal structures

Ya'acov Peterzil, Anand Pillay[2] & Sergei Starchenko[3]

1 Introduction

In this survey article we wish to describe our classification, or identification, of those (definably) simple groups G which are definable in O-minimal structures. Our basic result says that any such group is a linear semialgebraic group over some real closed field R. An even sharper result says that the structure $(G, .)$ is *bi-interpretable* with either the field R or with the field $R(i)$, where i is the square root of -1. These results will appear in [8] and [9]. (It should be stated here that for technical reasons our hypothesis on the group G is that it is actually definable in M rather than interpretable. We, however, expect all the results to go through under only the interpretability hypothesis.)

Our results yield an O-minimal analogue of the well-known Cherlin conjecture. Recall that the (as yet unproved) Cherlin conjecture states any (noncommutative) simple group of finite Morley rank is a linear algebraic group over an algebraically closed field. It can be shown that if G is a simple group of finite Morley rank, then there is some strongly minimal set D such that G can be defined in D. Thus the finite Morley rank hypothesis in the Cherlin conjecture can be replaced by : G is definable in some strongly minimal structure. So it is quite natural to ask the O-minimal analogue: what are the simple groups definable in O-minimal structures? And this is what we have answered. In fact our second result (the bi-interpretability result mentioned above) clearly yields the full Cherlin conjecture for simple groups of finite Morley rank which happen to be definable in O-minimal structures. Any such group is a linear algebraic group over an algebraically closed field of characteristic 0. Moreover this yields even a model-theoretic *characterisation* of simple algebraic groups over algebraically closed fields of characteristic 0.

For the remainder of this introduction we shall define our terms and give

[1] Received September 10, 96; revised version November 21, 1996.
[2] Partially supported by an NSF grant.
[3] Partially supported by an NSF grant

212   Y. Peterzil, A. Pillay & S. Starchenko

clear statements of the results.

As a rule when we speak of definable sets (in a structure) we allow parameters in the defining formulas.

By an O-minimal structure we mean a structure $(M, <, ...)$ where $<$ is a dense linear ordering, and every definable (with parameters) subset of M is a finite union of points and open intervals. We should emphasize that the hypothesis is on subsets of M not M^n for $n > 1$. Also by an interval we mean something of the form (a, b) where $a, b \in M \cup \{\pm\infty\}$.

By a group definable in the structure M we mean a group G such that both the universe of $(G, \cdot)$ and the graph of the group operation $\cdot$ are subsets of M^n and M^{3n} respectively (some $n > 1$) which are definable in the structure M. We will say that G is definably simple in the sense of M if G is noncommutative and has no proper nontrivial normal subgroup definable in M. If we replace the latter condition by: G has no proper nontrivial normal subgroups definable in the structure $(G, \cdot)$ we say that G is definably simple in the sense of $(G, \cdot)$.

Now suppose that R is a real closed field. The term "semialgebraic" (with respect to R) means definable in the structure $(R, +, .)$. Of course the ordering on R is semialgebraic, and in fact all semialgebraic sets will be quantifier-free definable after adding a symbol for the ordering. By a semialgebraic linear group with respect to R we mean a subgroup G of $GL(n, R)$ (some n) which is semialgebraic, namely definable in $(R, +, .)$. Such a group is said to be semialgebraically connected, if G has no proper semialgebraic subgroups of finite index. If R happens to be the field of real numbers $\mathbf{R}$, then G is a semialgebraic linear Lie group, and semialgebraic connectedness of G is equivalent to topological connectedness (in the Euclidean topology). Returning to the general situation where R is real closed and G is a semialgebraic linear group with respect to R, G will have a Zariski closure G_1 in $GL(n, R)$, G_1 being the smallest subset of $GL(n, R)$ defined by polynominal equations, which contains G. G_1 will be the universe of a subgroup of $GL(n, R)$ (which we still call G_1) and G will be a finite index subgroup of G_1. (Explanation:the Zariski closure of a subgroup G of $GL(n, R)$ will always also be a subgroup. On the other hand, if G is also semialgebraic, then it comes equipped with a dimension, which turns out to be the same as the dimension of its Zariski closure. Basic properties of this notion of dimension imply that if G and G_1 are semialgebraic groups with the same dimension, and G is a subgroup of G_1, then G is an open subgroup of finite index in G_1.) Now $R(i) = K$ is an algebraically closed field, and the subset of $GL(n, K)$ defined by the zero set of all polynomials over R vanishing on G_1 will be a subgroup of $GL(n, K)$, H say. H is a linear algebraic group, defined over R, and $H \cap GL(n, R) = G_1$. In general, by a linear algebraic group H we mean a subgroup of $GL(n, K)$ defined by polynomial equations over K, where K is an algebraically closed field. If these polynomial equations can be chosen over a subfield L of K, we say (at least in the characteristic 0 case) that H is defined over L. By

$H(L)$ we will mean $H \cap GL(n, L)$, which we call the group of L-rational points of H. H is said to be L-simple, if H has no proper nontrivial normal *algebraic* subgroup defined over L. To be K-simple (K algebraically closed) is the same as being abstractly simple. However, the L-simplicity of H does not even guarantee the abstract simplicity of $H(L)$. An example is given by taking $H = S0(3)$, the group of 3 by 3 orthogonal matrices of determinant 1, identified with $S0(3, K)$ for some big algebraically closed field (of characteristic 0 say). This group is abstractly simple. However, if R is a nonarchimedean real closed field, then $H(R)$ ($= S0(3, R)$) is not abstractly simple, as its elements which are infinitely close to the identity form a normal subgroup. On the other hand, $S0(3, R)$ is semialgebraically simple. This explains why we make the assumption "definably simple" in place of "simple" in the following result.

Theorem 1.1 *Suppose M is an 0-minimal structure, and G is a group definable in M which is definably simple in the sense of M. Then there is a real closed field R definable in M, and an R-simple linear algebraic group H defined over R such that G is definably isomorphic (in M) to the semialgebraically connected component of $H(R)$. Otherwise said, G is definably isomorphic to a semialgebraic, semialgebraically connected linear group with respect to some real closed field R definable in M.*

Remark. There is a similar version if G is assumed only to be definably simple in the sense of $(G, \cdot)$. The conclusion is for now a bit weaker: H need not be connected, or R-simple, although its connected component is semisimple. G will be definably isomorphic to a finite index semialgebraic subgroup of $H(R)$.

A key concern of current model theory is the question of bi-interpretability. A structure M is said to be interpretable in a structure N if there is an isomorphic copy $f(M)$ of M say whose universe and basic relations are all definable sets in N^{eq}. Now if M and N are each interpretable in the other, witnessed by $f(M)$ and $g(N)$, then $f.g$ yields an isomorphism of N with $f(g(N))$, the latter being also definable in N. Similarly $g.f$ is an isomorphism of M with $g(f(M))$. M and N are said to be bi-interpretable if one can choose both f and g such that the isomorphism $f.g$ is definable in N and the isomorphism $g.f$ is definable in M. Bi-interpretability means that the structures are essentially the same. An important observation of Poizat ([12]) is that if G is a simple algebraic group (with respect to a given algebraically closed field K), then the structures $(G, .)$ and $(K, +, .)$ are bi-interpretable. This is a model-theoretic version (and easily implies) a simple case of Borel-Tits theory: any abstract group automorphism of a simple algebraic group over an algebraically closed field is a composition of a field automorphism of K with a quasi-isomorphism of algebraic groups (i.e. an isomorphism definable in $(K, +, .)$). Much less trivial is the question of abstract automorphisms of groups of the form $G(k)$ where G is a k-simple algebraic group defined over an arbitrary field k. This is the concern of general Borel-Tits theory ([2]). Our second result bears on this question in the case where k is a real closed field. Note that if R is real closed then the algebraic closure of R is obtained by adjoining i, the square root of -1. Moreover clearly $R(i)$ is definable in R. We will have two fundamental cases for (definably) simple groups definable in the field R. An example of the first case is $PSL(2, R)$, which will be bi-interpretable with R. An example of the second case is $PSL(2, R(i))$ which will be bi-interpretable with $R(i)$ (and not with R). Actually in the proof the first case has 2 subcases: on the one hand, groups like $PSL(2, R)$, which for $R = \mathbf{R}$ are noncompact, and and the other hand groups like $SO(3, R)$ which for $R = \mathbf{R}$ are compact. The first (isotropic) subcase fits into Borel-Tits theory, whereas the second (anisotropic) case was studied by Weisfeiler [14]. It should be said, however, that our treatment is independent of either [2] or [14]

One would have liked to prove the next theorem by Theorem 1.1 and inspection. However we were not sure what to inspect and where to find it.

Theorem 1.2 *Let G be an infinite group which is definably simple in the sense of $(G, \cdot)$, and which is definable in some O-minimal structure. Then there is a real closed field R such that $(G, \cdot)$ is bi-interpretable either with $(R, +, .)$ or with $(R(i), +, .)$. In particular any simple Lie group is bi-interpretable with the field of reals or the field of complexes.*

With Theorem 1.2, one sees that definable simplicity of G in the sense of $(G, \cdot)$ is equivalent to definable simplicity in the ambient O-minimal structure. Theorem 1.2 also has bearing on Cherlin's original conjecture for it implies that bi-interpretability with R corresponds to the group being unstable.

Corollary 1.3 *Suppose G is an infinite simple group. Then the following are equivalent:*
(i) $Th(G, \cdot)$ is stable and $(G, \cdot)$ is definable in some O-minimal structure,
(ii) G is an algebraic group over some algebraically closed field of characteristic 0.

It should be said that the problem of proving O-minimal analogues of Cherlin's conjecture (i.e. for Lie groups) was originally raised by the second author, and the solution for groups of small dimension was given in [6].

In the next section we discuss the general ideology of the proof, and mention some aspects which may be of independent interest.

2 Outline of the proof.

Let us first discuss Poizat's original strategy for proving Cherlin's conjecture. Let G be an infinite simple group of finite Morley rank. His idea was to find an infinite solvable nonnilpotent definable subgroup B of G, and then use a result of Zilber which gives an infinite definable field K. K, having finite Morley rank too, has to be algebraically closed. The idea was then to show that the structure induced on K from $(G, \cdot)$ is just the field structure, then to use some model theory to show that $(G, \cdot)$ is definable back in $(K, +, .)$, then to conclude, by a version of Weil's Theorem, that G is an algebraic group over K. The existence of solvable nonnilpotent subgroups of a simple group of finite Morley rank, is the "no bad groups" hypothesis. The truth of this is still unknown, although it is not unlikely that there is a counterexample. The second thing one needs to know is that if K is an algebraically closed field definable in a structure M of finite Morley rank, then no structure on K is induced from M other than the field structure. This is a strong version of the "no bads fields" hypothesis. This turns out to be false: Hrushovski constructed counteramples. (It should be said that Cherlin and others have now a program to prove Cherlin's conjecture under the "no bad groups" hypothesis and a weaker version of the "no bad fields" hypothesis.) So the Poizat strategy falls through. But note that one of the main points was to define a field and show the field to be well-behaved.

Let us now pass to the O-minimal situation. Let M be an O-minimal structure. Sometimes it is convenient to assume M to be saturated. There is a general theory of definable sets and functions in O-minimal structures ([11],[3],[13]), yielding a notion of dimension and independence. Essentially algebraic closure on M is a pregeometry, and the associated dimension of definable sets has a topological significance: if $X \subset M^n$ then $dim(X)$ is the greatest r such that some projection of X on M^r contains an open set (in the product topology). Also definable functions are piecewise continuous.

Now let G be an infinite definably simple (in the sense of M) group definable in M. It turns out that the Poizat strategy is much more successful here. However, this is made possible by a theorem of the first and last authors ([7]) which shows the existence of a definable real closed field in an O-minimal structure M under some reasonable assumption on its complexity. In fact their result yields a definable real closed field in a neighbourhood U of a point $a \in M$, whenever there is a family F of germs of definable functions at a with $dim(F) > 1$. The noncommutativity of G (together with quite a bit of work) enables one to find a point $a \in M$ closely related to G satisying the above complexity hypothesis, yielding a definable real closed field R. Some additional work, using definable simplicity of G, is required to prove that some definable subset X of G of maximal dimension is contained in $dcl(R)$. Now R may have additional structure (other than semialgebraic structure) induced on it from M. However, what O-minimality yields is that definable (in M) functions from R^n to R are piecewise continuously differentiable (in fact piecewise C^k for abitrarily large k). Note that R is a densely ordered field so differentiability makes sense. Methods from [10] allow us to equip G with a "definable group manifold" structure over M such that moreover, some open neighbouhood U of the identity is identified with an open subset of R^n for some n, and that the group operation on this neighbourhood is continuously differentiable. At this point, a version of the classical adjoint representation comes into play: for any $g \in G$, the conjugation by g map $Inn_g : G \rightarrow G$ is continuously differentiable on a neighbourhood of the identity of G. Its differential is an element of $GL(n, R)$. G being centreless, this yields a definable embedding of G in $GL(n, R)$. We now have G as a *definable* subgroup of $GL(n, R)$, but we would like it to be a *semialgebraic* subgroup. For this, one has to develop some classical Lie theory (the relation between Lie groups and Lie algebras) in the "O-minimal expansions of real closed fields" context. This works quite successfully. We define a Lie algebra structure on the tangent space at the identity of G in a natural fashion, to obtain $L(G)$ the Lie algebra of G. This is an object definable in $(R, +, .)$. From semisimplicity of G (no definable normal abelian subgroups) we conclude semisimplicity of $L(G)$ (no proper abelian ideals). Properties of semisimple Lie algebras over *arbitrary* fields of characteristic 0 now imply that the (semialgebraic) group of automorphisms of $L(G)$ (linear transformations preserving the Lie brackets), has the same dimension as $L(G)$ and thus as G. The adjoint representation now yields an isomorphism of G with a finite index subgroup of $Aut(L(G))$ which must be semialgebraic. This completes the sketch of the proof of Theorem 1.1.

We now say a few words about the proof of Theorem 1.2. As already remarked, Poizat had already proved that if G is a simple algebraic group over an algebraically closed field (i.e. $G = G(K)$) then $(G, \cdot)$ and $(K, +, .)$ are bi-interpretable. The main point was to find an infinite field K' definable in $(G, \cdot)$, and then to show that (in the structure $(G, \cdot)$, G is contained in $dcl(K')$. Both these steps require some work when transferred to the real closed field setting. (We will here suppress the problem of working with definable simplicity in the sense of $(G, \cdot)$ rather than in the sense of M (or R).) So the situation now will be: R is a real closed field, H is an R-simple linear algebraic group defined over R, and G is the semialgebraic connected component of $H(R)$. The first problem is to find an infinite field definable in $(G, \cdot)$. At the outset we have a division into two cases: (i) H is anisotropic over R and (ii) H is isotropic over R. Condition (i) means that H has *no* (infinite) R-definable algebraic subgroup which is R-isomorphic to some product of copies of the multiplicative group. A typical example is when $G = S0(3, \mathbf{R})$, and corresponds in this classical case to G being *compact*. In this case, with a little work, the results on definability in compact Lie groups from [5] transfer to R, to give a 1-dimensional field definable (actually interpretable) in $(G, \cdot)$. In case (ii), where H is R-isotropic, we can fabricate, using the theory of parabolic subgroups from [1], a connected solvable centreless R-subgroup B of H. $B(R)$ will be enveloped in a solvable nonnilpotent subgroup of G definable in $(G, \cdot)$, and various methods enable one to define in $(G, \cdot)$ an infinite field K. (An O-minimal version of Zilber's Indecomposability Theorem is used here, as well as at other points of the proof.) Now the field K is definable in the real closed field $(R, +, \cdot)$, and is known to be semialgebraically isomorphic to R or to $R(i)$. In the first case $dim(K) = 1$ and there are methods, using simplicity of G, for showing that G is (in $(G, \cdot)$) contained in $dcl(K)$, so we finish. In the second case, we can consider the structure N consisting of $(K, +, .)$ equipped with all structure induced from $(G, \cdot)$, as a semialgebraic expansion of $(R(i), +, .)$. Results of Marker [4] imply that either a 1-dimensional field is definable in N, or N is simply an algebraically closed field with some constants. In the first subcase, we have a 1-dimensional real closed field definable in $(G, \cdot)$ and can proceed as before. In the second subcase, K with all its induced structure is strongly minimal. The original Zilber Indecomposability Theorem (or rather its proof), together with simplicity of G implies that $(G, \cdot)$ is a structure of finite Morley rank, and it is then easy to show that this structure is the definable closure of K.

This completes the outline, as well as the paper.

3 References

[1] A. Borel, Linear Algebraic Groups, Springer-Verlag, 1991.

[2] A. Borel and J. Tits, Homomorphismes "abstraits" de groupes alge-briques simples, Annals of Math. 97 (1973), 499-571.

[3] J. Knight, A. Pillay and C. Steinhorn, Definable sets in ordered structures II, Transactions of A.M.S. 295 (1986), 593-605.

[4] D. Marker, Semialgebraic expansions of C, Transactions of A.M.S. 320 (1990).

[5] A. Nesin and A. Pillay, Some model theory of compact Lie groups, Transactions of A.M.S. 326 (1991), 453-463.

[6] A. Nesin, A. Pillay and V. Rasenj, Groups of dimension two and three over O-minimal structures, Annals of Pure and Applied Logic 53 (1991), 279-296.

[7] Y. Peterzil and S. Starchenko, A trichotomy theorem for O-minimal structures, submitted to Journals of London Math. Soc.

[8] Y. Peterzil, A. Pillay and S. Starchenko, Definably simple groups in O-minimal structures, submitted for publication.

[9] Y. Peterzil, A. Pillay and S. Starchenko, Simple algebraic and semial-gebraic groups over real closed fields, submitted for publication.

[10] A. Pillay, On groups and fields definable in O-minimal structures, Journal of Pure and Applied Algebra 53 (1988), 239-255.

[11] A. Pillay and C. Steinhorn, Definable sets in ordered structures I, Transactions of A.M.S. 295 (1986), 565-592.

[12] B. Poizat, MM Borel, Tits, Zilber et le general nonsense, Journal of Symbolic Logic 53 (1988), 124-131.

[13] L. van den Dries, Tame topology and O-minimal structures, book, Cambridge University Press, to appear.

[14] B. Weisfeiler, On abstract homomorphisms of anisotropic algebraic groups over real closed fields, Journal of Algebra 60 (1979), 485-519.

Authors addresses

Ya'acov Peterzil
Department of Mathehmatics
University of Haifa
Haifa, Israel
email: kobi@mathcs2.haifa.ac.il

Anand Pillay
Department of Mathematics
University of Illinois
1409 West Green Street
Urbana, IL 61801
USA
email: pillay@math.uiuc.edu

Sergei Starchenko
Department of Mathematics
Vanderbilt University
Nashville, TN, 37240,
USA
email: starchen@math.vanderbilt.edu

Two-Dimensional Temporal Logic

Mark Reynolds[2]

ABSTRACT Two-dimensional combinations of temporal and modal logics
have been studied for some time for their logical properties and their applications to natural language semantics and computer science. In this survey,
we briefly describe a variety of these logics, concentrating on the temporal-
temporal combinations, their properties and uses. We also look at some
more recent results using irreflexivity rules, tiling and mosaic techniques.

1 Introduction

We survey some recent results about various two-dimensional temporal logics and some similar modal-temporal logics. We look at their simple logical
properties and applications in computer science and artificial intelligence.
For a more general account of multi-dimensional modal logic see [MV97]
and for broader surveys of temporal logic see [GHR95].

The logics we are most concerned with are defined over frames consisting
of a cross product of simpler structures. Valuations of propositional atoms
will be made at ordered pairs and so truth of formulas is also evaluated
at ordered pairs in structures. The accessibility relations of the modalities
will be restricted by keeping one of the coordinates of the two-dimensions
constant.

Such logics may be of interest to those investigating natural language semantics, describing changes in temporal information contained in databases,
using interval temporal logics to describe the relationships between processes or states of extended duration, combining temporal logic with logics
of possibility, knowledge or belief, describing systems of parallel processes
or trying to find modal approximates to the first-order logic of two variable
symbols.

In this paper we will briefly look at axiomatizations for these logics using

---

[1]Received January 10,96; revised version May 9, 1997.

[2]The author would like to thank the researchers in algebraic logic at London University, Marcelo Finger, Alberto Zanardo, Yde Venema, and the referees for their helpful
suggestions. This work is partly supported by EPSRC grant GR/K54946.

Gabbay's irreflexivity rules, undecidability proofs using the tiling technique and we also describe an application to decidability questions of a technique which was initially used by Istvan Németi in the area of algebraic logic (see [Nem86])–the mosaic method.

2   Compass Logic

The most straight forwardly two-dimensional temporal logic is the compass logic introduced by Venema in [Ven90] and [Ven92]. The language contains four interrelated modal diamonds: $\diamondsuit$, $\diamondsuit$, $\diamondsuit$ and $\diamondsuit$. Structures for this language consist of two linear orders $(T_1, <_1)$ and $(T_2, <_2)$: we shall call such a pair a *rectangular frame*. Two-dimensional valuations for atoms are made at ordered pairs from $T_1 \times T_2$. We can think of $(T_1, <_1)$ as lying horizontally and $(T_2, <_2)$ as lying vertically on a cartesian grid. The truth of formulas is also defined at ordered pairs in a natural way: for example,

- $(T_1, <_1, T_2, <_2, g), t_1, t_2 \models p$ iff $p \in g(t_1, t_2)$

- $(T_1, <_1, T_2, <_2, g), t_1, t_2 \models \diamondsuit A$ iff there is $s_1 \in T_1$ such that $t_1 <_1 s_1$ and $(T_1, <_1, T_2, <_2, g), s_1, t_2 \models A$.

It is useful to define some abbreviations including the corresponding universal modalities and some boolean combinations of the basic modalities: for example,

$$\boxdot A \equiv \neg \diamondsuit \neg A$$
$$\Box A \equiv A \wedge \boxdot A \wedge \boxdot A \wedge \boxdot A \wedge \boxdot A$$
$$\diamondsuit A \equiv \diamondsuit A \vee A \vee \diamondsuit A$$

Notice that this notation extends the appealingly intuitive geographical analogy suggested by the notation for a modal two-dimensional logic in [Seg73]. The intuition further suggests another possible application of this logic: to the field of spatial reasoning. Although there are modal logics for spatial reasoning (such as the logic of convex hulls in [Ben96]), we know of no investigation of the use of modalities for compass directions in this field.

An example of the kind of statement one can make in the logic is

$$\diamondsuit \boxdot A \rightarrow \boxdot \diamondsuit A$$

which is actually a validity where by a validity we here mean a formula which is true at every ordered pair in every rectangular structure. A *satisfiable* formula, on the other hand, is a formula ϕ for which there exists some rectangular structure $\mathcal{T} = (T_1, <_1, T_2, <_2, g)$ and some pair (t_1, t_2) such that

$$\mathcal{T}, t_1, t_2 \models \phi.$$

Just to be clear, we also define semantic consequence by

$$\Gamma \models A$$

iff $\mathcal{T}, t_1, t_2 \models A$ whenever we have $\mathcal{T}, t_1, t_2 \models \gamma$ for all $\gamma \in \Gamma$.

Below we will consider the expressive power of this logic and give an axiomatization for it. We will also see that validity is undecidable.

3 Variations

Variations on this logic arise in the usual ways: semantically, we can restrict our attention to certain subsets of the set of rectangular structures; syntactically, we can consider other temporal operators; or we can combine such variations.

Examples of restrictions are to only consider frames $(T_1, <_1, T_2, <_2)$ (i) with each $(T_i, <_i)$ being dense, (ii) with $(T_1, <_1) = (T_2, <_2)$ or (iii) with each $(T_i, <_i)$ being the natural numbers order. The logics resulting from such restrictions will, of course, usually have more validities.

Considering the language, then a more expressive two-dimensional temporal logic can be obtained by using Kamp's *until* U and *since* S operators. In one dimensional temporal logic, *until* (technically, the strict version of *until*) is defined as follows:

- $\mathcal{T}, t \models U(A, B)$ iff there is some time s such that $t < s$ and $\mathcal{T}, s \models A$ and for all times r such that $t < r < s$ we have $\mathcal{T}, r \models B$.

Since is defined dually, i.e. with $>$ instead of $<$. For two dimensions we end up with four operators: U^h, S^h, U^v and S^v, with a pair for each of the horizontal and vertical ordering. The compass operators can easily be defined in this language: for example $\Diamond\, A \equiv U^v(A, \top)$.

Note that there are also less expressive versions of *until* and *since* (called *non-strict*). In two-dimensions the horizontal (easterly) non-strict *until* has defining clause:

- $(T_1, <_1, T_2, <_2, g), t_1, t_2 \models U_{ns}^h(A, B)$ iff there is some $s_1 \in T_1$ such that $t_1 \leq_1 s_1$ and $(T_1, <_1, T_2, <_2, g), s_1, t_2 \models A$ and for all $r_1 \in T_1$ such that $t_1 \leq_1 r_1 <_1 s_1$ we have $(T_1, <_1, T_2, <_2, g), r_1, t_2 \models B$.

The difference between strict and non-strict versions of *until* and *since* is sometimes important in applications where time is taken to be the integers or natural numbers. In such a situation we also have the "next-time" operators $\ominus$, $\oplus$, $\ominus$ and $\oplus$. The semantic clause for $\ominus$, for example, is

$$(\mathbb{N}, <, \mathbb{N}, <, g), n, m \models \ominus\, \alpha \text{ iff } (\mathbb{N}, <, \mathbb{N}, <, g), n + 1, m \models \alpha$$

while that for $\oplus$ is

$$(\mathbb{N}, <, \mathbb{N}, <, g), n, m \models \oplus\, \alpha \text{ iff } n > 0 \text{ and } (\mathbb{N}, <, \mathbb{N}, <, g), n, m - 1 \models \alpha.$$

It is not hard to show that the next-time operators are definable from the strict *until-since* operators but not from the non-strict ones.

When the two dimensions of time are based on the same linear order $(T, <)$ (we can call such structures *square*) we have a diagonal in our structures: $\{(t,t)|t \in T\}$. Then we can make the language even more expressive by including modal constants which represent being on the diagonal or on one side of it as opposed to the other. For example we can introduce δ as a constant which is only true on the diagonal. Being in the north-west half-plane is determined by the truth of the formula $\Diamond \delta$.

Harel in [Har83] has considered a two-dimensional logic with just $\Diamond$, $\Diamond$, $\bigcirc$ and $\ominus$ over natural numbers squares.

In square structures we can also follow Vlach and Åqvist(see [GHR94]) and introduce a *converse* modality J (here written $\otimes$) and *projection* modalities J_1 and J_2. The definitions are:

$$(T, <, T, <, g), t, u \models \otimes\alpha \text{ iff } (T, <, T, <, g), u, t \models \alpha$$
$$(T, <, T, <, g), t_1, t_2 \models J_i\alpha \text{ iff } (T, <, T, <, g), t_i, t_i \models \alpha$$

There are many other variations on these logics: we can also relax some of our semantical assumptions instead of restricting them (e.g. consider structures where each $(T_i, <_i)$ is not necessarily a linear irreflexive order), or reduce the expressiveness of the language (e.g. do not use the "past" operators such as $\Diamond$ and $\Diamond$) instead of increasing it.

There are also less neatly two-dimensional combinations of temporal logics in the literature. For example, there are the logics arising from general *Temporalizing* [FG92] and combining [FG96] techniques. Temporalizing allows the adding of a temporal logic on top of any other logic. Truth is evaluated in two-dimensional structures but only a restricted language is available– formulas with a horizontal modality nested inside a vertical one, say, are outlawed. Combining or Fibring techniques, on the other hand, allow the full two-dimensional language but also allow very complex models without commutativity of the two accessibility relations $<_1$ and $<_2$. Such structures are sometimes known as independent combinations of modal logics [Tho80]. These logics are used to investigate the preservation of various logical properties under combination logics. They can also sometimes be the only way of keeping combinations of logic decidable. For more recent results in this area see, for example, [KW91], [Spa93] and [GS97].

4   Expressive Completeness

With all the variations on temporal language even for a fixed semantic domain, questions of relative expressiveness and expressive completeness are bound to be raised. Many of these have been answered by Venema and de Rijke in [Ven90], [Ven92], [Ven94] and [dR93].

Expressive completeness for temporal languages (see [GHR94]) concerns the well-known translation ([vB84]) from temporal or modal formulas into formulas in a monadic first-order logic. A temporal logic is expressively complete if and only if the translation is reversible. Without going into details we just summarize that in the one-dimensional case there do exist expressively complete temporal logics (e.g. the logic with U and S was proved to be so by [Kam68] for Dedekind complete flows of time) while for two-dimensional structures, Venema (in [Ven90]) has shown that no finite set of operators is complete for the class of all square frames built from a dense linear order.

However, [Ven94] presents an expressively complete two-dimensional logic when we restrict our attention to flat structures, i.e. structures in which the valuation g is such that $p \in g(t, u)$ iff $p \in g(t, v)$ for any t, v, u.

5 An Axiomatization

Here we adapt the systems in [Ven90] (for an interval logic), [GHR94] (for Vlach and Åqvist's logic) and [Fin94] (for a two-dimensional until-since logic) to give a complete finite axiom system for the compass logic. There is a slightly different axiomatization for the compass logic in [MV97]. Recall that the semantics are defined over frames with any pair of linear orders.

The inference rules are modus ponens and temporal generalization:

$$\frac{A, A \to B}{B} \qquad \frac{A}{\square\, A} \qquad \frac{A}{\square\, A} \qquad \frac{A}{\boxminus\, A} \qquad \frac{A}{\boxminus\, A}$$

along with two versions of the Gabbay IRR rule ([Gab81]):

$$\frac{\Diamond\, q \wedge \square\, \neg \Diamond\, q \to A}{A} \qquad \text{provided the atom } q \text{ does not appear in } A$$

and the version with $\Diamond$ instead of $\Diamond$ and $\boxminus$ instead of $\square$.

The axiom schemas include the usual ones for linear temporal logic:

A1: all classical tautologies,

A2: $\square\, (A \to B) \to (\square\, A \to \square\, B)$

A3: $A \to \square\, \Diamond\, A$

A4: $\square\, A \to \square\, \square\, A$

A5: $\Diamond\, (\square\, A \wedge B \wedge C) \to \square\, (A \vee \Diamond\, B \vee C)$

with south facing versions of A2, A3 and A4 and east-west versions of all these and schemas C(SE), C(SW), C(NW), C(NE) to describe the interaction of the two dimensions:

C(SE): $\diamondsuit \, \diamondsuit \, A \leftrightarrow \diamondsuit \, \diamondsuit \, A$
e.t.c.

To see that the IRR rules are sound, just suppose that the atom q does not appear in the formula A but that $\psi = ((\diamondsuit \, q \wedge \boxdot \, \neg\diamondsuit \, \neg q) \to A)$ is valid. Now consider any rectangular structure $T = (T_1, <_1, T_2, <_2, g)$ and any $(t_1, t_2) \in T_1 \times T_2$. Define a new structure $T' = (T_1, <_1, T_2, <_2, h)$ with valuation $h : (T_1 \times T_2) \to 2^L$ via

$$q \in h(u_1, u_2) \text{ iff } u_2 = t_2$$
$$p \in h(u_1, u_2) \text{ iff } p \in g(u_1, u_2) \quad \text{for } p \neq q.$$

It is clear that $T', t_1, t_2 \models \diamondsuit \, q \wedge \boxdot \, \neg\diamondsuit \, \neg q$ and so, as ψ is valid, we have $T', t_1, t_2 \models A$. A simple induction shows that for formulas like A which do not contain the atom q we have $T, t_1, t_2 \models A$. Thus the first IRR rule is valid and it is straight forward to show that the whole axiom system is sound.

The completeness part of the proof is made easy by the presence of the IRR rules. Instead of considering the set of maximally consistent sets of formulas as in many traditional completeness proofs we can confine our attention to a subset of such sets called the IRR sets. For each of these sets there is some atom which indicates its "latitude" relative to other comparable IRR sets and there is some atom which indicates its "longitude". It is then straight forward to arrange the sets of formulas in a two-dimensional grid and prove we have built a model of a given consistent set of formulas.

It is an open question whether there are axiomatizations for compass logic without using IRR-style rules (as there are in the case of some logics of historical necessity [Zan90]) but it is thought likely that there are not. Note that, as we will see below, it is not possible to axiomatize compass logic at all if we restrict the linear orders to be the natural numbers time.

6   Applications to Natural Language Semantics

Much of the initial motivation for studying multi-dimensional temporal logics came from the study of tenses in natural language. In giving formal semantics to various constructs it was realised that evaluating the truths of complex expressions was not adequately done with reference to a single time of evaluation. Often, for example, evaluating the truth of a past tense expression is not a simple matter of finding a time in the past at which some subexpression is true. The subexpression may depend on that past point and the overall time of utterance for its truth. Extra dimensions –often just 1 extra dimension– are used for points of reference in the semantics of various perfect tenses (see [GHR94], [Gue78]) and in special temporal constructs such as "now" (see [Kam71]) and "then" (see [Vla73]).

For a simple example consider the statement

when he was gaoled he didn't know that he would be released
before now.

Although, the statement contains an epistemic modality to confuse the
matter slightly, it is clear that its truth at time s could be adequately
established by finding a time t in the past at which the following statement
is true:

he is being gaoled and he doesn't know that he will be released
before now

provided that we are careful about the use of the word "now" and suppose
that it refers to time s. Thus the truth of the latter statement obviously
can only be evaluated at a pair of time points: one t specifying when the
evaluation takes place and the other, s, specifying what time corresponds
to that word "now".

7 Applications to Databases

Another very important use of temporal logic is in dealing with databases
which make use of time. We call these *temporal databases*. Time can be relevant to a database in one or both of two different ways. Each change to the
contents of the database will be made at some time: we refer to this as the
transaction time of the database update. Databases often also store information about the time of events: we refer to the actual time of occurrence
of an event as its *valid* time. Depending on which of these uses is made of
time or on whether both approaches have a role to play, we can identify
several different types of temporal databases but what is common to all;
as with all systems which change over time, is that describing or reasoning
about their evolution is very conveniently done with temporal logic. With
both the forms of temporal information involved, it was thus suggested
in [Fin92], that describing the evolution of a temporal database is best
done with two-dimensional temporal logic. This is because, for example, at
a certain transaction time today, say, we might realize that our database
has not been kept up to date and we may add some data about an event
which occurred (at a valid time) last week. Thus a one-dimensional model
which represents this-morning's view of the history of the recorded world,
is changed, by the afternoon, into a new one-dimensional model by having
the state of its view about last week altered. A series of one-dimensional
models arranged from one day to the next is clearly a structure for a two-
dimensional temporal logic. This application has recently been developed
into a logical language for controlling temporal databases (see [FR97]). Furthermore, it has recently been shown ([FG92]) that the restricted kind of
logic needed for database applications is much more amenable to reasoning
with than the usually undecidable general two-dimensional logics.

8   Intervals

As described in [vB95], there are many and various motivations for using an interval temporal logic: these include philosophical considerations of time and events, natural language, processes in computations and planning problems. For constraint problems in planning it may be sufficient to just consider networks of intervals with each pair related by some subset of the 13 possible basic relations (see [All81] for details). However, for more sophisticated reasoning about intervals an interval temporal logic such as that in [HS86] is better suited.

The modalities in [HS86] include:

- $\langle A \rangle$: at some interval beginning immediately after the end of the current one,

- $\langle B \rangle$: at some interval during the current one, beginning when the current one begins, and

- $\langle E \rangle$: at some interval during the current one, ending when the current one ends

As suggested in [Ven90], it is rewarding to notice that interval temporal logics are closely related to two-dimensional temporal logics. We can use the compass language to describe interval structures. Then the interval logic is much the same as the compass logic with a diagonal but with half the points missing. These logics are intertranslatable.

Suppose that $(T, <)$ is a linear order. An interval structure over $(T, <)$ is obtained by adding a valuation for the atoms on intervals i.e. on pairs (s, t) where $s \leq t$ in T. Truth of formulas is also evaluated at intervals.

We turn this into (half a) 2-dimensional structure $(T, <, T, <, g)$ for the compass language with the diagonal constant δ by the following moves:

- truth of formulas and the valuation of atoms only takes place at pairs (s, t) with $s \leq t$,

- $p \in g(t_1, t_2)$ iff $t_1 \leq t_2$ and $p \in h(t_1, t_2)$

- $(T, <, T, <, g), s, t \models \delta$ iff $s = t$

Then interval properties can be expressed in compass logic:

- "all subintervals satisfy p"
  is $\boxminus \boxdot p$

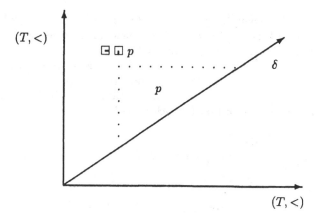

- "any interval strictly in the future satisfies p"
  is $\diamondsuit\ (\delta \wedge \boxdot\ \boxminus\ (\neg\diamondsuit\ \delta \rightarrow p))$

In fact Venema shows that all the formulas of Halpern and Shoham's logic can be easily translated: for example, $\langle B \rangle p$ simply translates to $\diamondsuit\ p$.

We know from a preceding section that the compass logic is not expressively complete. In fact, in [Ven91], Venema describes a useful construct which is not able to be expressed in the above interval logics. This is the two-place *chop* operator C which is defined by

$$\mathcal{T}, t, u \models \phi C \psi \text{ iff } \exists v \text{ such that } t \leq v \wedge v \leq u \text{ and}$$
$$\mathcal{T}, t, v \models \phi \text{ and } \mathcal{T}, v, u \models \psi$$

This operator would seem to have application in natural language semantics (as "and-then"), program semantics (as sequential composition), planning and be a generally useful composition construct in two-dimensional logics.

9   Undecidability via Tiling

The truly two-dimensional temporal logics we have met above are all undecidable. This is usually quite straight forward to show using the domino or tiling technique. The technique was first applied to two-dimensional logic and other logics of programs in [Har83] where many different tiling techniques are used to establish various levels of undecidability of the logics. Other techniques such as coding of Turing machine runs (see [HS86] or [HV89]) can be used but the tiling approach is very natural in this context. In [MR96] it is shown that the compass logic itself is undecidable. The authors use tiling but the proof is not completely straight forward. A much better demonstration of tiling in action can be gained from considering the two-dimensional temporal logic X^2 with "next" operators $\ominus$, $\oslash$, $\ominus$

and ⓙ as well as the compass operators used to describe structures where both dimensions of time are the natural numbers. This is very similar to the two-dimensional logic actually considered in [Har83].

We fix some denumerable set C of *colours*. *Tiles* are 4-tuples of colours and we define four projection maps so that each tile $\tau = (\text{left}(\tau), \text{right}(\tau), \text{up}(\tau), \text{down}(\tau))$. We say of a finite set T of tiles that T tiles $\mathbb{N} \times \mathbb{N}$ iff there is a map $\rho : \mathbb{N} \times \mathbb{N} \rightarrow T$ such that for each $i, j \in \mathbb{N}$,

- $\text{up}(\rho(i, j)) = \text{down}(\rho(i, j + 1))$

- and $\text{right}(\rho(i, j)) = \text{left}(\rho(i + 1, j))$.

The tiling problem for $\mathbb{N} \times \mathbb{N}$ is:

- given a finite set T of tiles, does T tile $\mathbb{N} \times \mathbb{N}$?

In [Rob71] it is shown that the tiling problem is co-r.e.-complete: and hence undecidable.

It is now straight forward to use this result to show the undecidability of the logic X^2.

Given a finite set T of tiles we define (recursively) a formula ϕ_T of the logic such that the satisfiability of ϕ_T is equivalent to the tiling of $\mathbb{N} \times \mathbb{N}$ by T. This will be clear. Since deciding validity is just deciding satisfiability of negations, this shows that validity in X^2 is r.e.-hard.

The formula ϕ_T, which uses the elements of T as propositional atoms is simply the conjunction of the following:

$$\neg\ominus \top \wedge \neg\textcircled{\tiny{$\uparrow$}} \top$$
$$\square \bigvee_{\tau \in T} \tau$$
$$\square \bigwedge_{\tau \neq \tau'} \neg(\tau \wedge \tau')$$
$$\square \bigwedge_{\text{up}(\tau) \neq \text{down}(\tau')} \neg(\tau \wedge \textcircled{\tiny{$\uparrow$}} \tau')$$
$$\square \bigwedge_{\text{right}(\tau) \neq \text{left}(\tau')} \neg(\tau \wedge \ominus \tau')$$

Such tiling proofs tend to be more complicated for other two-dimensional temporal logics. When the underlying flows are not necessarily the natural numbers then we must use the temporal logic to code in a discrete sub-flow. When the logic does not have next-time or *until* operators then the coding gets more complicated again. See [MR96] for details.

Note that we could have easily followed Harel and modified this proof using a different tiling problem to show that the logic X^2 actually has a Π_1^1-hard validity problem. This implies that the logic is not recursively axiomatisable. In fact, when the underlying linear orders are restricted to be natural numbers, integers or reals then many of the two-dimensional and interval logics we have seen are non-axiomatisable. When the full class of linear orders are available then this modified tiling problem, involving the infinite repetition of a certain tile, is not able to be encoded in the two-dimensional logic and we can only prove undecidability.

10   Combinations of Temporal and Modal Logic

Since they are similar we consider a few combined temporal-modal logics which exhibit a two-dimensional character. There is a survey of many of these logics in [Tho84].

The simplest of these is perhaps the logic which we will call $FP\Diamond$ (following [Rey98]). Models consist of a two-dimensional valuation on a cross product of a linear order and a set. The temporal operators F and P operate along the linear dimension perpendicularly to a modal $S5$ $\Diamond$ operator on the set.

$(U, T, <, g), u, t \models FA$ iff $\exists s \in T$ such that $t < s$ and $(U, T, <, g), u, s \models A$
$(U, T, <, g), u, t \models \Diamond A$ iff $\exists v \in U$ such that $(U, T, <, g), v, t \models A$
etc.

The logic was briefly mentioned in [Tho84] as being not very interesting. However, we will look at some of its niceties in the next section. It is also a logic which appears as a special case of many and various other combined temporal logics.

$FP\Diamond$ logic is a restriction of the Synchronized Ockhamist branching-time logic of [DZ94]. The semantical structures (called $T \times W$ frames by [Tho84]) for this logic involve the cross product of a linear order $(T, <)$ and a set W along with equivalence relations $\sim_t$ on W for each $t \in T$. The equivalence relations must satisfy the property that $w \sim_t w'$ and $t' < t$ implies $w \sim_{t'} w'$. The order $(T, <)$ represents time and the elements of the set W represents alternative histories. The $\sim_t$-class containing w can be used to represent the histories which are possible from the point of view of the world (t, w). Thus the modality $\Diamond_2$ defined by

$(T, <, W, \{\sim_t\}, g), t, w \models \Diamond_2 \alpha$ iff $\exists w' \in W$ such that $w \sim_t w'$
    and $(T, <, W, \{\sim_t\}, g), t, w' \models \alpha$

represents the idea of "at this time in some history which is currently considered possible". The modality $\Diamond_1$ defined by

$(T, <, W, \{\sim_t\}, g), t, w \models \Diamond_1 \alpha$ iff $\exists w' \in W$ such that
    $(T, <, W, \{\sim_t\}, g), t, w' \models \alpha$

represents the idea of "at this time in some history". It is this latter modality which extends the non-temporal modality in the logic $FP\Diamond$. Logics very similar to the synchronized logic form bases for logics of agency [BP90] and causation [vK93]. There are axiomatizations of such logic in [vKar] and [DZ96]. It seems to be an open problem whether this logic is decidable.

Logics of historical necessity or Ockhamist logics are closely related examples of a combined logic. They are not a neatly two-dimensional logics but we do have a modal logic of possibility in some sense orthogonal to a linear temporal logic. These logics are obtained by removing the $\Diamond_1$ modality from the Synchronized logic above. They are described in [Bur79] while there are axiomatizations in [Zan85], [Zan96] and [GHR94]. A special case of this logic is proved decidable in [GS85].

Many combinations of time and other modalities arise from formal investigations into how knowledge (or belief) changes over time. These logics are usually designed for reasoning about systems of multiple agents. See [FHMV95] for a comprehensive survey. A temporal-epistemic logic for n agents will use n knowledge modalities. Thus the versions which are of relevance to us here are simple ones, formalizing the changes in knowledge of one lone agent who knows about the world and her or his own knowledge. $S5$ is commonly taken to be the non-temporal logic of knowledge appropriate for one agent. So we can formalize the semantics of the temporal-epistemic logic using a two-dimensional frame very similar in general form to those for Synchronized historical necessity. However, the accessibility relation for the knowledge modality does not have to be restricted to being between worlds (t, w) and (t', w') with $t = t'$. In the case with time being the natural numbers these logics are well studied. In [FHMV95] there is an EXPSPACE-complete decision procedure and complete axiomatization for a logic like $FP\Diamond$ but with natural numbers time: this is a temporal-epsitemic logic of one agent who doesn't forget, doesn't learn and who knows the time.

We have mentioned that $FP\Diamond$ logic has applications to systems of parallel processes. There has been some work in developing two-dimensional logic for such applications. In [RS85], for example, we find a logic combining temporal and spatial modalities. Once again the temporal dimension is the natural numbers and we have the other dimension based on a set of processes. However, there is a set of names for links which may or may not connect one process to another. The language uses *until* in the temporal direction but has a spatial modality for each link as well as one for the transitive closure of all links. This leads to a highly undecidable, unaxiomatisable logic. In [SG87] on the other hand, we have a similar logic but without the linking modalities. There is just the one existential spatial modality as in $FP\Diamond$. With *until* as the temporal connective and the natural numbers as time, this logic is stated to be EXPSPACE-complete.

There are a wealth of two-dimensional non-temporal modal logics which have been investigated. As we will see in the next section some of the results and techniques have also some application to temporal logics. One of the most fruitful areas here has been the investigation of modal versions of first-order logic and their cylindric algebra counterparts. If we look at first-order logic with no function symbols, relations of arity only 1 or 2 and only two variable symbols then we can regard the existential quantifier as a modal operator and come with a two-dimensional modal logic which is the same as that in [Seg73]. There is a very recent account of this area in [AvBN97]. In [VM95], a similar modal logic is studied but its modal semantics is generalized from the first-order motivation. This logic is proved decidable and the proof is an example of the mosaic method which we now turn to.

11   Decidability via The Mosaic Method

The logic $FP\Diamond$ is worth a closer look. This is not just because it may have some application to systems of unbounded numbers of processes computing in parallel. It is also interesting that the mosaic method can be used to show the decidability of validity in this logic.

¿From its beginnings in [Nem86], the mosaic method has been increasingly used in proving decidability and completeness for various multi-modal logics. It is well explained in [VM95] where it is used to prove completeness and decidability of the logic LC_n.

Often, completeness and decidability proofs proceed in a step-by-step manner adding one point at a time to eventually build a model of a satisfiable formula. In the mosaic method we instead try to find a set of small pieces of a model which satisfies a certain closure property. This will be enough to guarantee that the small pieces can be put together to form a model. The actual putting together can either be done by a very simple step-by-step operation (as in [VM95]) or (as shown recently in [HHM+96]), we might be able to use new techniques (of Herwig and Hrushovski) to immediately find the model.

In using the mosaic method to give a decidability proof we need to define mosaics appropriate for the logic and define closure properties (dependent on a given formula) for a finite set of mosaics so that the existence of such a set of a certain size will be equivalent to the existence of a model for the formula.

The logic $FP\Diamond$ is an interesting candidate for a decidability proof via the mosaic method because, as shown in [Rey98], the logic does not have the finite model property. This shows that the finite set of mosaics with the closure property is not just a finite model in disguise.

Here is a brief summary of the decidability proof presented in [Rey98]. Suppose that we are interested in the satisfiability of the formula ϕ. Recall that the structures consist of a set U, a linearly ordered set $(T, <)$ and a valuation $g : (U \times T) \to 2^L$ for the atoms in L. The accessibility relation for F is $\{((u,t),(u,s))|t < s\}$ while that for $\Diamond$ is $\{((u,t),(v,t))\}$. We use the usual symbols G, H and $\Box$ for the universal modalities defined from F, P and $\Diamond$ respectively.

The object playing the part of a mosaic for this logic is called a *segment* and consists of a finite set X with a pair $\mu(x)$ and $\nu(x)$ of sets of subformulas (or negations of subformulas) of ϕ for each $x \in X$. We also impose some conditions on the sets. Let us try to motivate these. We want the segment to represent the small piece of a model $\mathcal{T} = (U, T, <, g)$ consisting of the set of pairs of sets of subformulas of ϕ which one obtains by choosing some s and t from T with $s < t$ and looking at

$$\{(A, B)| \quad \text{there is some } u \in U \text{ such that}$$
$$\alpha \in A \text{ iff } T, u, s \models \alpha \text{ and } \beta \in B \text{ iff } T, u, t \models \beta\}$$

In other words, slice the structure at two different times and look at the sets of formulas which are true on the slices on corresponding U lines.

Note that because we are interested only in subformulas of ϕ there will be only a finite number of pairs in any segment. The conditions that we impose on a segment are 10 in number but quite straight forward given the motivation. For example,

- we require the sets $\mu(x)$ and $\nu(x)$ to be maximally boolean consistent

- if $G\alpha \in \mu(x)$ we require $\alpha \in \nu(x)$ and $G\alpha \in \nu(x)$

- if $\Box\alpha \in \mu(x)$ we require α to be in each $\mu(y)$

- and, importantly, if $\Diamond\alpha \in \mu(x)$ we require there to be some $y \in X$ such that $\alpha \in \mu(y)$.

This is (part of) the definition of just a single segment. We then have to find some closure conditions on a finite set of such segments which will guarantee that they can be (copied as many times as necessary and) put together to build a model of ϕ. These closure properties are such as to allow us to show how to construct a model of ϕ in a step-by-step manner reminiscent of the construction in [Bur82] where defects are cured successively, thus gradually building the model. For example, if (X, μ, ν) is in our set of mosaics and $F\alpha \in \nu(x)$ then we would want there to be a mosaic (X', μ', ν') also in the set which contains some $x' \in X'$ with $\alpha \in \nu'(x')$ and which is such that it could be glued consistently after the first mosaic. Note that the gluing process sometimes involves multiplying copies of various U lines and it is in this way that we can end up with U being an infinite set.

At the eventual (possibly infinite) end of this building process we have a structure in which the labels of formulas on points are equal to the subformulas of ϕ which are true at that point in the model. Provided that we started the process with a segment with ϕ itself in a label then we will end with a model of ϕ.

The decision procedure is thus to check through the finite (but large) number of possible sets of segments for any that satisfy the closure properties. Provided we show also that any satisfiable formula has such a set of segments then the construction described above guarantees that we have a correct procedure.

It is worth mentioning that the complexity of the decision problem for this logic is an open problem: it is nexptime hard but the procedure described above is double exponential.

12 REFERENCES

[All81]   J. F. Allen. An interval based representation of temporal knowledge. In *Proc. 7th IJCAI*, 1981.

[AvBN97] H. Andréka, J. van Benthem, and I. Németi. Modal logics and bounded first-order fragments. *Journal of Philosophical Logic*, to appear, 1997.

[Ben96]   B. Bennett. Modal logics for qualitative spatial reasoning. *Journal of the Interest Group in Pure and Applied Logic (IGPL)*, 4, 1996.

[BP90]   N. Belnap and M. Perloff. Seeing to it that: a canonical form of agentives. In H. Kyburg, R. Loui, and G. Carlson, editors, *Knowledge representation and defeasible reasonings*, pages 175–199. Kluwer Academic Publisher, 1990.

[Bur79]   J. P. Burgess. Logic and time. *J. Symbolic Logic*, 44:566–582, 1979.

[Bur82]   J. P. Burgess. Axioms for tense logic I: 'since' and 'until'. *Notre Dame J. Formal Logic*, 23(2):367–374, 1982.

[dR93]   M. de Rijke. *Extending Modal Logics*. PhD thesis, University of Amsterdam, 1993.

[DZ94]   M. Di Maio and A. Zanardo. Synchronized histories in prior-thomason representation of branching time. In D. Gabbay and H. Ohlbach, editors, *Temporal Logic, Proceedings of ICTL '94*, number 827 in LNAI, pages 265–282. Springer-Verlag, 1994.

[DZ96]   M. Di Maio and A. Zanardo. A Gabbay-rule free axiomatization of $T \times W$ validity. Technical report, Dip. Matematica, Univ. Padova, 1996.

[FG92]   M. Finger and D. M. Gabbay. Adding a Temporal Dimension to a Logic System. *Journal of Logic Language and Information*, 1:203–233, 1992.

[FG96]   M. Finger and D. Gabbay. Combining Temporal Logic Systems. *Notre Dame Journal of Formal Logic*, 37(2):204–232, 1996. Special Issue on Combining Logics.

[FHMV95] R. Fagin, J. Halpern, Y. Moses, and M. Vardi. *Reasoning about Knowledge*. The MIT Press, 1995.

[Fin92]   M. Finger. Handling database updates in two-dimensional temporal logic. *J. of Applied Non-Classical Logic*, 2:201 – 224, 1992.

[Fin94]   M. Finger. *Changing the Past: Database Applications of Two-Dimensional Temporal Logics.* PhD thesis, Imperial College, 1994.

[FR97]   M. Finger and M. Reynolds. Imperative history: two-dimensional executable temporal logic. In H.J. Ohlbach and U. Reyle, editors, *Logic, Language and Reasoning.* Kluwer, 1997. To be published.

[Gab81]   D. M. Gabbay. An irreflexivity lemma with applications to axiomatizations of conditions on tense frames. In U. Monnich, editor, *Aspects of Philosophical Logic*, pages 67–89. Reidel, Dordrecht, 1981.

[GHR94]   D. Gabbay, I. Hodkinson, and M. Reynolds. *Temporal Logic: Mathematical Foundations and Computational Aspects, Volume 1.* Oxford University Press, 1994.

[GHR95]   D.M. Gabbay, C.J. Hogger, and J.A. Robinson, editors. *The Handbook of Logic in Artificial Intelligence and Logic Programming, vol. 4.* Oxford University Press, 1995.

[GS85]   Y. Gurevich and S. Shelah. The decision problem for branching time logic. *J. of Symbolic Logic*, 50:668–681, 1985.

[GS97]   D. Gabbay and V. Shehtman. Combining modal logics, 1997. submitted.

[Gue78]   F. Guenthner. Tense logic, time schemes and the tenses of English. In F. Guenthner and S. Schmidt, editors, *Formal Semantics and Pragmatics for Natural Languages.* Reidel, Dordrecht, 1978.

[Har83]   D. Harel. Recurring dominoes: Making the highly undecidable highly understandable. In *Conference on Foundations of Computing Theory*, pages 177–194. Springer, Berlin, 1983. Lec. Notes in Comp. Sci. 158.

[HHM⁺96] R. Hirsch, I. Hodkinson, M. Marx, S. Mikulás, and M. Reynolds. Mosaics and step-by-step, 1996. submitted.

[HS86]   J. Halpern and Y. Shoham. A propositional modal logic of time intervals. In *Proceedings, Symposium on Logic in Computer Science.* IEEE, Boston, 1986.

[HV89]   J. Halpern and M. Vardi. The complexity of reasoning about knowledge and time. i. lower bounds. *Journal of Computer and System Science*, 38:195–237, 1989.

[Kam68]   H. Kamp. *Tense logic and the theory of linear order.* PhD thesis, University of California, Los Angeles, 1968.

[Kam71]   H. Kamp. Formal properties of 'now'. *Theoria,* 37:237–273, 1971.

[KW91]   Marcus Kracht and Frank Wolter. Properties of independently axiomatizable bimodal logics. *J. Symbolic Logic,* 56(4):1469–1485, 1991.

[MR96]   M. Marx and M. Reynolds. Undecidability of compass logic, 1996. Submitted.

[MV97]   M. Marx and Y. Venema. *Multi-Dimensional Modal Logic.* Kluwer, 1997.

[Nem86]   I. Németi. *Free Algebras and Decidability in Algebraic Logic.* PhD thesis, Hungarian Academy of Sciences, Budapest, 1986. In Hungarian.

[Rey98]   M. Reynolds. A decidable of logic of parallelism, 1998. To appear in *Notre Dame Journal of Formal Logic.*

[Rob71]   R. Robinson. Undecidability and nonperiodicity for tilings of the plane. *Inventiones Math,* 12:177–209, 1971.

[RS85]   J. Reif and A. Sistla. A multiprocess network logic with temporal and spatial modalities. *Journal of Computer and System Sciences,* 30:41–53, 1985.

[Seg73]   K. Segerberg. Two-dimensional modal logic. *J. Philosophical Logic,* 2:77–96, 1973.

[SG87]   A. Sistla and S. German. Reasoning with many processes. In *Proceedings of Second IEEE Symposium on Logic in Computer Science,* pages 138–152, 1987.

[Spa93]   E. Spaan. *Complexity of Modal Logics.* PhD thesis, University of Amsterdam, 1993.

[Tho80]   S. Thomason. Independent propositional modal logics. *Studia Logica,* 39:143–144, 1980.

[Tho84]   R. Thomason. Combinations of tense and modality. In D. Gabbay and F. Guenthner, editors, *Handbook of Philosophical Logic, Vol II: Extensions of Classical Logic,* pages 135–165. Reidel, Dordrecht, 1984.

236   M. Reynolds

[vB84]   J. van Benthem. Correspondence theory. In D. M. Gabbay and F. Guenthner, editors, *Handbook of Philosophical Logic*, pages 167–247. D. Reidel, 1984.

[vB95]   J. van Benthem. Temporal logic. In D. Gabbay, editor, *Handbook of Logic in Artificial Intelligence and Logic Programming*, volume 4. Oxford University Press, 1995.

[Ven90] Y. Venema. Expressiveness and completeness of an interval tense logic. *Notre Dame J. Formal Logic*, 31:529–547, 1990.

[Ven91] Y. Venema. A logic with the chop operator. *J. Logic Comput.*, 1:453–476, 1991.

[Ven92] Y. Venema. *Many-dimensional modal logic*. PhD thesis, University of Amsterdam, 1992.

[Ven94] Y. Venema. Completeness through flatness in two-dimensional temporal logic. In D. Gabbay and H.-J. Ohlbach, editors, *Temporal Logic, First International Conference, ICTL '94, Bonn, Germany, July 11-14, 1994, Proceedings*, volume 827 of *Lecture Notes in A.I.*, pages 149–164. Springer-Verlag, 1994.

[vK93]   F. von Kutschera. Causation. *Journal of Philosophical Logic*, pages 563–588, 1993.

[vKar]   F. von Kutschera. $T \times W$ completeness. *Journal of Philosophical Logic*, to appear.

[Vla73] F. Vlach. *Now and then: A. formal study in the logic of tense anaphora*. PhD thesis, UCLA, 1973.

[VM95]   Y. Venema and M. Marx. A modal logic of relations. Technical Report IR-396, FWI, University of Amsterdam, 1995.

[Zan85] A. Zanardo. A finite axiomatization of the set of strongly valid Ockamist formulas. *J. of Philosophical Logic*, 14:447–468, 1985.

[Zan90] A. Zanardo. Axiomatization of 'Peircean' branching-time logic. *Studia Logica*, 49:183–195, 1990.

[Zan96] A. Zanardo. Branching-time logic with quantification over branches: the point of view of modal logic. *J. Symbolic Logic*, 61:1–39, 1996.

Author address

Department of Computer Science,
King's College, Strand,
London, WC2R 2LS, United Kingdom
email: markr@dcs.kcl.ac.uk

Rather Classless, Highly Saturated Models of Peano Arithmetic

J.H. Schmerl

Every saturated model of Peano Arithmetic having cardinality λ has 2^λ classes. Therefore, no saturated model of PA is rather classless. In other words, if $\kappa = \lambda$, then there are no rather classless, λ-saturated models of PA having cardinality κ. However, as long as λ is regular and $\kappa > \lambda$, there are no obstacles to the existence of rather classless, λ-saturated models of PA of cardinality κ other then there being no λ-saturated models of PA of cardinality κ at all. This is the content of the following theorem, which is the main result of this paper.

Theorem *If λ is regular, $\mathcal{N} \models$ PA is λ-saturated and $\lambda < |N|$, then there is a rather classless, λ-saturated $\mathcal{M} \succ \mathcal{N}$ such that $|M| = |N|$.*

The first rather classless, highly saturated models of Peano Arithmetic were constructed by Keisler [5]. His general theorem, specialized to models of PA, yields that whenever T is a consistent completion of PA, $\lambda^{<\lambda} = \lambda \geq \aleph_1$, and the combinatorial principle $\Diamond_{\lambda^+}$ holds, then there are rather classless, λ-saturated models of T of cardinality λ^+ (which, moreover, are λ^+-like). More rather classless, highly saturated models of PA can be obtained from a general theorem of Shelah (Theorem 12 of [8]) which, when specialized to models of PA, yields the following: If T is a consistent completion of PA, κ is the successor of a regular cardinal, and λ is a regular cardinal such that $\aleph_1 \leq \lambda < \kappa = \kappa^{<\lambda}$, then T has a rather classless, λ-saturated model of cardinality κ.

Kaufmann [3], assuming the combinatorial principle $\Diamond$, proved that there are $\aleph_1$-like, rather classless, recursively saturated models of each consistent completion of PA. Subsequently, this dependence on $\Diamond$ was eliminated by Shelah [8].

Rather classless, recursively saturated models of PA of each uncountable cardinality were constructed in Schmerl [7]. These models could be made

---

[1] Received september 26, 1996; revised version september 22, 1997.

$\aleph_0$-saturated by requiring that their standard systems contain all subsets of ω. The following, which is essentially Corollary 3.5 of [7], is that part of our main theorem in which $\lambda = \aleph_0$.

Theorem 1 *Let* $\mathcal{N} \models$ PA, *and suppose* $\kappa \geq |N| + 2^{\aleph_0}$. *Then there is a rather classless,* $\aleph_0$-*saturated* $\mathcal{M} \succ \mathcal{N}$ *of cardinality* κ.

Our aim is to generalize Theorem 1 by pushing up the amount of saturation. We do this in Theorem 2. It is well known that for singular λ, a model of PA is λ-saturated iff it is λ^+-saturated. Thus, we need consider λ-saturation only for regular λ. It is also well known that if $\mathcal{N}$ has cardinality κ and is λ-saturated, then $\kappa^{<\lambda} = \kappa$. Thus, the following theorem is easily seen to be an optimal result. Notice that Shelah's previously mentioned theorem handles the instance of Theorem 2 in which κ is the successor of a regular cardinal.

Theorem 2 *Suppose* $\mathcal{N} \models$ PA, *and suppose* λ *is regular,* $\aleph_1 \leq \lambda < \kappa = \kappa^{<\lambda}$ *and* $\kappa \geq |N|$. *Then there is a rather classless,* λ-*saturated* $\mathcal{M} \succ \mathcal{N}$ *of cardinality* κ.

To prove Theorem 2 we will obtain a rather classless, λ-saturated model $\mathcal{M}$ as the union of a continuous chain $\langle \mathcal{M}_\alpha : \alpha < \lambda^+ \rangle$ of models of T, where each model in the chain is an elementary end extension of each of the previous models. For an ordinal $\alpha < \lambda^+$, if α is not a limit ordinal or if $\mathrm{cf}(\alpha) \geq \lambda$, then $\mathcal{M}_\alpha$ will be λ-saturated. We will start with a λ-saturated model $\mathcal{M}_0 \succ \mathcal{N}$ having cardinality κ and descending cofinality λ^+. Two types of constructions of elementary end extensions will be used, resulting in model $\mathcal{M}_{\alpha+1}$ which is λ-generated over $\mathcal{M}_\alpha$. The type of construction which is used when $\mathcal{M}_\alpha$ is not λ-saturated (that is, when α is a limit ordinal and $\mathrm{cf}(\alpha) < \lambda$) is discussed in §3. When $\mathcal{M}_\alpha$ is λ-saturated, we will use a construction involving compatible sequences of satisfaction classes as described in §2. The proof of Theorem 2 will be completed in §4. Some preliminaries will be given in §1. Some open questions appear in §5.

1 Preliminaries

Consult Kaye [4] as a background reference to models of Peano Arithmetic.

We let $\mathcal{L}_{\mathsf{PA}} = \{+, \cdot, 0, 1, \leq\}$ be the language of Peano Arithmetic. We assume that the logic has term-building capabilities. For any language $\mathcal{L} \supseteq \mathcal{L}_{PA}$, we let PA*($\mathcal{L}$) (or just PA* if $\mathcal{L}$ is understood) be the extension of PA by all instances of the induction scheme in the language $\mathcal{L}$. In this paper it will always be understood that $\mathcal{L}$ is finite.

Consider a model $\mathcal{M} \models$ PA*. For $b \in M$, the set $[0, b] = \{x \in M : \mathcal{M} \models x \leq b\}$ is an initial segment of $\mathcal{M}$. A *cut* is an initial segment not of the form $[0, b]$. If I is a cut, then its *cofinality* $\mathrm{cf}(I)$ is the least cardinal κ such that some cofinal $X \subseteq I$ has cardinality κ. Similarly, $\mathrm{cf}(\mathcal{M})$ is the least cardinal κ such that some cofinal $X \subseteq M$ has cardinality κ. The *descending cofinality* of $\mathcal{M}$, denoted by $dcf(\mathcal{M})$, is the least cardinality κ of some set $X \subseteq M$ of nonstandard elements such that for each nonstandard $a \in M$ there is $b \in X$ for which $\mathcal{M} \models b < a$. Alternatively, if $c \in M$ is nonstandard, then $dcf(\mathcal{M}) = cf(\{x \in M : \mathcal{M} \models x + n \leq c \text{ for each } n \in \omega\})$.

The following easy lemma can be proved by a union of chains argument.

Lemma 1.1 *Let $\mathcal{N}$ be a model of* PA*. *Let μ, κ, λ be cardinals such that μ and λ are regular, $\kappa^{<\lambda} = \kappa \geq \mu \geq \lambda > \aleph_0$, and $|N| \leq \kappa$. Then there is a λ-saturated $\mathcal{M} \succ \mathcal{N}$ such that $|M| = \kappa$ and $dcf(\mathcal{M}) = \mu$.* $\qquad\square$

We let $Def(\mathcal{M})$ be the set of all subsets of M which are definable in $\mathcal{M}$ allowing parameters. A subset $X \subseteq M$ is a *class* of $\mathcal{M}$ iff $X \cap [0, b] \in Def(\mathcal{M})$ for each $b \in M$. If each class of $\mathcal{M}$ is in $Def(\mathcal{M})$, then we say that $\mathcal{M}$ is *rather classless*.

Let $\mathcal{M} \models$ PA*($\mathcal{L}$) and let $b \in M$. We will refer to a (partial) satisfaction class $S \subseteq M$ as a $\sum_b$-satisfaction class if $(\mathcal{M}, S) \models$ PA*($\mathcal{L} \cup \{S\}$) and S decides satisfaction for just the $\sum_b$-formulas (from the point of view of $\mathcal{M}$). If S is a $\sum_b$-satisfaction class and $a \leq b$, then $S|a$ is the unique subset of S which is a $\sum_a$-satisfaction class. The following proposition contains some well known facts about satisfaction classes.

Proposition 1.2 *Let $\mathcal{M} \models$ PA*, $n < \omega$ and $a, b \in M$.*
(1) *If S is a $\sum_{b+n}$-satisfaction class for $\mathcal{M}$, then $S \in Def((\mathcal{M}, S|b))$.*
(2) *If S is a $\sum_{a+b}$ satisfaction class for $\mathcal{M}$, then there is $S_0 \in Def((\mathcal{M}, S))$ which is a $\sum_b$-satisfaction class for $(\mathcal{M}, S|a)$.* $\qquad\square$

A *satisfaction class* is just a $\sum_b$-satisfaction class for some nonstandard b. It is well known that if $\mathcal{M} \models$ PA* is countable, then $\mathcal{M}$ has a satisfaction class iff it is recursively saturated. The next proposition shows that certain uncountable models also have satisfaction classes.

Proposition 1.3 *Let λ be a regular cardinal. Suppose that $\mathcal{M} \models$ PA* is λ-saturated and that $cf(\mathcal{M}) = \lambda$. Then $\mathcal{M}$ has a satisfaction class.*

Proof: Let $\mathcal{N}_0 \equiv \mathcal{M}$ be countable and recursively saturated, and let S_0 be a satisfaction class for $\mathcal{N}_0$. Let $(\mathcal{N}, S) \equiv (\mathcal{N}_0, S_0)$ be such that $cf(\mathcal{N}) = |N| = \lambda$. Then S is a satisfaction class for $\mathcal{N}$. Using the λ-saturation of $\mathcal{M}$, we can easily get a cofinal, elementary embedding of $\mathcal{N}$ into $\mathcal{M}$. Thus, we can assume that $\mathcal{N} \prec^{cf} \mathcal{M}$. Then, according to Theorem 1.2 of [7], there is $S' \subseteq M$ such that $(\mathcal{N}, S) \prec (\mathcal{M}, S')$, and therefore S' is a satisfaction class for $\mathcal{M}$. $\qquad\square$

If $\mathcal{M} \models$ PA* and $X \subseteq M$, then $\mathcal{M}$ is *generated* by X if $\mathcal{M}$ has no proper elementary substructures containing X. If $\mathcal{M}$ is generated by a set of cardinality at most λ, then $\mathcal{M}$ is λ-generated. If $A \subseteq M$, then $\mathcal{M}$ is

λ-generated over A if for some $X \subseteq M$, $|X| \leq \lambda$ and $\mathcal{M}$ is generated by $A \cup X$. If $\mathcal{M} \prec \mathcal{N}$ then $\mathcal{N}$ is a λ-generated elementary extension of $\mathcal{M}$ if $\mathcal{N}$ is λ-generated over M.

Lemma 1.4 *Suppose* $\mathcal{N} \prec^{end} \mathcal{M} \models \mathsf{PA}^*$, $b \in N$, $cf(\mathcal{N}) \leq \lambda$, *and* $\mathcal{M}$ *is* λ-generated over $[0, b]$. *Then* $\mathcal{N}$ *is* λ-generated over $[0, b]$.

Proof: Let $A \cup [0, b]$ generate $\mathcal{M}$, where $|A| \leq \lambda$; and let $C \subseteq N$ be cofinal, where $|C| \leq \lambda$. For each term $t(x, y)$ and elements $a \in A$ and $c \in C$ there is $d \leq c^c$ such that

$$\mathcal{M} \models \forall y < c \ (t(a, y) < c \to (d)_y = t(a, y)).$$

Each such d is in N, and there are at most λ of them. Clearly, the set of all such d generates $\mathcal{N}$ over $[0, b]$. $\Box$

2   Compatible Sequences

In this section we will discuss compatible sequences of satisfaction classes and of definable types. Let $\mathcal{M} \models \mathsf{PA}^*$ and let $I \subseteq M$ be a cut. A sequence $\langle S_k : k \in I \rangle$ is a *compatible sequence of satisfaction classes* (or CSSC) *for* $\mathcal{M}$ if for each $k \in I$, S_k is a $\sum_k$-satisfaction class and, whenever $j < k \in I$, then $S_j \subseteq S_k$. It is just as easy to get CSSC's as it is to get just satisfaction classes. For, let S be a $\sum_b$-satisfaction class for $\mathcal{M}$, let $S_k = S|k$ for each $k \leq b$, and let $I \subseteq [0, b]$ be a cut of $\mathcal{M}$. Then $\langle S_k : k \in I \rangle$ is a CSSC.

Recall the notion of definable types and their properties as developed by Gaifman [1]. Let T be a completion of $\mathsf{PA}^*(\mathcal{L})$. A type $p(x)$ of T is definable if it is unbounded (that is, for each $\theta(x)$ in $p(x)$, the sentence $\forall w \exists x > w \theta(x)$ is in T) and for each $\mathcal{L}$-formula $\varphi(x, u)$ there is an $\mathcal{L}$-formula $\sigma_\varphi(u)$ such that whenever t is a constant $\mathcal{L}$-term, then $\varphi(x, t) \in p(x)$ iff $\sigma_\varphi(t) \in T$. If $\mathcal{M} \models T$ and $p(x)$ is a definable type of T, then $p(x)$ can be extended to a type $p^M(x) = \{\varphi(x, a) : \varphi(x, u) \text{ is an } \mathcal{L}\text{-formula}, a \in M, \text{ and } \mathcal{M} \models \sigma_\varphi(a)\}$. If $\mathcal{M} \prec \mathcal{N}$ and $\mathcal{N}$ is generated by c over $\mathcal{M}$ and c realizes $p^M(x)$, then we say that $\mathcal{N}$ is a $p(x)$-extension of $\mathcal{M}$ generated by c. Each $p(x)$-extension $\mathcal{N}$ of $\mathcal{M}$ is conservative; that is, whenever $X \in \mathrm{Def}(\mathcal{N})$, then $X \cap M \in \mathrm{Def}(\mathcal{M})$.

Now let $\mathcal{M} \models \mathsf{PA}^*(\mathcal{L})$ and let $\langle S_k : k \in I \rangle$ be a CSSC for $\mathcal{M}$. For each $k \in I$, let $T_k = \mathrm{Th}((\mathcal{M}, S_k))$, which is a theory in the language $\mathcal{L}_k = \mathcal{L} \cup \{S_k\}$. We say that $\langle p_k(x) : k \in I \rangle$ is a *compatible sequence of definable types* (or CSDT) (*relative to* $\langle S_k : k \in I \rangle$ *and* $\mathcal{M}$) if the following two conditions hold:

(1) For each $k \in I$, $p_k(x)$ is a definable type of T_k. (Thus, for each $\mathcal{L}_k$-formula $\varphi(x, u)$ there is an $\mathcal{L}_k$-formula $\sigma_\varphi^k(u)$ such that whenever t is a constant $\mathcal{L}_k$-term, then $\varphi(x, t) \in p_k(x)$ iff $\sigma_\varphi^k(t) \in T_k$.)

(2) For $j < k \in I$, if $\psi(x)$ is an $\mathcal{L}_j$-formula and $\varphi(x, u)$ is the $\mathcal{L}_k$-formula derived from $\psi(x)$ by replacing each occurrence of S_j by $S_k|u$, then $\psi(x) \in p_j(x)$ iff $(\mathcal{M}, S_k) \models \sigma_\varphi^k(j)$.

The formula $\varphi(x, u)$, which is syntactically derived from $\psi(x)$, is related to $\psi(x)$ in the following semantic way: for any $a \in M$, $(\mathcal{M}, S_j) \models \psi(a)$ iff $(\mathcal{M}, S_k) \models \varphi(a, j)$.

Here is the point of the preceding definition of a CSDT. Suppose $\langle S_k : k \in I \rangle$ is a CSSC for $\mathcal{M}$ and $\langle p_k(x) : k \in I \rangle$ is a CSDT. Suppose $j < k \in I$. Let $(\mathcal{N}, S_k')$ be an elementary extension of $(\mathcal{M}, S_k)$ and let $c \in N$ realize $p_k^M(x)$ over $(\mathcal{M}, S_k)$. Then c (as an element of $(\mathcal{N}, S_k'|j)$) realizes $p_j^M(x)$ over $(\mathcal{M}, S_j)$.

So, now let $(\mathcal{N}_k, S_k')$ be a $p_k(x)$-extension of $(\mathcal{M}, S_k)$ generated by c_k, for each $k \in I$. If $j < k \in I$, then there is a unique elementary embedding $f_{jk} : (\mathcal{N}_j, S_j'|i)_{i \leq j} \to (\mathcal{N}_k, S_k'|i)_{i \leq j}$ which is the identity on M and for which $f(c_j) = c_k$. It follows from the uniqueness that these embeddings are compatible; that is, $f_{jk} \circ f_{ij} = f_{ik}$. Therefore, we can assume, without loss of generality, that $(\mathcal{N}_j, S_j'|i)_{i \leq j} \prec (\mathcal{N}_k, S_k'|i)_{i \leq j}$ and $c_j = c_k$ when $j < k \in I$. Let $(\mathcal{N}, R_k)_{k \in I} = \bigcup \{ (\mathcal{N}_k, S_k'|i)_{i \leq k} : k \in I \}$ and let $c = c_k$ (for any $k \in I$). We will refer to $(\mathcal{N}, R_k)_{k \in I}$ as the *canonical* $\langle p_k(x) : k \in I \rangle$-*extension* of $(\mathcal{M}, S_k)_{k \in I}$ generated by c. If the specific CSDT is not important, we will refer to $(\mathcal{N}, R_k)_{k \in I}$ as a canonical extension of $(\mathcal{M}, S_k)_{k \in I}$.

The following easily proved lemma contains the main properties of canonical extensions.

Lemma 2.1 *Suppose* $(\mathcal{N}, R_k)_{k \in I}$ *is a canonical extension of* $(\mathcal{M}, S_k)_{k \in I}$. *Then* $(\mathcal{N}, R_k)_{k \in I}$ *is a conservative extension of* $(\mathcal{M}, S_k)_{k \in I}$. *If* $\mathrm{cf}(I) = \lambda$, *then* $\mathcal{N}$ *is a* λ-*generated extension of* $\mathcal{M}$.

Proof: That the extension is conservative follows from the fact that $(\mathcal{N}, R_k)_{k \in I}$ is the union of an elementary chain each member of which is a conservative extension of some reduct of $(\mathcal{M}, S_k)_{k \in I}$.

Let $A \subseteq I$ be cofinal such that $|A| = \lambda$, and let $c \in N$ generate the canonical extension. Now let $C \subseteq N$ be such that $|C| = \lambda$, $c \in C$ and for each term $t(x, u)$ in the language of $(\mathcal{M}, S_k)_{k \in A}$ there is $d \in C$ such that $(\mathcal{N}, R_k)_{k \in I} \models \forall x \leq c((d)_x = t(x, c))$. Clearly, $(\mathcal{N}, R_k)_{k \in I}$ is generated by $C \cup M$. $\qquad\square$

It still needs to be demonstrated that CSDT's exist. That is the content of the next lemma, which, in fact, will show something stronger, namely that compatible sequences of minimal types exist. For each $n < \omega$ we say that a type $p(x)$ of T is n-indiscernible if for any n-ary formula $\psi(x_0, x_1, \ldots, x_{n-1})$ there is a formula $\varphi(x)$ in $p(x)$ such that either the sentence

$$\forall \bar{x}[(\varphi(x_0) \wedge \varphi(x_1) \wedge \cdots \wedge \varphi(x_{n-1}) \wedge x_0 < x_1 < \cdots < x_{n-1}) \to \psi(\bar{x})]$$

242   James H. Schmerl

or the sentence

$$\forall \bar{x}[(\varphi(x_0) \wedge \varphi(x_1) \wedge \cdots \wedge \varphi(x_{n-1}) \wedge x_0 < x_1 < \cdots < x_{n-1}) \rightarrow \neg \psi(\bar{x})]$$

is in T. Recall (see [6]) that the type $p(x)$ is minimal iff it is unbounded and n-indiscernible for each $n < \omega$. Also, $p(x)$ is minimal iff it is unbounded and 2-indiscernible. Minimal types are definable.

Lemma 2.2 *Let $\mathcal{M} \models$ PA* and let $I \subseteq M$ be a cut. Suppose $\langle S_k : k \in I \rangle$ is a CSSC. Then there exists a CSDT $\langle p_k(x) : k \in I \rangle$.*

Proof: We begin with a well known observation (see [2]) concerning the effectiveness of Ramsey's Theorem: If $k < \omega$ and $R_0, R_1, \cdots, R_k \subseteq \omega \times \omega$ are recursive binary relations, then there is an infinite Δ_3-set $H \subseteq \omega$ such that H is homogeneous for each R_i (that is, if $i < k$ then either whenever $x, y \in H$ and $x < y$ then $\langle x, y \rangle \in R_i$ or whenever $x, y \in H$ and $x < y$ then $\langle x, y \rangle \notin R_i$). This observation easily relativizes, yielding: If $k, n < \omega$, $X \subseteq \omega$ is an infinite $\sum_n$-set and $R_0, R_1, \cdots, R_k \subseteq \omega \times \omega$ are binary $\sum_n$-relations, then there is an infinite $\sum_{n+4}$-set $H \subseteq X$ which is homogeneous for each R_i. This statement is formalizable and provable as a scheme in PA*. Moreover, if $k \in I$, then the following sentence can be formalized and shown to hold in $(\mathcal{M}, S_k)$: if $j+4 \leq k$ and $X \subseteq M$ is an unbounded $\sum_j$-set, then there is an unbounded $\sum_{j+4}$-set $Y \subseteq X$ which is homogeneous for the first j binary $\sum_j$-relations $R \subseteq M^2$. Therefore, in $(\mathcal{M}, S_k)$ we can formally define the sequence $\langle H_j : j \leq k$ and $j \equiv 1 \pmod 4 \rangle$, where $H_1 = M$ and H_{4i+5} is the first unbounded $\sum_{4i+5}$-set $Y \subseteq H_{4i+1}$ which is homogeneous for each of the first $4i + 1$ binary $\sum_{4i+1}$-relations.

Notice that H_{4i+1} is independent of k (as long as $4i + 1 \leq k \in I$). Now "fill in" the sequence $\langle H_k : k \in I \rangle$ by setting $H_k = H_{4i+1}$, where $4i \leq k \leq 4i + 3$.

Clearly, for any $k \in I$ and $n < \omega$, H_{k+n} is definable in $(\mathcal{M}, S_k)$ without using parameters. Let $p_k(x)$ be the type consisting of all formulas $\varphi(x)$ in the language $\mathcal{L}_k$ such that for some $n < \omega$, $(\mathcal{M}, S_k) \models \forall x (x \in H_{k+n} \rightarrow \varphi(x))$. It is now an easy matter to check that $\langle p_k(x) : k \in I \rangle$ is a CSDT for $\langle S_k : k \in I \rangle$, completing the proof of the lemma. □

The next lemma shows how to construct λ-saturated, proper elementary end extensions of some λ-saturated models.

Lemma 2.3 *Let $\lambda \geq \aleph_1$ be a regular cardinal, let $\mathcal{M} \models$ PA* be λ-saturated, let $I \subseteq M$ be a cut for which $cf(I) \geq \lambda$, and let $\langle S_k : k \in I \rangle$ be a CSSC. If $(\mathcal{N}, R_k)_{k \in I}$ is a canonical extension of $(\mathcal{M}, S_k)_{k \in I}$, then $\mathcal{N}$ is λ-saturated.*

Proof: First, we show that $cf(\mathcal{N}) = cf(I) \geq \lambda$. This is an immediate consequence of Proposition 1.2(2) since whenever $m \in I$ there is $k \in I$ such that $m + n < k$, for each $n < \omega$, and for any such k, there is a function

$g \in \mathrm{Def}((\mathcal{M}, S_k))$ such that for any function $f \in \mathrm{Def}((\mathcal{M}, S_m))$, $(\mathcal{M}, S_k) \models \exists w \forall x > w(f(x) < g(x))$.

Next, we show that $\mathcal{N}$ is λ-saturated. Let $\mu < \lambda$ and let $\{A_\alpha : \alpha < \mu\}$ be a collection of nonempty, definable subsets of $\mathcal{N}$ which is closed under finite intersections. We need to show $\bigcap\{A_\alpha : \alpha < \mu\} \neq \emptyset$. Since $\mathrm{cf}(\mathcal{N}) \geq \lambda$, we can assume that each A_α is bounded. Since $\mathrm{cf}(I) \geq \lambda$, we can get $k \in I$ large enough so that $A_\alpha \in N_k$ for each $\alpha < \mu$. (Refer to the notation in the definition of canonical extension.) Then, for each $\alpha < \mu$, there is an $\mathcal{L}_k$-term $t_\alpha(u, v)$ and an element $b_\alpha \in M$ such that $(\mathcal{N}_k, S'_k) \models A_\alpha = t_\alpha(b_\alpha, c)$. Since $\mathcal{M}$ is λ-saturated, there is $d \in M$ such that $(d)_{b_\alpha} = t_\alpha$ for each $\alpha < \mu$. Let $m \in I$ be such that $k + n < m$ for each $n < \omega$. By Proposition 1.2(2), there is a $\sum_k$-satisfaction class for $(\mathcal{N}_k, S'_k)$ in $\mathrm{Def}((\mathcal{N}_m, S'_m))$, so that there is an $\mathcal{L}_m$-term $t(u, v, w)$ such that for each $\alpha < \mu$, $(\mathcal{N}_m, S'_m) \models A_\alpha = t(b_\alpha, c, d)$. Hence, there is $e \in N$ such that $A_\alpha = (e)_{b_\alpha}$ for each $\alpha < \mu$.

Let $b \in M$ be such that $b_\alpha < b$ for each $\alpha < \mu$. Let $E \in \mathrm{Def}(\mathcal{N})$ be the set of $f \in N$ such that for each definable $F \subseteq [0, b]$, $\emptyset \neq \bigcap\{(e)_x : x \in F\}$ iff $\emptyset \neq \bigcap\{(f)_x : x \in F\}$. In particular, $a \in E$. Clearly $E \cap M \neq \emptyset$, so let $f \in E \cap M$. By the λ-saturation of $\mathcal{M}$, there is $r \in M$ such that $r \in \bigcap\{(f)_{b_\alpha} : \alpha < \mu\}$. For each $e' \in E$, there is r' such that for each $x < b$, $r' \in (e')_x$ iff $r \in (f)_x$. Thus, there is s such that for each $x < b$, $s \in (e)_x$ iff $r \in (f)_x$, so in particular, $s \in \bigcap\{(e)_{b_\alpha} : \alpha < \mu\}$. Thus, $\bigcap\{A_\alpha : \alpha < \mu\} \neq \emptyset$. $\qquad\square$

The next corollary will not be explicitly used but is mentioned for its independent interest.

Corollary 2.4 *If $\mathcal{M}$ is λ-saturated and $\mathrm{cf}(\mathcal{M}) = \lambda$, then $\mathcal{M}$ has a λ-saturated, λ-generated elementary end extension.*

Proof: By Lemma 1.3, $\mathcal{M}$ has a $\sum_b$-satisfaction class S for some nonstandard b. Since $\mathcal{M}$ is λ-saturated, it has a cut I such that $\mathrm{cf}(I) = \lambda$ and $b \notin I$. Then $\langle S|k : k \in I \rangle$ is a CSSC. By Lemma 2.2 there is a CSDT. Now apply Lemmas 2.1 and 2.3 to get the λ-saturated, λ-generated elementary end extension. $\qquad\square$

3   Extending bdd λ-saturated models

In this section we present a simple lemma which shows that some bdd λ-saturated models have elementary end extensions which are λ-saturated.

First, we define bdd λ-saturation. Let λ be a regular cardinal and let $\mathcal{M} \models \mathrm{PA}^*(\mathcal{L})$. Then $\mathcal{M}$ is *bdd λ-saturated* if for each $b \in M$ there is a λ-saturated $\mathcal{N} \prec^{\mathrm{end}} \mathcal{M}$ such that $b \in N$. If $\mathcal{M}$ is λ-saturated, then $\mathcal{M}$ is bdd λ-saturated. Indeed, $\mathcal{M}$ is λ-saturated iff $\mathcal{M}$ is bdd λ-saturated and $\mathrm{cf}(\mathcal{M}) \geq \lambda$. If $\mathcal{M}$ is bdd λ-saturated and $\mathrm{cf}(\mathcal{M}) = \kappa$, then a *filtration* for $\mathcal{M}$ is a continuous sequence $\langle \mathcal{M}_\alpha : \alpha < \kappa \rangle$ whose union is $\mathcal{M}$ such that

$\mathcal{M}_\alpha \prec^{\mathrm{end}} \mathcal{M}_\beta \prec^{\mathrm{end}} \mathcal{M}$ whenever $\alpha < \beta < \kappa$ and $\mathcal{M}_\alpha$ is λ-saturated unless $\omega \leq \mathrm{cf}(\alpha) < \lambda$. Every bdd λ-saturated $\mathcal{M}$ has a filtration.

Lemma 3.1 *Let $\lambda \geq \aleph_1$ be regular. Suppose $\mathcal{M}$ is a bdd λ-saturated model of* PA* *such that $cf(\mathcal{M}) < \lambda$ and $\mathcal{M}$ is λ-generated over an initial segment. Then $\mathcal{M}$ has a λ-saturated, λ-generated elementary end extension.*

Proof: Let $\mathrm{cf}(\mathcal{M}) = \kappa < \lambda$, and let $b \in M$ be such that $\mathcal{M}$ is λ-generated over $[0, b]$. Because it is bdd λ-saturated, $\mathcal{M}$ has a filtration $\langle \mathcal{M}_\alpha : \alpha < \kappa \rangle$ such that that $b \in M_0$. Then $\mathcal{M}_0$ has a filtration $\langle \mathcal{N}_\beta : \beta < \lambda \rangle$ with $b \in N_0$. By Lemma 1.4, $\mathcal{M}_0$ is λ-generated over $[0, b]$

Let $C \subseteq M$ be a cofinal subset of M such that $|C| = \kappa < \lambda$. More specifically, let $C = \{c_\alpha : \alpha < \kappa\}$ where $c_\alpha \in M_{\alpha+1} \backslash M_\alpha$. By the λ-saturation of $\mathcal{M}_0$ there is a function $f_0 : C \to M_0$ which is elementary over $[0, b]$ and $f_0(c_\alpha) \in N_{\alpha+1} \backslash N_\alpha$ for each $\alpha < \kappa$. By Lemma 1.4, $\mathcal{N}_\kappa$ is λ-generated over $[0, b]$. Let $X \subseteq M$ be such that $|X| \leq \lambda$ and $X \cup [0, b]$ generates $\mathcal{M}$, and let $Y \subseteq N_\kappa$ be such that $Y \cup [0, b]$ generates $\mathcal{N}_\kappa$ and $|Y| \leq \lambda$. By a back-and-forth argument, we can extend f_0 to f, which is elementary over $[0, b]$ such that $X \subseteq D = \mathrm{dom} f$ and $Y \subseteq f``D$. Since X and Y generate $\mathcal{M}$ and $\mathcal{N}_\kappa$, respectively, over $[0, b]$, we can extend f to an isomorphism of $\mathcal{M}$ and $\mathcal{N}_\kappa$. Since $\mathcal{M}_0$ is λ-generated over $[0, b]$, it certainly is a λ-generated extension of $\mathcal{N}_\kappa$. Thus, $\mathcal{N}_\kappa$ has a λ-saturated, λ-generated elementary end extension (namely $\mathcal{M}_0$), then so does $\mathcal{M}$. $\quad\square$

4   Proving Theorem 2

The proof of Theorem 2 will be completed in this section.

Let $\mathcal{M}_0 \succ \mathcal{N}$ be λ-saturated such that $\mathrm{dcf}(\mathcal{M}_0) = \lambda^+$ and $|M_0| = \kappa$. Moreover, we want $\mathcal{M}_0$ to have a $\sum_b$-satisfaction class S for some nonstandard $b \in M_0$. (To get such a model $\mathcal{M}_0$, apply Lemma 1.1 to the theory T together with the sentences asserting that S is a satisfaction class.) Let $\langle I_\alpha : \alpha < \lambda^+ \rangle$ be a sequence of cuts of $\mathcal{M}_0$ such that:

(1) if $\alpha < \lambda^+$, then $I_\alpha \subseteq [0, b]$ and $\mathrm{cf}(I_\alpha) = \lambda$;

(2) if $\alpha < \beta < \lambda^+$, then $I_\beta \subseteq I_\alpha$ and $I_\beta \neq I_\alpha$;

(3) $\omega = \bigcap \{I_\alpha : \alpha < \lambda^+\}$.

Let $S_k^0 = S|k$ for each $k \in I_0$, and then let $\mathcal{M}_0^* = (\mathcal{M}_0, S_k^0)_{k \in I_0}$. Notice that $\langle S_k^0 : k \in I_0 \rangle$ is a CSSC.

Construct a sequence $\langle \mathcal{M}_\alpha^* : \alpha < \lambda^+ \rangle$ of structures $\mathcal{M}_\alpha^* = (\mathcal{M}_\alpha, S_k^\alpha)_{k \in I_\alpha}$ such that for each $\alpha < \lambda^+$, the following hold:

(4) if $\alpha = 0$, α is a successor ordinal or $\mathrm{cf}(\alpha) = \lambda$, then $\mathcal{M}_{\alpha+1}^*$ is a canonical extension of $(\mathcal{M}_\alpha, S_k^\alpha)_{k \in I_{\alpha+1}}$;

(5) if $\aleph_0 \leq \mathrm{cf}(\alpha) < \lambda$, then $\mathcal{M}^*_{\alpha+1}$ is an elementary end extension of $(\mathcal{M}_\alpha, S^\alpha_k)_{k \in I_{\alpha+1}}$ and $\mathcal{M}_{\alpha+1}$ is a λ-saturated, λ-generated extension of $\mathcal{M}_\alpha$;

(6) if α is a limit ordinal, then $\mathcal{M}^*_\alpha = \bigcup\{(\mathcal{M}_\beta, S^\beta_k)_{k \in I_\alpha} : \beta < \alpha\}$.

We will construct this sequence by recursion. Suppose that $\gamma < \lambda^+$ and that we already have $\langle \mathcal{M}^*_\alpha : \alpha < \gamma \rangle$ such that for each $\alpha < \gamma$, (4)–(6) hold. If γ is a limit ordinal, then let $\mathcal{M}^*_\gamma = \bigcup\{(\mathcal{M}_\alpha, S^\alpha_k)_{k \in I_\gamma} : \alpha < \gamma\}$. If $\gamma = \alpha + 1$ and either $\alpha = 0$ or α is a successor ordinal or $\mathrm{cf}(\alpha) = \lambda$, then apply Lemmas 2.1, 2.2 and 2.3 to get $\mathcal{M}^*_\gamma$. If $\gamma = \alpha+1$ and $\aleph_0 \leq \mathrm{cf}(\alpha) < \lambda$, let $j \in I_\alpha \backslash I_\gamma$ and apply Lemma 3.1 to the structure $(\mathcal{M}_\alpha, S^\alpha_j)$, getting $(\mathcal{M}_\gamma, S^\gamma_j)$, and then letting $\mathcal{M}^*_\gamma = (\mathcal{M}_\gamma, S^\gamma_j | k)_{k \in I_\gamma}$.

Now let $\mathcal{M} = \bigcup\{\mathcal{M}_\alpha : \alpha < \lambda^+\}$. Clearly, $\mathcal{M}$ is λ-saturated and $|M| = \kappa$. Suppose $X \subseteq M$ is a class and that $X \notin \mathrm{Def}(\mathcal{M})$. By an argument like the one in the proof of Lemma 3.1 of [7], there is some $\alpha < \lambda^+$ such that $X \cap M_\alpha \notin \mathrm{Def}(\mathcal{M}_\alpha)$. Then, by Lemma 2.4 of [7], there is $k \in I_\alpha$ such that $X \cap M_\alpha \notin \mathrm{Def}((\mathcal{M}_\alpha, S^\alpha_k))$. Let $\beta < \lambda^+$ be such that $\alpha < \beta$, $\mathrm{cf}(\beta) = \lambda$ and $k \notin I_\beta$. Then $X \cap M_\beta \in \mathrm{Def}(\mathcal{M}_\beta)$, so for some $j \in I_\beta$, $X \cap M_\beta \in \mathrm{Def}((\mathcal{M}_\beta, S^\beta_j))$. Since $\mathcal{M}^*_\alpha \prec \mathcal{M}^*_\beta$, then $X \cap M_\alpha \in \mathrm{Def}((\mathcal{M}_\alpha, S^\alpha_j)) \subseteq \mathrm{Def}((\mathcal{M}_\alpha, S^\alpha_k))$, entailing a contradiction.

This completes the proof of Theorem 2.   □

5   Questions

The model $\mathcal{M}$ which was constructed in the proof of Theorem 2 has cofinality λ^+. With a little more effort, we can get $\mathrm{cf}(\mathcal{M}) = \lambda^{++}$ provided, of course, that $\lambda^{++} \leq \kappa$. This suggests the following question.

Question 5.1 For which cardinals κ, λ, μ do there exist rather classless, λ-saturated models $\mathcal{M} \models \mathrm{PA}$ such that $|M| = \kappa$ and $\mathrm{cf}(\mathcal{M}) = \mu$?

In [7], assuming $V = L$, for each $\kappa \geq \aleph_2$, such that $\mathrm{cf}(\kappa) > \aleph_0$ and κ is not weakly compact, we constructed a rather classless, κ-like $\aleph_0$-saturated model of PA. By the theorem of Keisler [5], if $\kappa = \lambda^+$ and λ is regular, then we can get a rather classless, κ-like λ-saturated model of PA. This suggests the following question.

Question 5.2 (Assume $V = L$.) For which cardinals κ, λ do there exist rather classless, κ-like, λ-saturated models of PA?

6   REFERENCES

[1] Gaifman, H., Models and types of Peano's arithmetic, Ann. Math. Logic **9** (1976), 223–306.

[2] Jockusch, C.G., Ramsey's theorem and recursion theory, J. Symb. Logic **37** (1972), 268-280.

[3] Kaufmann, M., A rather classless model, Proc. Amer. Math. Soc. **62** (1977), 330-333.

[4] Kaye, Richard, *Models of Peano Arithmetic*, Oxford University Press, Oxford, 1991.

[5] Keisler, H.J., Models with tree structures, in: *Proceedings of the Tarski Symposium* (eds. L. Henkin et al.), American Math. Soc., 1974, Providence, RI, 331-348.

[6] Kossak, R., Kotlarski, H., Schmerl, J.H., On maximal subgroups of the automorphism group of a countable recursively saturated model of PA, Ann. Pure Appl. Logic **65** (1993), 125-148.

[7] Schmerl, J.H., Recursively saturated, rather classless models of Peano Arithmetic, in: *Logic Year 1979-80* (eds. M. Lerman et al.), Lecture Notes in Mathematics **859**, Springer-Verlag, Berlin, 1981, pp. 268-282.

[8] Shelah, S., Models with second order properties II. Trees with no undefined branches, Annals Math. Logic **14** (1978), 73-87.

Author address

Department of Mathematics,
University of Connecticut
Storrs, CT 06269, USA
email: schmerl@math.uconn.edu

Incompleteness theorems and S_2^i versus S_2^{i+1}

Gaisi Takeuti

In [2], S. Buss introduced the systems of Bounded Arithmetic for $S_2^i (i = 0, 1, 2, \ldots)$ which have a close relationship to classes in polynomial hierarchy. As Buss stated in the introduction of his book, one of the most important problems on Bounded Arithmetic is the separation problems on $S_2^i (i = 1, 2, 3, \ldots)$ i.e., the problems to show $S_2^i \neq S_2^{i+1} (i = 1, 2, 3, \ldots)$. We believe that the separation problems of S_2^i and the separation problems of classes of the polynomial hierarchy are the same problem in the sense that the difficulty of these two problems comes from the same source. We also believe that the solution of one of them will lead to the solution of the other problem.

This idea is partially supported by the following theorem in [4].

Theorem. (Krajíček-Pudlák-Takeuti) *If $S_2^i = S_2^{i+1}$, then $\Sigma_{i+2}^P = \Pi_{i+2}^P$.*

Very often, a stronger theory is shown to be strictly stronger than a weaker theory by proving that the stronger theory proves the consistency of the weaker system. In [2], S. Buss proved that the second incompleteness theorem also holds for S_2^i. However this method does not work for the separations of S_2^i since the theorem of Wilkie and Paris in [6] immediately implies that $S_2 = \bigcup_i S_2^i$ does not prove the consistency of S_2^o. The reason for this phenomenon is that the consistency here is the consistency of the theory with unbounded quantifiers. The expressing power of unbounded quantifiers is too strong to be handled by Bounded Arithmetic. The ordinary consistency is totally inadequate for Bounded Arithmetic. We need some more delicate consistency.

In [5], we introduced a delicate notion of proof in S_2^i and delicate notions of consistency of S_2^i and Gödel sentences of S_2^i. Using them we proved that a Gödel sentence of S_2^i is provable in $S_{2,n}^{i+1}$ though it is not provable in S_2^i, therefore $S_{2,n}^{i+1} \neq S_2^i$ holds in the language of S_2^i. S_2^{i+1} is a limit of $S_{2,n}^{i+1}$ if n goes to ∞ but this result does not imply $S_2^{i+1} \neq S_2^i$.

---

[1]Received August 96; revised version December 1996.

In this paper we further develop the idea of [5] and propose many conjectures on delicate consistency and delicate Gödel sentences of S_2^i which imply $S_2^{i+1} \neq S_2^i$. We believe that a little more advance on the knowledge of these consistencies and Gödel sentences would prove $S_2^i \neq S_2^{i+1}$ and $P \neq NP$.

1   The formalized terms

We define $|a|_n$ for $n = 0, 1, 2, \ldots$ by $|a|_0 = a$ and $|a|_{n+1} = |\,|a|_n\,|$. In [5], we expand the language of S_2^i by introducing $a \dot{-} b$, $\max(a, b), \ldots, \beta(i, w)$. In this paper we further expand the language by adding finitely many function symbols whose intended meanings are polynomial time computable functions.

Let $\tilde{n}$ be the Gödel number of a formalized term in the language of S_2^i with only free variables $\ulcorner a_1 \urcorner, \ldots, \ulcorner a_n \urcorner$ where $n = 0, 1, 2, \ldots$. In this case, we often denote $\tilde{n}$ by $\ulcorner t(a_1, \ldots, a_n) \urcorner$ though we cannot find a term $t(a_1, \ldots, a_n)$ in general. Let $v(\ulcorner t(a_1, \ldots, a_n) \urcorner, b_1, \ldots, b_n)$ be the value which $\ulcorner t(a_1, \ldots, a_n) \urcorner$ represents when $\ulcorner a_i \urcorner$ represents b_i respectively. Let $\vec{a}$ and $\vec{b}$ express $a_1, \ldots, a_n$ and $b_1, \ldots, b_n$ respectively. In [5], we proved that $\exp(\ulcorner t(\vec{a}) \urcorner |b|^{|\ulcorner t(\vec{a}) \urcorner|})$ is a bound of $v(\ulcorner t(\vec{a}) \urcorner, \vec{b})$ where $\exp(a) = 2^a$ and b is the maximum of $\vec{b}$ and 2 and also that

$$v(\ulcorner t \urcorner) \leq \exp(\ulcorner t \urcorner)$$

if t is the Gödel number of a closed term.

Let f be a new polynomial time computable function whose function symbol is in the language and $f(a) \leq a \# \ldots \# a$, where the number of a is n. Let $t(a) = f(t_0(a))$. We are going to prove

$$v(\ulcorner t(a) \urcorner, b) \leq \exp(\ulcorner t(a) \urcorner |b|^{|\ulcorner t(a) \urcorner|})$$

under the assumption $v(\ulcorner t_0(a) \urcorner, b) \leq \exp(\ulcorner t_0(a) \urcorner |b|^{|\ulcorner t_0(a) \urcorner|})$

$$
\begin{aligned}
v(\ulcorner t(a) \urcorner, b) &\leq v(\ulcorner t_0(a) \urcorner, b) \# \ldots \# v(\ulcorner t_0(a) \urcorner, b) \\
&\leq \exp(\ulcorner t_0(a) \urcorner |b|^{|\ulcorner t_0(a) \urcorner|}) \# \ldots \# \exp(\ulcorner t_0(a) \urcorner |b|^{\ulcorner t_0(a) \urcorner}) \\
&\leq \exp((\ulcorner t_0(a) \urcorner |b|^{|\ulcorner t_0(a) \urcorner|} + 1)^n) \\
&\leq \exp(c(\ulcorner t_0(a) \urcorner |b|^{|\ulcorner t_0(a) \urcorner|})^n) \\
&\quad \text{for some constant } c \\
&\leq \exp(\ulcorner t(a) \urcorner \cdot |b|^{|\ulcorner t(a) \urcorner|})
\end{aligned}
$$

where we define $\ulcorner f \urcorner$ to be sufficiently large and $f(a)$ to be an abbreviation of $f(a, \ldots, a)$. More precisely we define $\ulcorner f(t) \urcorner$ to be the sequence number of ($\ulcorner f \urcorner, \ulcorner t \urcorner, \ldots, \ulcorner t \urcorner$) where $\ulcorner t \urcorner$ occurs n times for a fixed number n so that the above calculation goes through. It is very easy to find such $\ulcorner f \urcorner$ and n for every polynomial time computable function f.

2   Truth definition

We say 'a is n-small' if there exists x such that $a \leq |x|_n$. We say 'a is small' if a is 1-small. In this section we always assume $i > 0$. Let u satisfy the following condition. $|u|_2$ is greater than the Gödel number $\ulcorner t(\vec{a}) \urcorner$ of a term $t(\vec{a})$ in the language of S_2^i with only free variables $\vec{a}$. The length of $\vec{a}$ is less than $|\ulcorner t(\vec{a}) \urcorner|$ and it is 3-small. Let $\vec{b}$ be a sequence with the same length as $\vec{a}$. As before we define $b = \max(2, \vec{b})$. As in Section 1, we define $v(\ulcorner t(\vec{a}) \urcorner, \vec{b})$ and the following holds.

$$v(\ulcorner t(\vec{a}) \urcorner, \vec{b}) \leq \exp(|u|_2 |b|^{|u|_3}).$$

Here $\exp(|u|_2 |b|^{|u|_3})$ is a Σ_1^b-definable function in S_2^1 when b is small.

The system S_2^i is defined by the following axioms and inferences.

(a) Basic axioms.

The language of S_2^i consists of $\leq$ and finitely many function symbols which express polynomial time computable functions.

The defining axioms of the functions in the language and the predicate $\leq$. All these axioms are included in the form of initial sequents without logical symbols.

(b) Σ_i^b - PIND

$$\frac{A(\llcorner \tfrac{1}{2} a \lrcorner), \Gamma \to \Delta, A(a)}{A(0), \Gamma \to \Delta, A(t)}$$

where $A(0)$ is Σ_i^b and a satisfies an eigenvariable condition. We further extend a) by introducing finitely many forms of initial sequents without logical symbols. E.g.

$$\longrightarrow |s| \leq s.$$

This saves unnecessary use of induction in order to prove some necessary properties. Here the following must be satisfied.

1.   The number of the forms of initial sequents must be finite.

2.   The initial sequent thus introduced must be valid and has no occurrences of logical symbols. As a stronger case of this type of extension, later we also consider the following S_2^i under the assumption of $P = NP$.

If $P = NP$, there exists an NP-complete predicate $\exists x \leq t(a)A(x,a)$ with a sharply bounded $A(x,a)$ and a polynomial time computable function f satisfying the following condition

$$\exists x \leq t(a)A(x,a) \rightarrow f(a) \leq t(a) \wedge A(f(a),a).$$

If we introduce finitely many polynomial time computable functions, then we can assume that $A(x,a)$ is an atomic formula and $P = NP$ can be expressed by the following forms of initial sequents without logical symbols

$$s' \leq t(s), A(s', t(s)) \rightarrow f(s) \leq t(s)$$
$$s' \leq t(s), A(s', t(s)) \rightarrow A(f(s), s).$$

Our theory developed later also holds for extended S_2^i in this way and will be used to find conjectures to prove $P \neq NP$. The outline of this plan goes as follows.

$$P = NP \rightarrow \Sigma_i^b = \Sigma_{i+1}^b \dashrightarrow S_2^i = S_2^{i+1}.$$

Therefore $S_2^i \neq S_2^{i+1} \dashrightarrow P \neq NP$. Here $\dashrightarrow$ holds if S_2^i and S_2^{i+1} are extended S_2^i and S_2^{i+1} discussed in the above.

In [5], it is proved that $v(\ulcorner t(\vec{a}) \urcorner, \vec{b})$ is definable under the assumption that $\exp(|u|_2|b|^{|u|_3})$ is definable where $\ulcorner t(\vec{a}) \urcorner \leq |u|_2$ and $b = \max(2, \vec{b})$. In the same way, we can show that $v(\ulcorner t(\vec{a}) \urcorner, \vec{b})$ is Σ_1^b-definable in S_2^1 and satisfies the following conditions if $\ulcorner t(\vec{a}) \urcorner \leq |u|_2$ and b is small.

1. If $\ulcorner f(t_1(\vec{a}), \ldots, t_k(\vec{a})) \urcorner$ is a subterm of $\ulcorner t(\vec{a}) \urcorner$, then

$$v(\ulcorner f(t_1(\vec{a}), \ldots, t_k(\vec{a}) \urcorner, \vec{b}) = f(v(\ulcorner t_1(\vec{a}) \urcorner, \vec{b}), \ldots, v(\ulcorner t_n(\vec{a}) \urcorner, \vec{b})).$$

2. $v(\ulcorner 0 \urcorner, \vec{b}) = 0$ and $v(\ulcorner a_i \urcorner, \vec{b}) = b_i$.

All these properties are provable in S_2^1.

Let $\ulcorner t \urcorner$ be a formalized closed term and small. Then in the same way as above, $v(\ulcorner t \urcorner)$ is Σ_1^b-definable function in S_2^1 and satisfies the following conditions.

1. $v(\ulcorner 0 \urcorner) = 0$.

2. $v(\ulcorner f(t_1, \ldots, t_n) \urcorner) = f(v(\ulcorner t_1 \urcorner), \ldots, v(\ulcorner t_n \urcorner))$.

These properties are provable in S_2^1.

Now let u satisfy the following condition. $|u|_2$ is greater than the Gödel number $\ulcorner \varphi(\vec{a}) \urcorner$ of a quantifier free formula in the language of S_2^i with only free variables $\vec{a}$. Let $\vec{b}$ be a sequence with the same length as $\vec{a}$ and $b = \max(\vec{b}, 2)$ be small. In [5], the truth definition of $T_0(\ulcorner \varphi(\vec{a}) \urcorner, \vec{b})$ was defined by using $\exp(|u|_2|b|^{|u|_3})$ is definable. The following properties were proved.

1. If $\ulcorner t_1 \leq t_2 \urcorner$ is a subformula of $\ulcorner \varphi(\vec{a}) \urcorner$, then

$$T_0(\ulcorner t_1 \leq t_2 \urcorner, \vec{b}) \text{ iff } v(\ulcorner t_1 \urcorner, \vec{b}) \leq v(\ulcorner t_2 \urcorner, \vec{b}).$$

2. If $\ulcorner t_1 = t_2 \urcorner$ is a subformula of $\ulcorner \varphi(\vec{a}) \urcorner$, then

$$T_0(\ulcorner t_1 = t_2 \urcorner, \vec{b}) \text{ iff } v(\ulcorner t_1 \urcorner, \vec{b}) = v(\ulcorner t_2 \urcorner, \vec{b}).$$

3. If $\ulcorner \psi_1 \wedge \psi_2 \urcorner$ is a subformula of $\ulcorner \varphi(\vec{a}) \urcorner$, then

$$T_0(\ulcorner \psi_1 \wedge \psi_2 \urcorner, \vec{b}) \text{ iff } T_0(\ulcorner \psi_1 \urcorner, \vec{b}) \wedge T_0(\ulcorner \psi_2 \urcorner, \vec{b}).$$

4. If $\ulcorner \psi_1 \vee \psi_2 \urcorner$ is a subformula of $\ulcorner \varphi(\vec{a}) \urcorner$, then

$$T_0(\ulcorner \psi_1 \vee \psi_2 \urcorner, \vec{b}) \text{ iff } T_0(\ulcorner \psi_1 \urcorner, \vec{b}) \vee T_0(\ulcorner \psi_2 \urcorner, \vec{b}).$$

5. If $\ulcorner \neg \psi \urcorner$ is a subformula of $\ulcorner \psi(\vec{a}) \urcorner$, then

$$T_0(\ulcorner \neg \psi \urcorner, \vec{b}) \text{ iff } \neg T_0(\ulcorner \psi \urcorner, \vec{b}).$$

$T_0(\ulcorner \varphi(\vec{a}) \urcorner, \vec{b})$ is Σ_1^b-definable in S_2^1 and all these properties are provable in S_2^1.

In the same way, we can Σ_1^b-define a truth definition $T_0(\ulcorner \varphi \urcorner)$ in S_2^1 if $\ulcorner \varphi \urcorner$ is the Gödel number of a quantifier free sentence in S_2^1 and $\ulcorner \varphi \urcorner$ is small and the following properties are provable in S_2^1.

1. If $\ulcorner t_1 \leq t_2 \urcorner$ is a subformula of $\ulcorner \psi \urcorner$, then

$$T_0(\ulcorner t_1 \leq t_2 \urcorner) \text{ iff } v(\ulcorner t_1 \urcorner) \leq v(\ulcorner t_2 \urcorner).$$

2. If $\ulcorner t_1 = t_2 \urcorner$ is a subformula of $\ulcorner \varphi \urcorner$, then

$$T_0(\ulcorner t_1 = t_2 \urcorner) \text{ iff } v(\ulcorner t_1 \urcorner) = v(\ulcorner t_2 \urcorner).$$

3. If $\ulcorner \psi_1 \wedge \psi_2 \urcorner$ is a subformula of $\ulcorner \psi \urcorner$, then

$$T_0(\ulcorner \psi_1 \wedge \psi_2 \urcorner) \text{ iff } T_0(\ulcorner \psi_1 \urcorner) \wedge T_0(\ulcorner \psi_2 \urcorner).$$

4. If $\ulcorner \psi_1 \vee \psi_2 \urcorner$ is a subformula of $\ulcorner \varphi \urcorner$, then

$$T_0(\ulcorner \psi_1 \vee \psi_2 \urcorner) \text{ iff } T_0(\ulcorner \psi_1 \urcorner) \vee T_0(\ulcorner \psi_2 \urcorner).$$

5. If $\ulcorner \neg \psi \urcorner$ is a subformula of $\ulcorner \psi \urcorner$, then

$$T_0(\ulcorner \neg \psi \urcorner) \text{ iff } \neg T_0(\ulcorner \psi \urcorner).$$

A formula $\varphi(\vec{a})$ in the language of S_2^1 is said to be a pure i-form if the following conditions are satisfied.

1. The only free variables in $\varphi(\vec{a})$ are $\vec{a}$.

2. $\varphi(\vec{a})$ is of the form

$$\exists x_1 \le t_1(\vec{a}) \forall x_2 \le t_2(\vec{a}, x_2) \ldots Q_i x_i \le t_i(\vec{a}, x_1, \ldots, x_{i-1})$$
$$Q_{i+1} x_{i+1} \le |t_{i+1}(\vec{a}, x_1, \ldots, x_i)| A(\vec{a}, x_1, \ldots, x_{i+1})$$

where Q_i is $\forall$ and Q_{i+1} is $\exists$ if i is even and Q_i is $\exists$ and Q_{i+1} is $\forall$ if i is odd and $A(\vec{a}, a_1, \ldots, a_{i+1})$ is a quantifier free formula in the language of S_2^1. The formula described in the above is denoted by

$$\exists x_1 \le t_1 \ldots Q_{i+1} x_{i+1} \le |t_{i+1}| A(\vec{a}, \vec{x}).$$

Since S_2^1 has $\beta(i, a)$, every Σ_i^p-formula with only free variables $\vec{a}$ is equivalent to a pure i-form. The formalized notion of pure i-form, i.e., "x is a Gödel number of pure i-form formula" is Δ_1^b with respect to S_2^1.

A formula $\varphi(\vec{a})$ in the language of S_2^1 is said to be of i-form if the following conditions are satisfied.

1. The only free variables in $\varphi(\vec{a})$ are $\vec{a}$.

2. $\varphi(\vec{a})$ is a subformula of a pure i-form formula, i.e., it is of the form

$$Q_j x_j \le t_j \ldots Q_{i+1} x_{i+1} \le |t_{i+1}| A(\vec{a}, \vec{x})$$

where $A(\vec{a}, \vec{x})$ is quantifier free and Q_k is $\forall$ if k is even and Q_k is $\exists$ if k is odd and t_k is of the form $t_k(\vec{a}, x_j, \ldots, x_{k-1})$. Quantifier free formulas and formulas of the form $Q_{i+1} x_{i+1} \le |t_{i+1}(\vec{a})| A(\vec{a}, x_{i+1})$ are included as special cases of i-form formulas.

If $\ulcorner \varphi(\vec{a}) \urcorner$ is the Gödel number of an i-form formula, then $\varphi(\vec{a})$ is of the form

$$Q_j x_j \le t_j \ldots Q_i x_i \le t_i Q_{i+1} x_{i+1} \le |t_{i+1}| A(\vec{a}, \vec{x})$$

and j is calculated from $\ulcorner \varphi(\vec{a}) \urcorner$.

We are going to define $\widetilde{T}(\tilde{u}, \ulcorner \varphi(\vec{a}) \urcorner, \vec{a})$, which is a truth definition of $\ulcorner \varphi(\vec{a}) \urcorner$ by assigning the value a_i to $\ulcorner a_i \urcorner$.

Later we will give a condition such that all the terms occurring in the computation of $\widetilde{T}(\tilde{u}, \ulcorner \varphi(\vec{a}) \urcorner, \vec{a})$ are bounded by $\tilde{u}$. Under this condition, $\widetilde{T}(\tilde{u}, \ulcorner \varphi(\vec{a}) \urcorner, \vec{a})$ is defined to be $\bigwedge_j$ (if $\ulcorner \varphi(\vec{a}) \urcorner$ is of the form

$$\ulcorner Q_j x_j \le t_j \ldots Q_{i+1} x_{i+1} \le |t_{i+1}| A(\vec{a}, \vec{x}) \urcorner,$$

then

$$Q_j x_j \leq v(\ulcorner t_j \urcorner, \vec{a}) \quad \ldots Q_{i+1} x_{i+1} \leq |v(\ulcorner t'_{i+1} \urcorner, \vec{a}, x_j, \ldots, x_i)$$
$$T_0(\ulcorner A(\vec{a}, b_j, \qquad \ldots, b_i \urcorner), \vec{a}, x_j, \ldots, x_i)),$$

where we denote $t_k(\vec{a}, b_j, \ldots, b_{k-1})$ by t'_k. Then $\widetilde{T}(\widetilde{u}, \ulcorner \varphi(\vec{a}) \urcorner, \vec{a})$ is Σ_i^b in S_2^1.

In [5], we defined strictly i-normal proof and i-normal proof. This notion is very useful to evaluate the proof in $S_{2,n}^{i+1}$. Now we need a stronger notion since we would like to replace $S_{2,n}^{i+1}$ by S_2^{i+1}. In this paper we call this stronger notion i-normal proof.

Definition. *A proof P in S_2^i is said to be strictly i-normal if the following conditions are satisfied.*

1. *Every formula in P is i-normal.*

2. *P is in free variable normal form.*

3. *Let $\vec{c}$ be all parameter variables in P and $\vec{b}$ be an enumeration of all other free variables in P satisfying the condition that if the elimination inference for b_i is below the elimination inference for b_j then $i < j$. There exists an assignment $t_i(\vec{c})$ for b_i satisfying the following conditions.*

   (a) $t_i(\vec{c})$ is a term in the language of S_2^1 and all function symbols occurring in $t_i(\vec{c})$ are function symbols of increasing functions.

   (b) If the elimination inference of b_i is

   $$\frac{A(\lfloor \tfrac{1}{2} b_i \rfloor), \Gamma \to \Delta, A(b_i)}{A(0), \Gamma \to \Delta, A(t(b_1, \ldots, b_{i-1}, \vec{c}))}$$

   or

   $$\frac{b_i \leq t(b_1, \ldots, b_{i-1}, \vec{c}), A(b_i), \Gamma \to \Delta}{\exists x \leq t(b_1, \ldots, b_{i-1}, \vec{c}) A(x), \Gamma \to \Delta}$$

   or

   $$\frac{b_i \leq t(b_1, \ldots, b_{i-1}, \vec{c}), \Gamma \to \Delta, A(b_i)}{\Gamma \to \Delta, \forall x \leq t(b_1, \ldots, b_{i-1}, \vec{c}) A(x)}$$

   then $a_1 \leq t_1(\vec{c}), \ldots, a_{i-1} \leq t_{i-1}(\vec{c}) \to t(a_1, \ldots, a_{i-1}, \vec{c}) \leq t_i(\vec{c})$ is provable without using logical inference, induction, or any free variables other than $a_1, \ldots, a_{i-1}$ and $\vec{c}$. All the information for condition 3) is called a supplementary proof. The proof P includes all these supplementary proofs.

4. *A sequence $\ldots, t(t_1(\vec{c}), \ldots, t_i(\vec{c}), \vec{c}), \ldots$ is provided where $t(b_1, \ldots, b_i, \vec{c})$ ranges over all terms in the proof.*

Precisely P together with all supplementary proofs and the sequence of terms described in 4) is called strictly i-normal proof. Let $\Gamma \to \Delta$ be provable in S_2^i, where all formulas in Γ and Δ are i-normal. Then we first make a free cut free proof of $\Gamma \to \Delta$ in S_2^i and then we can easily make a strictly i-normal proof of $\Gamma \to \Delta$.

Let $\varphi(a)$ be an i-form formula. Then a proof P of $\forall x \neg \varphi(x)$ is said to be i-normal if P is obtained from a strictly i-normal proof P_0 in the following way.

$$\vdots \quad P_0$$

$$\varphi(a) \quad \to$$

$$\overline{\qquad \to \neg \varphi(a) \qquad}$$

$$\overline{\qquad \to \forall x \neg \varphi(x) \qquad}$$

We denote the formalized notion "w is a strictly i-normal proof of $\ulcorner \Gamma \to \Delta \urcorner$" and "$w$ is an i-normal proof of $\ulcorner \forall x \neg \varphi(x) \urcorner$" by $Prf^i(w, \ulcorner \Gamma \to \Delta \urcorner)$ and $Prf^i(w, \ulcorner \forall x \neg \varphi(x) \urcorner)$ respectively."

In §2 Lemma 3 in [5], we proved the following theorem.

Theorem. *Let $\varphi(c)$ be an i-normal formula with only free variable c. Then*

$$S_{2,n}^{i+1} \vdash Prf^i(|w|_n, \ulcorner \varphi(c) \urcorner) \to \varphi(c).$$

The key point of the proof of this theorem is that if $\ulcorner t(\vec{a}) \urcorner$ is n-small, i.e., of the form $|u|_n$, then $|b|^{|\ulcorner t(\vec{a}) \urcorner|}$ is small in $S_{2,n}^1$. Therefore the bound $\exp(\ulcorner t(\vec{a}) \urcorner |b|^{|\ulcorner t(\vec{a}) \urcorner|})$ can be expressed in $S_{2,n}^1$ and $v(\ulcorner t(\vec{a}) \urcorner, \vec{b})$ can be expressed in $S_{2,n}^1$.

We add several remarks on the theorem. In $Prf^i(|w|_n, \ulcorner \varphi(c) \urcorner)$, c in $\ulcorner \varphi(c) \urcorner$ is a variable, therefore we might write the theorem

$$S_{2,n}^{i+1} \vdash Prf^i(|w|_n, \ulcorner \varphi(a) \urcorner) \to \varphi(c).$$

We define I_k by the following as usual

$$I_0 = 0$$
$$I_{2k+1} = I_{2k} + 1$$
$$I_{2(k+1)} = 2 \cdot I_{k+1}$$

where 2 is $1 + 1$. Then $v(\ulcorner t(a_1, \ldots, a_n) \urcorner, b_1, \ldots, b_n) = v(\ulcorner t(I_{b_1}, \ldots, I_{b_n}) \urcorner)$. Therefore we have the following theorem with the same proof.

Theorem. *Let $\varphi(c)$ be an i-normal formula with only free variable c. Then*

$$S_{2,n}^{i+1} \vdash Prf^i(|w|_n, \ulcorner\varphi(I_c)\urcorner) \to \varphi(c).$$

Now we come back to our present case in S_2^{i+1}. Then $v(\ulcorner t(\vec{a})\urcorner, \vec{b})$ is definable if $\ulcorner t(\vec{a})\urcorner$ is 2 small and b is small and $v(\ulcorner t\urcorner)$ is definable if $\ulcorner t\urcorner$ is the Gödel number of a closed term and $\ulcorner t\urcorner$ is small.

These two conditions give the following two theorems which are proved in the same way as in the proof of §2 Lemma 3 in [5].

Theorem 1. *Let $\Gamma(c) \to \Delta(c)$ be a sequent with only free variable c and all formulas in $\Gamma(c)$ and $\Delta(c)$ be i-normal. Then*

$$S_2^{i+1} \vdash c \le |d|, Prf^i(|w|_2, \ulcorner\Gamma(a) \to \Delta(a)\urcorner), \Gamma(c) \to \Delta(c).$$

Theorem 2. *Let $\Gamma \to \Delta$ be a sequent and all formulas in Γ and Δ be i-normal sentences. Then*

$$S_2^{i+1} \vdash Prf^i(|w|, \ulcorner\Gamma \to \Delta\urcorner), \Gamma \to \Delta.$$

We have the following corollaries.

Corollary 3. *Let $\varphi(a)$ be an i-normal formula in which a is only free variable. Then*

$$S_2^{i+1} \vdash \exists w Prf^i(|w|_2, \ulcorner\forall x\neg\varphi(x)\urcorner) \to \forall x\neg\varphi(|x|).$$

Corollary 4.
$$S_2^{i+1} \vdash \neg Prf^i(|w|, \ulcorner\to\urcorner).$$

Remark. The following theorem is obtained by the same method with the proof of Corollary 3.

Theorem. *Let $\varphi(a)$ be an i-normal formula in which a is only free variable. Then*

$$S_2^1 \vdash \exists w Prf^i(|w|_3, \ulcorner\forall x\neg\varphi(x)\urcorner) \to \forall x\neg\varphi(|x|_2).$$

3   Gödel sentences

See §7.5 in [2] for the general theory of Gödel sentences in Bounded Arithmetic. We define Gödel sentences φ_k^i satisfying

$$S_2^1 \vdash \varphi_k^i \longleftrightarrow \forall x \neg Prf^i(|x|_k, \ulcorner \varphi_k^i \urcorner).$$

From the definition of φ_k^i, φ_k^i is of the form $\forall x \neg G_k^i(|x|_k)$ where G_k^i is an i-form formula and we have

$$S_2^1 \vdash G_k^i(a) \longleftrightarrow Prf^i(a, \ulcorner \varphi_k^i \urcorner).$$

The following properties on Gödel sentences are proved by the standard argument.

Theorem 5. S_2^i does not prove $\varphi_k^i (k = 0, 1, 2, \ldots)$ and $S_2^1 \vdash \varphi_0^i \longleftrightarrow \forall x \neg Prf^i(x, \ulcorner \rightarrow \urcorner)$ therefore $S_2^i \nvdash \forall x \neg Prf^i(x, \ulcorner \rightarrow \urcorner)$.

The following natural question arises here. Is $\forall x \neg Prf^i(x, \ulcorner \rightarrow \urcorner)$ provable in S_2^{i+1}? If the answer is yes, then we have $S_2^i \neq S_2^{i+1}$ and $P \neq NP$. But we believe the answer is no in the following reason.

Let $S_2^{-\infty}$ be the equational theory involving equations $s = t$, where s and t are closed terms in the Buss' original language of S_2^1 with the natural rules of the function symbols. Therefore $S_2^{-\infty}$ does not have any free variables, any logical symbols or any inductions.

As is stated in [3], we conjecture

$$S_2 \nvdash Con(S_2^{-\infty})$$

where $Con(S_2^{-\infty})$ is the consistency of $S_2^{-\infty}$. $S_2^{-\infty}$ is an extremely weak system and therefore $Con(S_2^{-\infty})$ is much weaker than $\forall x \neg Prf^0(x, \ulcorner \rightarrow \urcorner)$. Therefore our conjecture implies

$$S_2^i \nvdash \forall x \neg Prf^0(x, \ulcorner \rightarrow \urcorner)$$

and we believe that there is no hope to prove

$$S_2 \vdash \forall x \neg Prf^i(x, \ulcorner \rightarrow \urcorner).$$

On the other hand, we have the following comjecture.

Conjecture 6. $S_2^i \nvdash \forall x \neg Prf^i(|x|, \ulcorner \rightarrow \urcorner)$.

This conjecture together with Corollary 4 certainly implies $S_2^i \neq S_2^{i+1}$ and $P \neq NP$. However it should be noted that $S_2^1 \vdash \forall x \neg Prf^i(|x|_2, \ulcorner \rightarrow \urcorner)$.

Remark. For the feasibility of our conjecture we would like to discuss its relation with Baker-Gill-Solovay's result in [1].

First our system S_2^i must satisfy the following basic conditions as we stated before.

(a) The predicate constants are only $\leq$ and $=$.

(b) The number of function constants are finite and all the function constants express polynomial time computable functions.

(c) Extra true axioms are allowed to be used only when they can be expressed by finitely many initial sequents without logical symbols.

These basic conditions cannot accept Baker-Gill-Soloray type relativizations. Moreover we discuss the comparison of the typical Baker-Gill-Soloray case and our case.

Let $\widetilde{S_2^i}$ be S_2^i+ (Axiom on PSPACE–complete predicate) and $\widetilde{P}$ and $\widetilde{NP}$ be the class of P and NP in the language of $\widetilde{S_2^i}$. Then we have $\widetilde{P} = \widetilde{NP}$. We consider a similar stronger example $\widetilde{\widetilde{S_2^i}} = S_2^i+$ (Axiom on exponential function) and let $\widetilde{\widetilde{P}}$ and $\widetilde{\widetilde{NP}}$ be P and NP formulated in the language of $\widetilde{\widetilde{S_2^i}}$. Then $\widetilde{\widetilde{P}}=\widetilde{\widetilde{NP}}$. In this case of $\widetilde{\widetilde{S_2^i}}$, the situation is totally opposite to our case of conjecture i.e.,

$$\widetilde{\widetilde{S_2^{i+1}}} \nvdash \forall x \neg \, \widetilde{\widetilde{Prf^i}} \, (|x|, \ulcorner \to \urcorner)$$

in the place of $S_2^{i+1} \vdash \forall x \neg Prf^i(|x|, \ulcorner \to \urcorner)$ and $\widetilde{\widetilde{S_2^i}} \nvdash \forall x \neg \, \widetilde{\widetilde{Prf^i}} \, (|x|, \ulcorner \to \urcorner)$ is trivial in the place of our conjecture $S_2^i \nvdash \forall x \neg Prf^i(|x|, \ulcorner \to \urcorner)$. Therefore Baker-Gill-Solovay's result has no relation with the feasibility of our conjecture.

Theorem 7. *We have for* $k > 2$

$$S_2^{i+1} \vdash \forall x \neg G_k^i(|x|_{k+1})$$

especially $S_2^{i+1} \vdash \forall x \neg G^i(|x|_3)$.

Proof. By the definition of φ_k^i, we have

$$S_2^{i+1} \vdash \neg \varphi_k^i \to \exists x Prf^i(|x|_k, \ulcorner \forall x \neg G_k^i(|x|_k) \urcorner).$$

From this and Corollary 3 follows

$$S_2^{i+1} \vdash \neg \varphi_k^i \to \forall x \neg G_k^i(|x|_{k+1}).$$

Therefore $S_2^{i+1} \vdash \forall x \neg G_k^i(|x|_k) \vee \forall x \neg G_k^i(|x|_{k+1})$. Since $G_k^i(a)$ is equivalent to a form $Prf^i(a, \ulcorner \varphi_k^i \urcorner)$, we have $S_2^{i+1} \vdash \forall x \neg G_k^i(|x|_{k+1})$. $\square$

Corollary 8. *For* $k \geq 2$ *we have*

$$S_2^{i+1} \vdash \forall x \neg Prf^i(|x|_{k+1}, \ulcorner \varphi_k^i \urcorner)$$

especially $S_2^{i+1} \vdash \forall x \neg Prf^i(|x|_3, \ulcorner \varphi_2^i \urcorner)$.

Conjecture 9.
$$S_2^i \nvdash \forall x \neg Prf^i(|x|_3, \ulcorner \varphi_2^i \urcorner).$$

Obviously this conjecture implies $S_2^i \neq S_2^{i+1}$ and $P \neq NP$. It should be noted that for $k \geq 3$, $S_2^1 \vdash \forall x \neg G_k^i(|x|_{k+2})$ i.e., $S_2^1 \vdash \forall x \neg Prf^i(|x|_{k+2}, \ulcorner \varphi_k^i \urcorner)$.

Theorem 10. *We have*
$$S_2^{i+1} \vdash \varphi_2^i \vee \forall x \neg Prf^i(|x|_2, \ulcorner \forall x \neg G_2^i(|x|) \urcorner).$$

Proof. We discuss this inside of S_2^{i+1}. Suppose
$$\neg \forall x \neg Prf^i(|x|_2, \ulcorner \forall x \neg G_2^i(|x|) \urcorner)$$
i.e, $\exists x Prf^i(|x|_2, \ulcorner \forall x \neg G_2^i(|x|) \urcorner)$. Then from Corollary 3 follows
$$\forall x \neg G_2^i(|x|_2) \text{ i.e., } \varphi_2^i.$$

□

Conjecture 11. *The following statement holds.*
$$S_2^{i+1} \vdash \forall x \neg Prf^i(|x|_2, \ulcorner \forall x \neg G_2^i(|x|) \urcorner) \to \varphi_2^i$$

It is easily seen that this conjecture also implies $S_2^i \neq S_2^{i+1}$ and $P \neq NP$.

Theorem 12. *If $S_2^{i+1} \vdash \varphi_k^i \to \forall x \neg A(x)$ and $k \geq 3$ and $A(x)$ is i-normal, then*
$$S_2^{i+1} \vdash \forall x \neg A(|x|).$$

Proof. $S_2^{i+1} \vdash \varphi_k^i \to \forall x \neg A(x)$ implies that there exists a constant size proof of the following form

$$\vdots$$

$$\cdots \to \cdots Prf^i(|t|_k, \ulcorner \psi_k^i \urcorner)$$
$$\overline{\neg Prf^i(|t|_k, \ulcorner \psi_k^i \urcorner), \cdots \to \cdots}$$
$$\overline{\forall x \neg Prf^i(|x|_k, \ulcorner \psi_k^i \urcorner), \cdots \to \cdots}$$
$$\vdots$$
$$\overline{A(a), \forall x \neg Prf^i(|x|_k, \ulcorner \psi_k^i \urcorner), \cdots \to \cdots}$$

Let us denote the size of this proof by a constant c_0. Now we discuss inside of S_2^{i+1} and suppose $\neg\varphi_k^i$ i.e., $\exists x Prf^i(|x|_k, \ulcorner\forall x \neg Prf^i(|x|_k, \ulcorner\psi_k^i\urcorner)\urcorner)$. Then there exists a proof $|x_0|_k$ of

$$Prf^i(|b|_k, \ulcorner\psi_k^i\urcorner) \to \quad .$$

Using the previous proof of size c_0 and this proof $|x_0|_k$ and making many cut of the following form, we get a proof of

$$A(a) \to$$

i.e.,

$$\vdots$$

$$\frac{\cdots \; \to \cdots, Prf^i(|t|_k, \ulcorner\psi_k^i\urcorner)Prf^i(|t|_k, \ulcorner\psi_k^i\urcorner) \to}{\cdots \; \to \cdots}$$

$$\vdots$$

$$\overline{}$$

$$A(a) \to$$

Since the whole procedure is polynomial time computable from two proofs, the size of the new proof is not greater than $t(|x_0|_k)$ for some term t.

Since $2 < k$, we have

$$\exists z Prf^i(|z|_2, \ulcorner\forall x \neg A(x)\urcorner)$$

and we proved

$$S_2^{i+1} \vdash \neg\varphi_k^i \to \exists z Prf^i(|z|_2, \ulcorner\forall x \neg A(x)\urcorner).$$

Then by Corollary 3 we have

$$S_2^{i+1} \vdash \neg\varphi_k^i \to \forall x \neg A(|x|).$$

Therefore we have

$$S_2^{i+1} \vdash \forall x \neg A(x) \lor \forall x \neg A(|x|)$$

i.e.,

$$S_2^{i+1} \vdash \forall x \neg A(|x|).$$

$\square$

Corollary 13. *If $S_2^{i+1} \vdash \varphi_k^i \to \forall x > Prf^i(|x|_{k-1}, \ulcorner\varphi_k^i\urcorner)$ and $k > 2$, then $S_2^{i+1} \vdash \varphi_k^i$ and therefore $S_2^{i+1} \vdash \forall x \neg Prf^i(|x|_{k-1}, \ulcorner\varphi_k^i\urcorner)$ and $S_2^{i+1} \neq S_2^i$ and $P \neq NP$.*

Proof. By Theorem 12, we have $S_2^{i+1} \vdash \forall x \neg Prf^i(|x|_k, \ulcorner\varphi_k^i\urcorner)$ i.e., $S_2^{i+1} \vdash \varphi_k^i$ therefore $S_2^{i+1} \vdash \forall x \neg Prf^i(|x|_{k-1}, \ulcorner\varphi_k^i\urcorner)$ and $S_2^{i+1} \neq S_2^i$ and $P \neq NP$. $\Box$

Theorem 14. *For $k \geq 1$ we have*

$$S_2^{i+1} \vdash \forall x \neg Prf^i(|x|_{k+1}, \ulcorner\forall x \neg G_k^i(|x|_{k-1})\urcorner).$$

Proof. We have

$$S_2^{i+1} \vdash \exists x Prf^i(|x|_2, \ulcorner\forall x \neg G_k^i(|x|_{k-1})\urcorner) \to \forall x \neg G_k^i(|x|_k).$$

On the other hand, we have

$$S_2^1 \vdash \exists x Prf^i(|x|_{k+1}, \ulcorner\forall x \neg G_k^i(|x|_{k-1})\urcorner)$$
$$\to \exists x Prf^i(|x|_k, \ulcorner\forall x \neg G_k^i(|x|_k)\urcorner)$$

since a proof of $\forall x \neg G_k^i(|x|_k)$ is obtained from a proof of $\forall x \neg G_k^i(|x|_{k-1})$ by a polynomial time computable operation. Therefore we have

$$S_2^{i+1} \vdash \quad \neg(\exists x Prf^i(|x|_2, \ulcorner\forall x \neg G_k^i(|x|_{k-1})\urcorner)$$
$$\wedge \exists x Prf^i(|x|_{k+1}, \ulcorner\forall x \neg G_k^i(|x|_{k-1})\urcorner)$$

since $\exists x Prf^i(|x|_k, \ulcorner\forall x \neg G_k^i(|x|_k)\urcorner)$ is equivalent to $\neg\varphi_k^i$ and $\forall x \neg G_k^i(|x|_k)$ is φ_k^i. Therefore we have

$$S_2^{i+1} \vdash \neg\exists x Prf^i(|x|_{k+1}, \ulcorner\forall x \neg G_k^i(|x|_{k-1})\urcorner).$$

$\Box$

Conclusion. At this moment, there are many mysteries regarding the nature of Gödel sentences in Bounded Arithmetic. We strongly believe that $S_2^i \neq S_2^{i+1}$ and $P \neq NP$ would be proved if our knowledge on these Gödel sentences could be improved. We list several problems on Gödel sentences.

1. Is $S_2^{i+1} \vdash \varphi_k^i$ true? We conjecture that this is false for $k = 0, 1$.

2. Is $S_2^{i+1} \vdash \forall x \neg G_0^i(|x|)$ true? Is $S_2^{i+1} \vdash \forall x \neg G_1^i(|x|_2)$ true? At this moment we have no idea about them.

3. Is $S_2^{i+1} \vdash \forall x \neg Prf^i(|x|_k, \ulcorner\forall x \neg G_k^i(|x|_{k-1})\urcorner)$ true for some k?

Especially is
$S_2^{i+1} \vdash \forall x \neg Prf^i(|x|, \ulcorner\forall x \neg G_1^i(x)\urcorner)$ true?

4 REFERENCES

[1] T. Baker, J. Gill & R. Solovay, Relativization of the $P =?NP$ question, SIAM Journal of Computing 4 (1975), 431–442.

[2] S. Buss, Bounded Arithmetic, Bibliopolis, Napoli, (1986)

[3] P. Clote & J. Krajíček, Problem 1.2.9 in p 5 in Open problems, Arithmetic, proof theory and computational complexity, Oxford, (1993)

[4] J. Krajíček, P. Pudlák & G. Takeuti, Bounded Arithmetic and the Polynomial Hierachy, Annals of Pure and Applied Logic, vol. 52 (1991), 143–153

[5] G. Takeuti, Bounded Arithmetic and Truth Definition, Annals of Pure and Applied Logic, vol. 39 (1988), 75–104.

[6] A. Wilkie & J. Paris, On the scheme of induction for bounded arithmetic formulas, Proc. of an ASL Conference in Manchester, (1986) North Holland.

Author address

Department of Mathematics
University of Illinois
1409 W Green Street
Urbana, Il 61801 USA
email: takeuti@symcom.math.uiuc.edu

Printed in the United States
By Bookmasters